From Scaling and Perturbation to Stability

This book aims not only to teach technical methods, but also to cultivate an "asymptotic way of thinking" that emphasizes scaling, dominant balances, approximation, and physical insight as essential tools in mathematical modeling. **From Scaling and Perturbation to Stability** is designed to help students extract meaningful approximations from complex mathematical models before seeking exact solutions.

The author takes a modeling-first, intuition-driven approach. The book begins with dimensional analysis to emphasize physical reasoning before formal solution techniques are introduced. The development proceeds from simple to complex: from scalar problems to systems, from regular to singular perturbations, and from linear to nonlinear stability.

The presentation is unified and systematic. Methods are introduced in a coherent framework, repeatedly connected to physical interpretation and to the limitations of approximations. The text reflects extensive classroom testing through a two-semester sequence, and its structure is shaped by what has proven most effective for student learning.

Distinguishing Features

- **Extensive, carefully graded exercises**, ranging from introductory to challenging.
- **Strong emphasis on physical intuition and modeling**, not merely formal manipulation.
- **Unified treatment** of perturbation methods, asymptotic analysis, integrals, and dynamical systems.
- **Integration of stability and bifurcation theory** into an asymptotic framework.
- **Clear connection between theory and applications**, with numerous worked examples drawn from science and engineering.

This text provides a coherent, example-driven introduction to asymptotic and perturbation methods for students in applied mathematics, science, and engineering.

Other books by **Youssef N. Raffoul**, published by CRC Press: ***Fourier Series and Boundary Value Problems with Engineering Applications*** (2025) and ***Applied Mathematics for Scientists and Engineers*** (2023).

Textbooks in Mathematics
Series editors:
Al Boggess, Kenneth H. Rosen

Math Anxiety How to Beat It
Brian Cafarella

Lectures on Differential Geometry with Maple
Mayer Humi

Numerical Analysis for Engineers
Methods and Applications
Bilal M. Ayyub and Richard H. McCuen

An Invitation to Real Analysis
Andrew D. Hwang

Fourier Series and Boundary Value Problems with Engineering Applications
Youssef N. Raffoul

A Course in Real Analysis, Second Edition
Hugo D. Junghenn

Real and Functional Analysis
Kenneth Kuttler

Mathematics for Artificial Intelligence
Jane Hawkins

Elements of Topology, Second Edition
Theory and Practice
Steven G. Krantz

Convex Analysis, Second Edition
Steven G. Krantz

A Second Course in Complex Analysis
Peter V. Dovbush and Steven G. Krantz

Real and Complex Analysis, Second Edition
Volume 1
Christopher Apelian and Steve Surace

Elementary Differential Equations, Third Edition
Charles E. Roberts Jr. and Miklós Bóna

A Tour through Graph Theory, Second Edition
Karin R. Saoub

From Scaling and Perturbation to Stability
Youssef N. Raffoul

https://www.routledge.com/Textbooks-in-Mathematics/book-series/CANDHTEX-BOOMTH

From Scaling and Perturbation to Stability

Youssef N. Raffoul

CRC Press
Taylor & Francis Group
Boca Raton London New York

CRC Press is an imprint of the
Taylor & Francis Group, an **informa** business
A CHAPMAN & HALL BOOK

Designed cover image: Youssef N. Raffoul

First edition published 2027
by CRC Press
2385 NW Executive Center Drive, Suite 320, Boca Raton FL 33431

and by CRC Press
4 Park Square, Milton Park, Abingdon, Oxon, OX14 4RN

CRC Press is an imprint of Taylor & Francis Group, LLC

ISBN: 978-1-041-37684-2 (hbk)
ISBN: 978-1-041-37685-9 (pbk)
ISBN: 978-1-042-00922-0 (ebk)

DOI: 10.1201/9781042009220

Typeset in Nimbus Roman font
by KnowledgeWorks Global Ltd.

Publisher's note: This book has been prepared from camera-ready copy provided by the authors.

February 04, 1956 – January 25, 2026

In loving memory of my beloved brother, Melhem Naim Raffoul. Words cannot fully express the depth of our loss. You were the pillar of the Naim family, an engineer, a man of strength, wisdom, and unwavering support. Your encouragement was a constant source of inspiration in my life. I will carry your memory with me always. Rest in peace.

Contents

Preface

Applied mathematics occupies a central role in modern science, engineering, and technology. The development of mathematical models capable of describing physical, biological, and engineering phenomena often leads naturally to differential equations containing small or large parameters, multiple time scales, nonlinear interactions, and stability questions. In many important applications, exact solutions are either unavailable or too complicated to provide meaningful qualitative insight. Consequently, asymptotic and perturbation methods have become indispensable tools for understanding the behavior of mathematical models and for obtaining approximations that reveal the underlying structure of complex systems.

This book grew out of many years of teaching applied mathematics to upper-level undergraduate and beginning graduate students, with the aim of presenting a coherent and example-driven introduction to scaling, asymptotic, and perturbation methods as they arise naturally in science and engineering. The material has been thoroughly classroom tested and refined through repeated offerings of a two-semester sequence, and its organization reflects what has proven most effective for student learning.

The reader is expected to be comfortable with ordinary differential equations, linear algebra, and material at the level of an intermediate real analysis course. No prior familiarity with asymptotic methods, perturbation theory, or dynamical systems is required.

With this preparation, the book is appropriate for senior undergraduate students, beginning graduate students, and first-year graduate courses in applied mathematics, mathematics, physics, and engineering. The material is designed to support a two-semester course, although other organizations are possible. Chapters 1–5 form a natural first semester, covering dimensional analysis, scaling, asymptotic expansions, regular and singular perturbation methods, and boundary layer theory. Chapters 6–8 form a natural second semester devoted to systems of differential equations, matrix methods, stability theory, and nonlinear dynamics. Alternatively, instructors may select topics from several chapters to design a one-semester course tailored to the needs of their students.

The inclusion of stability theory is intentional. Many of the models developed in the early chapters lead naturally to systems of differential equations whose long term behavior must be understood. The later chapters therefore provide the tools needed to analyze the stability, qualitative behavior, and bifurcations of the very models that motivate the asymptotic and perturbation methods introduced at the beginning of the text.

The text opens with dimensional analysis and the Buckingham Pi theorem, emphasizing physical intuition and modeling from the outset. Through a wide range of illustrative examples, including blast waves, surface waves generated by a raindrop, relativistic motion, and gravitational systems, students learn how dimensionless parameters expose the fundamental structure of a problem before any detailed solution is attempted. These ideas establish the central theme of the book: identifying dominant balances and extracting meaningful approximations from complex mathematical models.

Building on this foundation, the book develops scaling arguments, asymptotic expansions, and perturbation methods for ordinary differential equations. Regular perturbation techniques are introduced alongside systematic procedures for selecting appropriate scalings. More advanced methods, including multiple scales, averaging, and the Poincaré–Lindstedt technique, are presented in a unified framework. Particular emphasis is placed on the physical interpretation of asymptotic results and on understanding the limitations of the approximations obtained.

Singular perturbation problems and boundary layer theory are treated in depth. The text covers the formation of boundary and interior layers, higher-order corrections, nonlinear boundary layers, and equations with variable coefficients. Numerous worked examples illustrate how matched asymptotic expansions resolve apparent contradictions between limiting equations and boundary conditions.

Asymptotic methods are then extended beyond differential equations to include asymptotic expansions of integrals. Laplace-type integrals, exponentially weighted integrals, repeated integration by parts, and the WKB approximation for singularly perturbed second-order equations are developed carefully, providing students with tools widely used in applied mathematics, physics, and engineering.

The book next turns to systems of ordinary differential equations, moving from scalar problems to coupled dynamics. After establishing basic preliminaries, the fundamental matrix is introduced as a central concept in the analysis of linear systems. Linear autonomous systems of the form $x' = Ax$ and nonautonomous systems $x' = A(t)x + g(t)$ are studied in detail, with particular emphasis on the matrix exponential e^{At} and its role in describing solution behavior. These chapters lay the groundwork for a systematic treatment of stability theory.

Subsequent chapters focus on the stability of linear systems, presenting precise definitions, illustrative examples, and systematic methods for determining stability in both time-dependent and autonomous settings. Phase–plane analysis of planar systems is used to connect algebraic criteria with geometric intuition and qualitative behavior.

The final chapters address stability and bifurcations in nonlinear systems. Scalar bifurcations, stability by linearization, and classical models such as epidemic and competition systems are used to illustrate how qualitative changes in dynamics arise as parameters vary. Limit cycles, planar bifurcations, and Lyapunov functions are developed as fundamental tools for analyzing long-term behavior, stability, and instability in nonlinear models.

A defining feature of this book is its extensive collection of exercises. Problems are carefully graded from introductory to challenging, making them effective tools for reinforcing concepts and developing technical skill. Early exercises focus on consolidating core ideas, while later problems encourage synthesis and independent thinking. This progression makes the book particularly well suited for a two-semester course sequence, while also allowing instructors the flexibility to select individual chapters or selected topics for specialized courses.

It is our hope that this text not only provides students with practical analytical techniques, but also fosters an asymptotic mode of thinking that emphasizes scaling, approximation, and physical insight alongside exact analysis.

Special thanks are extended to my graduate class of Spring 2026, with particular appreciation to Mr. Christian Hemsath.

Youssef N. Raffoul, Professor and Graduate Program Director
Department of Mathematics
University of Dayton
Dayton, Ohio
May, 2026

Author

Youssef N. Raffoul is a professor of mathematics at the University of Dayton, whose research focuses on the qualitative theory of differential, difference, integral, and functional dynamical systems, with particular emphasis on Volterra equations and dynamical systems on time scales. He earned his PhD in mathematics from Southern Illinois University at Carbondale in 1996, following earlier degrees from Indiana University and the University of Dayton. Over nearly three decades, he has made sustained contributions to the study of stability, boundedness, periodicity, and oscillation in continuous, discrete, stochastic, and hybrid systems, developing analytical techniques that connect functional analysis with applied scientific models. Since joining the University of Dayton in 1999, Dr. Raffoul has advanced to the rank of full professor and has played a central role in the department's graduate mission, including service as director of Graduate Programs in Mathematics. His research encompasses differential and difference equations, Volterra integro-differential systems, functional equations, and dynamical systems on time scales. He is especially known for his systematic use of Lyapunov functionals, fixed point theory, resolvent techniques, and variation-of-parameters methods. He has authored or coauthored more than 160 peer-reviewed research articles widely cited in the international mathematical literature. Raffoul is also the author of several influential monographs and advanced textbooks that bridge rigorous mathematical theory with scientific and engineering applications, including *Qualitative Theory of Volterra Difference Equations* (Springer), *Stability, Periodicity and Boundedness in Functional Dynamical Systems on Time Scales* (Springer), *Advanced Differential Equations* (Elsevier), *Applied Mathematics for Scientists and Engineers* (CRC Press), *Difference Equations and Applications* (Elsevier), and *Fourier Series and Boundary Value Problems with Engineering Applications* (CRC Press/Taylor & Francis). Beyond his research and authorship, he has contributed extensively as an editor, guest editor, conference organizer, and invited speaker, serving on the editorial boards of several international journals and delivering plenary and keynote lectures worldwide. His work has been recognized through numerous honors and distinctions, including the Lebanese Order of Merit (Silver Medal with Distinction), the honorary title Allama, a national Career in Science Excellence Award, and major University of Dayton awards for scholarly achievements. Throughout his career, Dr. Raffoul has remained deeply committed to mentoring students and fostering international collaboration while maintaining a strong emphasis on clarity, accessibility, and practical application in applied mathematics.

1

Dimensional Analysis

Dimensional analysis and scaling arguments provide some of the most powerful and economical tools in theoretical and applied science. Long before a detailed solution of governing equations is available, these methods allow us to identify the essential physical parameters, construct meaningful dimensionless groups, and reveal how physical systems respond to changes in scale. By exploiting the invariance of physical laws under changes of units, dimensional analysis offers deep insight into complex phenomena using minimal mathematical machinery.

This chapter begins with a concrete and classical example: the drag force acting on a sphere moving through a fluid. By examining the limits of low and high Reynolds numbers, we demonstrate how distinct physical regimes emerge naturally from scaling considerations. The Stokes regime, dominated by viscous effects, contrasts sharply with the inertial regime, where nonlinear fluid motion governs the dynamics. These limiting cases illustrate how dimensional reasoning not only predicts functional forms but also clarifies the range of validity of approximate models used in practice.

The central theoretical framework of the chapter is the Buckingham Pi theorem, which provides a systematic procedure for constructing dimensionless parameters from a given set of physical variables. Its power is illustrated through a range of examples, from classical fluid mechanics to problems of extraordinary historical and physical significance. In particular, we examine the dimensional analysis that underlies estimates of the energy released in the first atomic bomb detonation, highlighting how scaling arguments can extract quantitative information from limited data.

Further applications extend to wave phenomena, compressible flows, relativistic motion, and gravitational systems. Dimensionless numbers such as the Mach number emerge naturally as measures of competing physical effects, while their interpretation provides a unifying language across diverse areas of physics. The chapter concludes with exercises designed to strengthen intuition, encourage independent reasoning, and develop fluency in applying dimensional and scaling arguments to both familiar and unfamiliar problems.

Introduction to Dimensional Analysis

Dimensional analysis stands as one of the most powerful and elegant techniques in mathematical modeling, providing a systematic framework for simplifying complex

DOI: 10.1201/9781042009220-1

physical problems. By examining the dimensions of the quantities involved, we can reduce the number of variables, identify key dimensionless parameters, and uncover scaling laws that govern physical phenomena. This approach not only streamlines the analysis of differential equations but also reveals profound insights into the underlying structure of natural systems.

At the heart of dimensional analysis lies the **Buckingham Pi Theorem**, which offers a rigorous method for determining the complete set of independent dimensionless groups from the governing parameters. This theorem ensures that any physically meaningful equation can be expressed in terms of these dimensionless quantities, dramatically reducing the complexity of experimental and theoretical investigations. The effectiveness of the Pi Theorem is exemplified in classic problems such as:

- **Drag on a Sphere:** The force experienced by a sphere moving through a fluid depends on its velocity, size, and the fluid's density and viscosity. Dimensional analysis yields the drag coefficient as a function of the Reynolds number, a dimensionless parameter that characterizes flow regimes from laminar to turbulent.
- **Atomic Explosion of 1945:** G.I. Taylor used dimensional analysis to determine the energy released in the first nuclear detonation solely from photographs of the blast wave. By identifying the relevant dimensionless groups, he derived a scaling law relating the fireball radius to time and energy, demonstrating the predictive power of dimensional reasoning.
- **Reynolds Number:** This fundamental dimensionless quantity in fluid mechanics distinguishes between flow patterns and predicts the transition from smooth laminar flow to chaotic turbulence. It emerges naturally from non-dimensionalizing the Navier_Stokes equations and governs similarity in fluid systems across scales.

In this chapter, we will explore the principles of dimensional analysis, develop the Buckingham Pi Theorem, and in later chapters, apply these tools to scale differential equations across various domains from fluid dynamics to heat transfer and beyond. Through these applications, we will see how dimensionless groups not only simplify problems but also reveal universal behaviors, enabling engineers and scientists to extrapolate results from models to full-scale systems.

The central idea is that every physical quantity can be expressed in terms of a few **fundamental dimensions**, such as **mass** (M), **length** (L), and **time** (T). Other physical quantities such as velocity, force, energy, and power are then derived from these basic dimensions. For example,

$$[V] = LT^{-1}, \qquad [F] = MLT^{-2}, \qquad [E] = ML^2T^{-2}.$$

As for the pressure, we let A be the area of the surface. Then $[A] = L^2$, and as a consequence we have that

$$[P] = \frac{[F]}{[A]} = \frac{MLT^{-2}}{L^2} = ML^{-1}T^{-2}.$$

The **fundamental system of units** provides the foundation upon which all measurements and equations in physics and engineering are built. In the International System of Units (SI), the base units for mechanics are the kilogram (kg), the meter (m), and the second (s). These units correspond respectively to the fundamental dimensions of mass, length, and time. In other fields, additional base quantities are required: temperature (Θ) for thermodynamics, electric current (I) for electromagnetism, and luminous intensity or amount of substance in certain applied sciences. However, in the context of classical mechanics, most relationships can be adequately described using the triplet (M, L, T).

The **choice of fundamental units** is not arbitrary but guided by convenience, precision, and universality. For instance, in particle physics or relativistic mechanics, it is often advantageous to replace mass, length, and time with alternative natural units that simplify expressions and highlight physical symmetries. In systems where the speed of light c and Planck's constant h play central roles, these constants can be set to unity ($c = 1$, $h = 1$), effectively merging the dimensions of length and time into a single relativistic framework. This simplification illustrates the deep connection between the selection of units and the nature of the physical laws being described.

In **mechanical problems**, the fundamental dimensions (M, L, T) are sufficient to characterize the relationships among quantities such as displacement, velocity, acceleration, force, and energy. When extending these principles to **heat flow**, however, additional dimensional considerations enter the picture. Thermal conductivity, temperature difference, and heat flux introduce new derived dimensions. For example, the rate of heat transfer per unit area has dimensions MT^{-3}, while the **thermal diffusivity**, which combines conductivity, density, and specific heat, carries the dimensions L^2T^{-1}. Thus, even within a mechanical framework, thermal phenomena expand the scope of dimensional analysis to capture interactions between mechanical and thermal processes.

In **relativistic mechanics**, dimensional analysis takes on a more profound role. The dimensions of time and length become intimately related through the constant speed of light c, with one second corresponding to approximately 3×10^8 meters in distance. Energy and mass are similarly linked through Einstein's celebrated equation

$$E = mc^2,$$

showing that energy and mass are not separate quantities but share a unified dimensional basis. Consequently, the distinction between M, L, and T becomes less rigid, and physical laws are often expressed in systems of units (such as natural or geometric units) that reduce the number of independent dimensions.

Viscosity and the Reynolds Number.

When dealing with **fluid mechanics**, new physical quantities arise that further demonstrate the power of dimensional reasoning. The concept of **viscosity**, for example, reflects the internal friction of a fluid and has dimensions

$$[\mu] = ML^{-1}T^{-1}.$$

It measures how momentum diffuses through a fluid, much as thermal conductivity measures the diffusion of heat. The **Reynolds number**, a dimensionless parameter defined as

$$Re = \frac{\rho U L}{\mu},$$

where ρ is the fluid density, U a characteristic velocity, L a characteristic length, and μ the dynamic viscosity, plays a central role in describing the similarity of fluid flows. The Reynolds number determines whether a flow is likely to be laminar or turbulent. Systems that share the same Reynolds number are said to be **dynamically similar**, meaning that they exhibit comparable flow behavior even when their characteristic lengths, velocities, or viscosities differ substantially.

Remark 1 *Throughout this text, the notation*

$$A \approx B$$

means that A and B are approximately equal to not exactly equal, but close in numerical value. Thus, the statement

$$C \approx D$$

may indicate that the two quantities have the same order of magnitude or that one is an approximation of the other under some assumptions or simplifications.

Consider a rigid sphere of radius L translating through an incompressible fluid at constant speed U. The primary quantity of interest is the total drag force D exerted by the fluid on the sphere. The parameters that enter the problem are

$$D,\ \rho,\ \mu,\ U,\ L,$$

where ρ is the fluid density and μ its dynamic viscosity. The fundamental dimensions used here are mass, length, and time, denoted $M,\ L,\ T$. The dimensions are

$$[D] = MLT^{-2}, \quad [\rho] = ML^{-3}, \quad [\mu] = ML^{-1}T^{-1}, \quad [U] = LT^{-1}, \quad [L] = L.$$

Anticipating the Buckingham Pi theorem developed later in this chapter, we identify two independent dimensionless quantities. Namely, a convenient choice is the Reynolds number,

$$\mathrm{Re} = \frac{\rho U L}{\mu},$$

and the non-dimensional drag

$$\pi \equiv \frac{D}{\rho U^2 L^2}.$$

Thus the drag must have the form

$$D = \rho U^2 L^2 f(\mathrm{Re}) \tag{1.1}$$

for some dimensionless function $f(\mathrm{Re})$. This representation makes explicit that the flow regime (through Re) controls the scaling of the drag.

When viscous forces dominate inertia, the flow is in the Stokes regime. For a sphere of radius L (often denoted as a in the literature), the exact steady creeping-flow result is

$$D_{\text{Stokes}} = 6\pi\mu LU. \tag{1.2}$$

Expressed in the form (1.1),

$$\frac{D_{\text{Stokes}}}{\rho U^2 L^2} = \frac{6\pi\mu LU}{\rho U^2 L^2} = \frac{6\pi}{\text{Re}},$$

so that for $\text{Re} \ll 1$

$$f(\text{Re}) \sim \frac{6\pi}{\text{Re}}.$$

A first inertial correction (Oseen correction) yields the small-Re expansion

$$D = 6\pi\mu LU\left(1 + \frac{3}{8}\text{Re} + O(\text{Re}^2)\right),$$

which shows how weak inertia begins to alter the Stokes result.

At large Re viscous effects are confined to thin boundary layers and separated wakes; inertial pressure forces dominate. Empirically and from inviscid scaling, the drag takes the form of a dynamic pressure times a cross-sectional area,

$$D \approx \tfrac{1}{2}C_D(\text{Re})\,\rho U^2 A, \qquad A = \pi L^2, \tag{1.3}$$

so that

$$\frac{D}{\rho U^2 L^2} = \frac{1}{2}\pi C_D(\text{Re}).$$

For many practical ranges of Re (except close to the drag crisis), the drag coefficient C_D for a smooth sphere tends to an approximately constant value $C_D \approx 0.4$–0.5. Thus, $f(\text{Re})$ approaches a constant:

$$f(\text{Re}) \xrightarrow{\text{Re}\to\infty} \frac{1}{2}\pi C_D \approx 0.7 \text{ (order unity)}.$$

Drag crisis.

The behavior of $C_D(\text{Re})$ is not monotone. Around a critical Reynolds number (for a smooth sphere, typically $\text{Re} \sim 2 \times 10^5$), the drag coefficient undergoes a sudden decrease known as the *drag crisis*. This phenomenon occurs because the transition to turbulence within the boundary layer delays flow separation, reduces the size of the wake behind the sphere, and consequently reduces the pressure drag.

These results illustrate how dimensional analysis collapses the dependence of the drag onto the single parameter Re (when fluid compressibility and other physics are excluded) and how different asymptotic behaviors appear in different Re regimes. In summary, the framework of dimensional analysis anchored by the fundamental dimensions of mass, length, and time serves as the backbone for understanding and

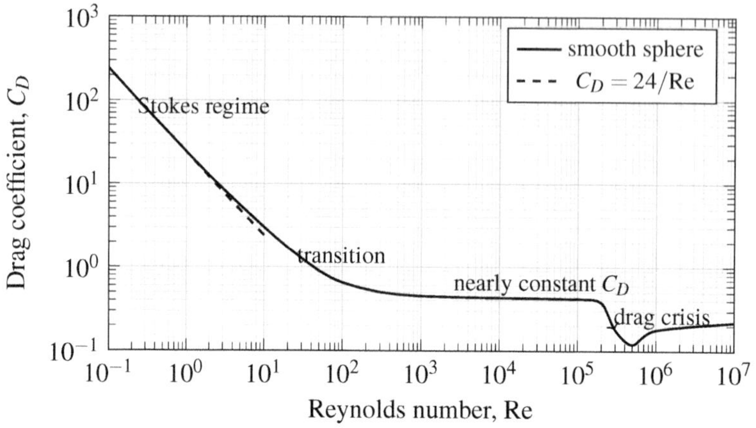

FIGURE 1.1: Qualitative dependence of the drag coefficient C_D on the Reynolds number Re for flow past a smooth sphere. At a small Reynolds numbers, Stokes' law gives $C_D \sim 24/\mathrm{Re}$. At a large Reynolds number, C_D is approximately constant over a wide range before the drag crisis occurs.

simplifying physical laws across disciplines. Whether dealing with mechanical motion, heat transfer, relativistic effects, or viscous flow, the same underlying logic applies: physical relationships must remain consistent and dimensionally homogeneous. This universality makes dimensional analysis not merely a mathematical convenience but a cornerstone of all physical reasoning.

We now state the following definition concerning a unit-free physical law, that is, one which is independent of the system of units chosen. Consider the physical law

$$f(Q_1, Q_2, \ldots, Q_n) = 0. \tag{1.4}$$

Definition 1.1 (Unit-Free (Dimensionally Homogeneous) Physical Law) *The physical law given by* (1.4) *is said to be* ***unit-free*** *or* ***dimensionally homogeneous*** *if every term in the equation has the same dimensions.*

Mathematically, the physical law given by (1.4) *is unit-free if for all choices of* $\gamma_i > 0$, $i = 1, \cdots n$, *we have*

$$f(\bar{Q}_1, \bar{Q}_2, \ldots, \bar{Q}_n) = 0,$$

if, and only if,

$$f(Q_1, Q_2, \ldots, Q_n) = 0,$$

where

$$\bar{Q}_1 = \gamma_1 Q_1, \; \bar{Q}_2 = \gamma_2 Q_2, \; \ldots \; , \bar{Q}_n = \gamma_n Q_n.$$

In other words, the validity of the law does not depend on the particular system of units used (e.g., SI, CGS), but only on the physical quantities involved. This ensures that the equation remains true regardless of the choice of fundamental units.

Example 1.1 *Consider Newton's second law of motion*

$$F = ma.$$

Define

$$f(F, m, a) = F - ma.$$

The dimensions of the physical quantities are

$$[F] = MLT^{-2}, \qquad [m] = M, \qquad [a] = LT^{-2}.$$

Hence,

$$[m][a] = M(LT^{-2}) = MLT^{-2} = [F].$$

Thus, both sides of the equation have the same dimensions, and the law is dimensionally homogeneous (unit-free).

To illustrate invariance under a change of units, let

$$\gamma_1 = \frac{1}{2.54}, \qquad \gamma_2 = 2.20462, \qquad \gamma_3 = \frac{1}{60},$$

corresponding respectively to changes in length, mass, and time units.

Define the transformed quantities

$$\bar{m} = \gamma_2 m, \qquad \bar{a} = \gamma_1 \gamma_3^{-2} a, \qquad \bar{F} = \gamma_2 \gamma_1 \gamma_3^{-2} F.$$

Then

$$\begin{aligned} f(\bar{F}, \bar{m}, \bar{a}) &= \bar{F} - \bar{m}\bar{a} \\ &= \gamma_2 \gamma_1 \gamma_3^{-2} F - (\gamma_2 m)(\gamma_1 \gamma_3^{-2} a) \\ &= \gamma_1 \gamma_2 \gamma_3^{-2} (F - ma) \\ &= \gamma_1 \gamma_2 \gamma_3^{-2} f(F, m, a). \end{aligned}$$

Therefore,

$$f(\bar{F}, \bar{m}, \bar{a}) = 0 \quad \textit{if and only if} \quad f(F, m, a) = 0.$$

Hence, Newton's law is independent of the particular system of units chosen. □

Example 1.2 *A piano wire has mass m, length l, and tension force F. Suppose the period of vibration is given by*

$$T = 2\pi \sqrt{\frac{ml}{F}}.$$

We use dimensional analysis to verify that the formula is dimensionally consistent.

Since

$$[m] = M, \quad [l] = L, \quad [F] = MLT^{-2}, \quad [2\pi] = 1,$$

we obtain

$$\left[2\pi\sqrt{\frac{ml}{F}}\right] = [2\pi]\left[\frac{ml}{F}\right]^{1/2}$$
$$= \left(\frac{ML}{MLT^{-2}}\right)^{1/2}$$
$$= (T^2)^{1/2}$$
$$= T.$$

Hence, the expression has the dimensions of time, confirming that the formula is dimensionally consistent. □

1.1 The Buckingham Pi Theorem

As we have mentioned before, in the study of physical phenomena, it is often desirable to reduce the number of variables governing a system by expressing them in terms of dimensionless parameters. This process, known as *dimensional analysis*, provides deep insight into the underlying physics and often reveals fundamental similarity laws between apparently different systems. Among the most powerful tools in dimensional analysis is the *Buckingham Pi Theorem*, which establishes a systematic procedure for forming independent dimensionless combinations of the variables involved in a physical problem. The theorem not only simplifies experimental and theoretical investigations but also aids in scaling laws, model testing, and data correlation. Before presenting the formal statement and proof of the theorem, we summarize the essential requirements that must be satisfied for its valid application.

In order to apply the Buckingham Pi Theorem correctly to a physical system,

$$f(q_1, q_2, \ldots, q_n) = 0, \tag{1.5}$$

several fundamental requirements must be satisfied. These ensure that the resulting dimensionless parameters (the π–groups) meaningfully describe the physics of the problem.

1. Dimensional Homogeneity

The physical law must be *dimensionally homogeneous*, meaning that every additive term in the equation has the same dimensions.

2. Independence of Variables

The variables $q_1, q_2, \ldots, q_n$ must be *independent physical quantities*. No one variable should be an explicit function of another; otherwise, the number of independent dimensionless parameters would be incorrectly counted.

3. Finite Number of Fundamental Dimensions

The system must involve only a finite and countable number k of independent fundamental dimensions such as mass (M), length (L), time (T), temperature (Θ), electric current (I), etc. These form a basis for expressing the dimensions of all the physical quantities in the problem.

4. Completeness of the Variable Set

All physical quantities that significantly affect the phenomenon must be included among the variables. If any important variable is omitted, the resulting dimensionless groups will not completely represent the behavior of the system.

5. Dimensional Expressibility

Each variable q_i must be expressible as a product of powers of the fundamental dimensions, i.e.,

$$[q_i] = M^{a_i} L^{b_i} T^{c_i} \Theta^{d_i} \cdots,$$

where $a_i, b_i, c_i, d_i, \ldots$ are real numbers. This allows the construction of the dimensional matrix used in the theorem.

6. Existence of a Functional Relationship

A continuous and single–valued functional relationship must exist among the variables,

$$f(q_1, q_2, \ldots, q_n) = 0 \quad \text{or equivalently} \quad q_n = \phi(q_1, q_2, \ldots, q_{n-1}),$$

ensuring that the variables can be grouped into dimensionless combinations.

These requirements ensure that the Buckingham Pi Theorem can be properly applied to yield $n-k$ independent dimensionless parameters that fully characterize the physical system.

Theorem 1.1 (Buckingham's Pi Theorem) *Suppose a physically meaningful equation involves n physical variables,*

$$f(q_1, q_2, \ldots, q_n) = 0,$$

and these variables involve k independent physical dimensions (such as mass, length, time, etc.). Then, the equation can be equivalently expressed as a relationship among

$$r = n - k$$

dimensionless products, called π-terms, *namely*

$$F(\pi_1, \pi_2, \ldots, \pi_r) = 0,$$

where each π_i is a dimensionless monomial, that is, a product of the original variables raised to certain real powers:

$$\pi_i = q_1^{a_{i1}} q_2^{a_{i2}} \cdots q_n^{a_{in}}, \quad i = 1, 2, \ldots, r.$$

Proof *Below, we provide a step-by-step outline of the proof. Let the k fundamental physical dimensions be denoted symbolically by $D_1, D_2, \ldots, D_k$ (for example, L, M, T, etc.). Each variable q_i has dimensions that can be expressed as a product of powers of these fundamentals:*

$$[q_i] = D_1^{a_{i1}} D_2^{a_{i2}} \cdots D_k^{a_{ik}}.$$

The exponents a_{ij} are real numbers (often integers or rationals).

Next we construct the dimensional matrix. To do so, we collect the exponents a_{ij} into an $n \times k$ dimensional matrix:

$$A = \begin{bmatrix} a_{11} & a_{12} & \cdots & a_{1k} \\ a_{21} & a_{22} & \cdots & a_{2k} \\ \vdots & \vdots & \ddots & \vdots \\ a_{n1} & a_{n2} & \cdots & a_{nk} \end{bmatrix}.$$

Each row corresponds to one of the physical variables q_i, and each column corresponds to one of the k fundamental dimensions. Consider a general monomial formed by the n variables:

$$\pi = q_1^{\alpha_1} q_2^{\alpha_2} \cdots q_n^{\alpha_n},$$

where $\alpha_1, \alpha_2, \ldots, \alpha_n$ are real exponents.

The dimensions of π are given by

$$[\pi] = [q_1]^{\alpha_1} [q_2]^{\alpha_2} \cdots [q_n]^{\alpha_n} = D_1^{b_1} D_2^{b_2} \cdots D_k^{b_k},$$

where each b_j is the total exponent of D_j in π, computed as

$$b_j = a_{1j}\alpha_1 + a_{2j}\alpha_2 + \cdots + a_{nj}\alpha_n.$$

The product π is dimensionless *if and only if*

$$b_j = 0 \quad \text{for all } j = 1, 2, \ldots, k.$$

In vector form, let $\alpha = (\alpha_1, \alpha_2, \ldots, \alpha_n)^T$. Then the above k equations can be written compactly as

$$A^T \alpha = 0.$$

Hence, the exponents that form dimensionless monomials correspond exactly to the vectors α in the null space of A^T. Recall that the null space of a matrix M is defined by

$$Null(M) = \{x \in \mathbb{R}^k : Mx = 0\},$$

that is, the set of all vectors that are mapped to the zero vector by M.

By the Rank–Nullity Theorem from linear algebra, we have

$$\dim(Null(A^T)) = n - \text{rank}(A^T).$$

Thus, there exist r linearly independent solutions $\alpha^{(1)}, \alpha^{(2)}, \ldots, \alpha^{(r)}$ *to* $A^T\alpha = 0$. *To construct independent* π*-terms, we notice that each solution vector* $\alpha^{(i)} = (\alpha_1^{(i)}, \ldots, \alpha_n^{(i)})^T$ *defines a distinct dimensionless monomial*

$$\pi_i = q_1^{\alpha_1^{(i)}} q_2^{\alpha_2^{(i)}} \cdots q_n^{\alpha_n^{(i)}}, \qquad i = 1, 2, \ldots, r.$$

Because the $\alpha^{(i)}$ *are linearly independent, the corresponding* π_i *are also independent (none can be written as a product or power of the others).*

Finally, the original dimensional equation

$$f(q_1, q_2, \ldots, q_n) = 0$$

must be valid in any system of units. If we change the fundamental units (for example, rescale L, M, or T by arbitrary factors), the numerical values of the variables q_i *change, but the functional relationship among dimensionless quantities must remain the same.*

Hence, the physical content of the relationship is entirely captured by dimensionless combinations of the variables, i.e., by $\pi_1, \pi_2, \ldots, \pi_r$. *Therefore, the original equation can be expressed equivalently as*

$$F(\pi_1, \pi_2, \ldots, \pi_r) = 0.$$

This completes the proof of the Buckingham Pi Theorem. □

We note that in some parts of the book, the **Buckingham Pi Theorem** is referred to simply as the **Pi Theorem.** We hope that this abbreviated terminology will not cause any confusion.

Example 1.3 *Consider the drag force D experienced by a sphere of characteristic length L moving through an incompressible viscous fluid at speed U. We assume the drag depends on the (listed) variables:*

$$D = f(\rho, \mu, L, U),$$

or, equivalently, a relation among the five variables

$$F(D, \rho, \mu, L, U) = 0,$$

where

$$[D] = MLT^{-2}, \quad [\rho] = ML^{-3}, \quad [\mu] = ML^{-1}T^{-1}, \quad [L] = L, \quad [U] = LT^{-1}.$$

Therefore, we have $n = 5$ variables and $k = 3$ fundamental dimensions (M, L, T), so by Buckingham we expect $n - k = 2$ independent dimensionless groups (two independent π-numbers). Below, we show two full derivations that produce the standard choices:

$$\pi_1 = \frac{D}{\rho U^2 L^2} \quad \text{(drag coefficient)}, \qquad \pi_2 = \frac{\rho U L}{\mu} \quad \text{(Reynolds number)}.$$

The dimension matrix A has rows corresponding to M, L, T, and columns to the variables:

$$\begin{aligned}
[D] &= M^1 L^1 T^{-2} &&\longrightarrow (1, 1, -2)^\top, \\
[\rho] &= M^1 L^{-3} T^0 &&\longrightarrow (1, -3, 0)^\top, \\
[\mu] &= M^1 L^{-1} T^{-1} &&\longrightarrow (1, -1, -1)^\top, \\
[L] &= M^0 L^1 T^0 &&\longrightarrow (0, 1, 0)^\top, \\
[U] &= M^0 L^1 T^{-1} &&\longrightarrow (0, 1, -1)^\top.
\end{aligned}$$

Thus (writing rows as M, L, T), the 3×5 matrix A is

$$A = \begin{matrix} \\ M \\ L \\ T \end{matrix} \begin{matrix} \begin{matrix} D & \rho & \mu & L & U \end{matrix} \\ \begin{bmatrix} 1 & 1 & 1 & 0 & 0 \\ 1 & -3 & -1 & 1 & 1 \\ -2 & 0 & -1 & 0 & -1 \end{bmatrix} \end{matrix}$$

A dimensionless monomial (product)

$$\pi = D^{\alpha_1} \rho^{\alpha_2} \mu^{\alpha_3} L^{\alpha_4} U^{\alpha_5}.$$

is dimensionless iff

$$A\alpha = 0, \qquad \text{where } \alpha = (\alpha_1, \ldots, \alpha_5)^\top.$$

We solve the linear homogeneous system $A\alpha = 0$. The rank of A is 3 (which equals the number of fundamental dimensions), hence the nullspace has dimension $5 - 3 = 2$ and we will obtain two linearly independent solution vectors by solving the system of equations

$$\begin{aligned}
\alpha_1 + \alpha_2 + \alpha_3 &= 0, \\
\alpha_1 - 3\alpha_2 - \alpha_3 + \alpha_4 + \alpha_5 &= 0, \\
-2\alpha_1 - \alpha_3 - \alpha_5 &= 0.
\end{aligned}$$

In the first equation, solve for α_2 in terms of α_1 and α_3 to obtain

$$\alpha_2 = -\alpha_1 - \alpha_3.$$

Next we substitute α_2 into the second equation of the original system, and after some simplifications, we arrive at

$$4\alpha_1 + 2\alpha_3 + \alpha_4 + \alpha_5 = 0. \tag{1.6}$$

Now we solve for α_5 in terms of α_1 and α_3 in the third equation of our original system and obtain

$$\alpha_5 = -2\alpha_1 - \alpha_3.$$

Then, substituting this α_5 into (1.6) *gives*

$$\alpha_4 = -2\alpha_1 - \alpha_3.$$

Collecting all variables in terms of α_1 and α_3 we have

$$\alpha_2 = -\alpha_1 - \alpha_3,$$
$$\alpha_4 = -2\alpha_1 - \alpha_3,$$
$$\alpha_5 = -2\alpha_1 - \alpha_3.$$

In vector notation, the above system is equivalent to

$$\begin{bmatrix} \alpha_1 \\ \alpha_2 \\ \alpha_3 \\ \alpha_4 \\ \alpha_5 \end{bmatrix} = \alpha_1 \begin{bmatrix} 1 \\ -1 \\ 0 \\ -2 \\ -2 \end{bmatrix} + \alpha_3 \begin{bmatrix} 0 \\ -1 \\ 1 \\ -1 \\ -1 \end{bmatrix}$$

Each α_j defines a dimensionless product. For example, if we set $\alpha_1 = -\frac{1}{2}$ and $\alpha_3 = 1$, we obtain $\alpha_2 = -\frac{1}{2}$, and $\alpha_4 = \alpha_5 = 0$. Similarly, if we set $\alpha_1 = -\frac{1}{2}$ and $\alpha_3 = 0$, we get $\alpha_2 = \frac{1}{2}$, and $\alpha_4 = \alpha_5 = 1$. Consequently, we have

$$\pi_1 = D^{-1/2}\rho^{-1/2}\mu^1 = \frac{\mu}{\sqrt{D\rho}}, \quad \pi_2 = D^{-1/2}\rho^{1/2}LU = \frac{\sqrt{\rho}\,LU}{\sqrt{D}}.$$

These two π's are valid dimensionless combinations but are somewhat 'non-standard' in appearance (they come from the particular basis chosen). We may replace them with any other two independent dimensionless combinations (spanning the same nullspace). In particular, the classical choices are obtained by taking different basis vectors. For example, by choosing $\alpha_1 = 1$, $\alpha_3 = 0$, and $\alpha_1 = 0$, $\alpha_3 = -1$, we arrive at convenient integer-exponent linearly independent vectors

$$\begin{bmatrix} \alpha_1 \\ \alpha_2 \\ \alpha_3 \\ \alpha_4 \\ \alpha_5 \end{bmatrix} = \begin{bmatrix} 1 \\ -1 \\ 0 \\ -2 \\ -2 \end{bmatrix}, \quad \begin{bmatrix} \alpha_1 \\ \alpha_2 \\ \alpha_3 \\ \alpha_4 \\ \alpha_5 \end{bmatrix} = \begin{bmatrix} 0 \\ 1 \\ -1 \\ 1 \\ 1 \end{bmatrix},$$

respectively. Under this scenario we get

$$\pi_1 = D\rho^{-1}L^{-2}U^{-2} = \frac{D}{\rho U^2 L^2}, \qquad \pi_2 = \rho\mu^{-1}LU = \frac{\rho UL}{\mu}.$$

Thus, the nullspace approach shows there are exactly two independent π-groups, and we may choose them to be the (dimensionless) drag coefficient and the Reynolds number:

$$\pi_1 = \frac{D}{\rho U^2 L^2}, \qquad \pi_2 = \frac{\rho U L}{\mu}.$$

Buckingham then implies that the functional relation among the original variables can be written as

$$\Phi(\pi_1, \pi_2) = 0$$

if and only if

$$\frac{D}{\rho U^2 L^2} = \Psi(\mathrm{Re}), \quad \textit{where } \mathrm{Re} = \frac{\rho U L}{\mu}.$$

In conclusion, from either method, we obtain two independent dimensionless groups, and hence the original physical relation can be written as

$$\Phi\left(\frac{D}{\rho U^2 L^2}, \frac{\rho U L}{\mu}\right) = 0,$$

or, equivalently,

$$\frac{D}{\rho U^2 L^2} = \Psi(\mathrm{Re}), \qquad \mathrm{Re} = \frac{\rho U L}{\mu}.$$

Interpretation: the (nondimensional) drag coefficient $C_D = \frac{D}{\rho U^2 L^2}$ is a (universal) function of the Reynolds number Re *alone (for this set of assumptions, such as incompressible flow, same body shape, etc.). Buckingham's theorem guarantees the reduction of variables; experiments or theory determine Ψ.* □

Remark 2
- *The linear algebra approach exhibits the nullspace of the dimension matrix; any two independent nullspace vectors give two valid π-groups. The repeating-variables method is a constructive way in obtaining a convenient set of π-groups (typically those with integer exponents and standard physical meaning).*
- *One may choose different repeating variables (e.g., μ, U, L or ρ, U, L), but the resulting set of π-groups will span the same nullspace (and hence be algebraically equivalent).*
- *Buckingham's theorem does not determine the function Ψ; it only reduces the number of independent variables from five to two in this example.*

We provide another application in the form of an example.

1.1.1 First Detonation of an Atomic Bomb

The first detonation of an atomic bomb occurred during the *Trinity Test* in New Mexico in 1945. Several years later, a sequence of photographs capturing the explosion was released to the public, each accompanied by a time stamp and a reference scale indicating size. Using only these published images, the British physicist Geoffrey Ingram Taylor (1886–1975) remarkably succeeded in estimating the total energy released by the explosion, which was still classified at the time. The natural question arises: *how can such photographs be used to deduce the magnitude of the energy released?* Taylor approached this question using the principles of *dimensional analysis*. He reasoned that the radius R of the expanding blast wave depends primarily on three quantities: the total energy released E, the elapsed time t since detonation, and the density of the surrounding air ρ. Assuming that the effects of gravity and other physical parameters are negligible in the early stages of the explosion, the relationship among these quantities must be expressible in the form

$$R = C\left(\frac{Et^2}{\rho}\right)^{1/5},$$

where C is a dimensionless constant determined experimentally or theoretically. From this expression, the total energy E can be inferred once $R(t)$ is measured from the photographs. By plotting the radius of the fireball against time and fitting the data to the above relation, Taylor was able to estimate the explosive yield of the Trinity device with remarkable accuracy, demonstrating the extraordinary power of dimensional reasoning in physics.

Below we provide the details. For our purposes, it will be sufficient to consider only three: time (T), length (L), and mass (M). For instance, density (ρ) has dimension mass over cubic length, and energy (E) has dimension mass times squared length divided by squared time. The atomic explosion can be represented as:

$$[R] = L, \quad [t] = T, \quad [\rho] = \frac{M}{L^3} = ML^{-3}, \quad [E] = \frac{ML^2}{T^2} = ML^2T^{-2},$$

since $E = \frac{1}{2}mv^2$. It is important to note that although different systems of units can be employed, the physical dimensions remain the same. As we have done in the previous example, the matrix A is given by

$$A = \begin{array}{c} \\ M \\ L \\ T \end{array} \begin{array}{c} \begin{matrix} t & R & \rho & E \end{matrix} \\ \begin{bmatrix} 0 & 0 & 1 & 1 \\ 0 & 1 & -3 & 2 \\ 1 & 0 & 0 & -2 \end{bmatrix} \end{array}$$

A dimensionless monomial (product)

$$\pi = t^{\alpha_1} R^{\alpha_2} \rho^{\alpha_3} E^{\alpha_4} \tag{1.7}$$

is dimensionless if, and only if,

$$A\alpha = 0 \text{ where } \alpha = (\alpha_1, \ldots, \alpha_4)^\top.$$

We solve the linear homogeneous system $A\alpha = 0$. The rank of A is 3 (which equals the number of fundamental dimensions), hence the nullspace has dimension $4-3=1$, and we will have only one π quantity. We have the system of equations

$$\alpha_3 + \alpha_4 = 0,$$

$$\alpha_2 - 3\alpha_3 + 2\alpha_4 = 0,$$

$$\alpha_1 - 2\alpha_4 = 0.$$

Before we embark on solving the system, it is worth mentioning that solving $A\alpha = 0$ is equivalent to the following. Since the quantity π is dimensionless, then we must have from (1.7) that

$$\begin{aligned} 1 &= [\pi] = [t^{\alpha_1} R^{\alpha_2} \rho^{\alpha_3} E^{\alpha_4}] \\ &= T^{\alpha_1} L^{\alpha_2} (ML^{-3})^{\alpha_3} (ML^2T^{-2})^{\alpha_4} \\ &= T^{\alpha_1 - 2\alpha_4} L^{\alpha_2 - 3\alpha_3 + 2\alpha_4} M^{\alpha_3 + \alpha_4}. \end{aligned}$$

Setting each exponent equal to zero, we obtain the same system of equations. We now proceed to solve the system of equations. Solving the first equation for α_3 in terms of α_4 and substituting into the second equation yields an equation for α_2, from which we obtain

$$\begin{bmatrix} \alpha_1 \\ \alpha_2 \\ \alpha_3 \\ \alpha_4 \end{bmatrix} = \alpha_4 \begin{bmatrix} 2 \\ -5 \\ -1 \\ 1 \end{bmatrix}$$

Setting $\alpha_4 = -1$, we obtain

$$\alpha_1 = -2, \quad \alpha_2 = 5, \quad \alpha_3 = 1.$$

Thus,

$$\pi = t^{-2} R^5 \rho E^{-1} = \frac{R^5 \rho}{t^2 E}.$$

By the Pi Theorem, the physical law is equivalent to a physical law involving only the dimension quantities; in this case

$$\Phi(\pi) = \Phi(\frac{R^5\rho}{t^2 E}) = 0,$$

where Φ is some function of a single variable. Because π is the only independent dimensionless quantity, this implies that

$$\frac{R^5 \rho}{t^2 E} = C,$$

or

$$R = C \left(\frac{Et^2}{\rho} \right)^{1/5},$$

where C is a dimensionless constant determined experimentally or theoretically. Surprisingly, the empirical values of the dimensionless constant have been found to lie in the range

$$0.9 \leq C \leq 1.1,$$

which makes it reasonable to assume $C = 1$ for practical purposes. Let us now estimate the energy released during the Trinity test. From the photos of the explosion, it can be observed that approximately 0.025 s after the explosion, the radius of the fireball was

$$R = 130 \text{ m}.$$

Furthermore, the density of air at 20°C is

$$\rho = 1.20 \text{ kg/m}^3.$$

Substituting these values into the derived formula yields an estimated energy

$$E = 7.13 \times 10^{13} \text{ J},$$

which corresponds to approximately 17.1 kilotons of TNT, which is remarkably close to Taylor's original result. In summary, we have obtained a reliable estimate of the energy released in a nuclear explosion without invoking any details of nuclear physics. Rather, this result emerged from insightful dimensional reasoning, a few simplifying assumptions, and the powerful guidance of the Buckingham Pi theorem. The usefulness of the Buckingham Pi theorem extends far beyond this simple illustration. In more advanced applications, it serves as an elegant and indispensable tool for analyzing thermodynamic, fluid dynamic, and gravitational systems. Often underappreciated, the theorem highlights profound links between mathematics and physics, revealing the underlying symmetries that govern the natural world.

1.1.2 Dimensional Analysis of Surface Waves Generated by a Raindrop

A small raindrop lands in a puddle, causing waves to propagate outward from the point where it lands. We wish to investigate the variables that affect the speed at which these waves propagate. Assume that the wave speed V depends on the following variables:

$$V, \rho, \lambda, h, \sigma,$$

where

- ρ is the density of the fluid,
- λ is the wavelength (distance between two successive crests),
- h is the wave height (amplitude),
- σ is the surface tension of the fluid.

Our goal is to determine the dimensionless groups that describe this problem.

The variables have the following dimensional representation,

$$[V] = LT^{-1}, \quad [\rho] = ML^{-3}, \quad [\lambda] = L, \quad [h] = L, \quad [\sigma] = MT^{-2}.$$

There are $n = 5$ variables and $k = 3$ fundamental dimensions (M, L, T), so according to Buckingham's Pi Theorem, we expect

$$n - k = 5 - 3 = 2$$

dimensionless groups. The repeated variables are ρ, λ, and σ, since they involve all three fundamental dimensions and are independent of one another. Let the first π–group involve V:

$$\pi_1 = V\rho^a\lambda^b\sigma^c.$$

Substituting the dimensions, we obtain

$$\begin{aligned} 1 &= [V\rho^a\lambda^b\sigma^c] \\ &= LT^{-1}(ML^{-3})^a L^b (MT^{-2})^c \\ &= M^{a+c}L^{1-3a+b}T^{-1-2c}. \end{aligned}$$

For π_1 to be dimensionless, the exponents of M, L, and T must vanish:

$$\begin{cases} a + c = 0, \\ 1 - 3a + b = 0, \\ -1 - 2c = 0. \end{cases}$$

Solving these equations gives

$$c = -\tfrac{1}{2}, \quad a = \tfrac{1}{2}, \quad b = \tfrac{1}{2}.$$

Therefore,

$$\pi_1 = V\rho^{1/2}\lambda^{1/2}\sigma^{-1/2} = V\sqrt{\frac{\rho\lambda}{\sigma}}.$$

For the formulation of the second π–group we involve h so that

$$\pi_2 = h\rho^\alpha\lambda^\beta\sigma^\gamma.$$

$$\begin{aligned} 1 &= [h\rho^\alpha\lambda^\beta\sigma^\gamma] \\ &= L(ML^{-3})^\alpha L^\beta (MT^{-2})^\gamma \\ &= M^{\alpha+\gamma}L^{1-3\alpha+\beta}T^{-2\gamma}. \end{aligned}$$

Setting the exponents of M, L, and T equal to zero:

$$\begin{cases} \alpha + \gamma = 0, \\ 1 - 3\alpha + \beta = 0, \\ -2\gamma = 0. \end{cases}$$

From the third equation, $\gamma = 0$, hence $\alpha = 0$, and $\beta = -1$. Thus

$$\pi_2 = \frac{h}{\lambda}.$$

By the Buckingham Pi Theorem, the physical relationship can be expressed as

$$\pi_1 = f(\pi_2),$$

that is,

$$V\sqrt{\frac{\rho\lambda}{\sigma}} = f\left(\frac{h}{\lambda}\right).$$

Equivalently,

$$V = \sqrt{\frac{\sigma}{\rho\lambda}}\,F\left(\frac{h}{\lambda}\right),$$

where F is a dimensionless function. The term $\sqrt{\dfrac{\sigma}{\rho\lambda}}$ represents the characteristic speed of capillary waves (balance between surface tension and inertia). The ratio $\dfrac{h}{\lambda}$ accounts for nonlinear or finite–amplitude effects. Alternatively,

$$\pi_1^2 = \frac{\rho V^2 \lambda}{\sigma}$$

is the **Weber number**, so the functional relation may be written as

$$\mathrm{We} = G\left(\frac{h}{\lambda}\right).$$

The **Weber number** (We) is a dimensionless quantity that expresses how surface tension effects depend on the relative length scale $\frac{h}{\lambda}$.

1.1.3 Mach Number

Example 1.4 *Consider the resistance to motion of a sphere of diameter D moving at velocity V in a compressible liquid of density ρ and bulk modulus of compression E. We wish to show that the dimensionless relationship between these quantities is a function of the Mach number. The given variables are*

$$R,\ D,\ V,\ \rho,\ E,$$

where R is the resistance (force) on the sphere, D its diameter, V the velocity, ρ the density of the liquid, and E the bulk modulus of compression (i.e., pressure). Thus, the dimensions are

$$[R] = MLT^{-2}, \quad [D] = L, \quad [V] = LT^{-1}, \quad [\rho] = ML^{-3}, \quad [E] = ML^{-1}T^{-2}.$$

The number of variables is $n = 5$ and the number of fundamental dimensions is $k = 3$. Hence, the number of independent dimensionless π–groups is

$$n - k = 5 - 3 = 2.$$

We choose the repeating variables ρ, V, and D, since they contain all three fundamental dimensions and are independent. Assume

$$\pi_1 = R\rho^a V^b D^c.$$

The overall dimensions are

$$\begin{aligned}[R]\,[\rho]^a[V]^b[D]^c &= (MLT^{-2})(ML^{-3})^a(LT^{-1})^b(L)^c \\ &= M^{1+a}L^{1-3a+b+c}T^{-2-b}.\end{aligned}$$

To make π_1 dimensionless, we equate the exponents of M, L, and T to zero. That is

$$\begin{cases} 1 + a = 0, \\ 1 - 3a + b + c = 0, \\ -2 - b = 0. \end{cases}$$

Solving gives

$$a = -1, \qquad b = -2, \qquad c = -2.$$

Hence

$$\pi_1 = R\rho^{-1}V^{-2}D^{-2},$$

or

$$\pi_1 = \frac{R}{\rho V^2 D^2}.$$

This represents a drag coefficient (up to a constant factor). Next, we construct $\boldsymbol{\pi_2}$ *containing E. That is, we assume*

$$\pi_2 = E\rho^\alpha V^\beta D^\gamma.$$

Compute the dimensions:

$$\begin{aligned}[E]\,[\rho]^\alpha[V]^\beta[D]^\gamma &= (ML^{-1}T^{-2})(ML^{-3})^\alpha(LT^{-1})^\beta(L)^\gamma \\ &= M^{1+\alpha}L^{-1-3\alpha+\beta+\gamma}T^{-2-\beta}.\end{aligned}$$

Setting exponents to zero:

$$\begin{cases} 1+\alpha=0, \\ -1-3\alpha+\beta+\gamma=0, \\ -2-\beta=0. \end{cases}$$

Solving yields

$$\alpha=-1, \qquad \beta=-2, \qquad \gamma=0.$$

Therefore

$$\pi_2 = E\rho^{-1}V^{-2} = \frac{E}{\rho V^2}.$$

By Buckingham's Pi Theorem, the variables are related as

$$\frac{R}{\rho V^2 D^2} = f\left(\frac{E}{\rho V^2}\right),$$

where f is a dimensionless function. The speed of sound in the compressible liquid is given by

$$c = \sqrt{\frac{E}{\rho}}.$$

Hence,

$$\frac{E}{\rho V^2} = \frac{c^2}{V^2} = \frac{1}{\mathrm{Ma}^2},$$

where

$$\mathrm{Ma} = \frac{V}{c}$$

is the Mach number.

Thus, the dimensionless relationship becomes

$$C_D \equiv \frac{R}{\rho V^2 D^2} = F(\mathrm{Ma}),$$

that is, the dimensionless resistance (or drag coefficient) depends only on the Mach number. □

In the next section, we offer an application related to general relativity.

1.1.4 Dimensionless Parameters in Relativistic Motion

Consider a particle of mass m moving at speed v under the laws of special relativity. The key quantities that may influence its relativistic energy E are

$$E = f(m, v, c),$$

where c is the speed of light. The dimensions of the quantities are

$$[E] = ML^2T^{-2}, \quad [m] = M, \quad [v] = LT^{-1}, \quad [c] = LT^{-1}.$$

Thus, the number of variables is $n = 4$, and the number of independent dimensions is $k = 3$ (M, L, T). By Buckingham's Pi theorem, there should be $n - k = 1$ dimensionless group.

Let us assume a general dimensionless π-term of the form

$$\pi_1 = E^a m^b v^e c^d,$$

from which we have

$$\pi_1 = (ML^2T^{-2})^a M^b (LT^{-1})^e (LT^{-1})^d.$$

For dimensional consistency:

$$M^{a+b} L^{2a+e+d} T^{-2a-e-d} = 1.$$

Equating the exponents of M, L, and T to zero gives

$$\begin{cases} a + b = 0, \\ 2a + e + d = 0, \\ -2a - e - d = 0. \end{cases}$$

The last two equations are identical, implying one degree of freedom. Solving gives

$$b = -a, \quad e = -2a - d.$$

Choosing $a = 1$ and $d = 0$, we get

$$\pi_1 = \frac{E}{mv^2}.$$

Choosing instead $d = -2a$, we obtain the dimensionless group

$$\pi_1 = \frac{E}{mc^2}.$$

This dimensionless ratio expresses the relativistic energy of the particle relative to its rest energy. In special relativity,

$$E = \gamma mc^2 \quad \text{where } \gamma = \frac{1}{\sqrt{1 - v^2/c^2}}.$$

Hence,

$$\pi_1 = \gamma,$$

the *Lorentz factor*, which emerges directly from the dimensional structure of the theory.

Example 1.5 *(Dimensionless group in gravitational systems)*

Let us apply the Buckingham π-Theorem to a relativistic gravitational system. Suppose we wish to understand how the gravitational field around a massive object depends on its physical parameters. For a static, spherically symmetric field, the relevant variables are

$$r,\ m,\ G,\ c,$$

where

- *r is the radial distance from the center of the gravitating body,*
- *m is the mass of the central object,*
- *G is Newton's gravitational constant,*
- *c is the speed of light in vacuum.*

The total number of variables is $n = 4$, and the number of fundamental dimensions (M, L, T) is $k = 3$. Hence, the Buckingham π-Theorem predicts $n - k = 1$ dimensionless group.

We look for a dimensionless combination of the form

$$\pi_1 = G^a m^b c^d r^e.$$

The dimensions of the variables are

$$[G] = L^3 M^{-1} T^{-2}, \qquad [c] = LT^{-1}, \qquad [m] = M, \qquad [r] = L.$$

Thus,

$$[G^a m^b c^d r^e] = L^{3a+d+e} M^{-a+b} T^{-2a-d}.$$

For π_1 to be dimensionless, we require

$$\begin{cases} 3a + d + e - 0, \\ -a + b = 0, \\ -2a - d = 0. \end{cases}$$

Solving,

$$b = a, \qquad d = -2a, \qquad e = -a,$$

and therefore, up to a constant power,

$$\pi_1 = \frac{Gm}{rc^2}.$$

This dimensionless parameter measures the strength of the gravitational field in relativistic units.

If

$$\frac{Gm}{rc^2} \ll 1,$$

the field is weak, and Newtonian gravity is a good approximation. If

$$\frac{Gm}{rc^2} \sim 1,$$

relativistic effects become important. In general relativity, this occurs near the Schwarzschild radius

$$r_s = \frac{2Gm}{c^2},$$

which corresponds to the event horizon of a non-rotating black hole.

The quantity

$$\frac{Gm}{rc^2}$$

can be interpreted as the ratio of a characteristic gravitational energy scale Gm^2/r to the relativistic rest energy mc^2. Thus, the Buckingham π-Theorem identifies the natural small parameter that determines when relativistic corrections to Newtonian gravity become significant. □

1.1.5 Dimensions of Useful Quantities in Mechanics

Below, is a list of the dimensions of useful quantities in mechanics. All come from the definition of the relevant quantity.

$$\begin{aligned}
[\text{mass}] &= M,\\
[\text{length}] &= L,\\
[\text{time}] &= T,\\
[\text{area}] &= [\text{length} \times \text{length}] = L^2,\\
[\text{volume}] &= [\text{area} \times \text{length}] = L^3,\\
[\text{density}] &= [\text{mass}/\text{volume}] = \frac{M}{L^3} = ML^{-3},\\
[\text{velocity}] &= [\text{length}/\text{time}] = \frac{L}{T} = LT^{-1},\\
[\text{acceleration}] &= [\text{velocity}/\text{time}] = \frac{LT^{-1}}{T} = LT^{-2},\\
[\text{force}] &= [\text{mass} \times \text{acceleration}] = MLT^{-2},\\
[\text{moment of force, torque}] &= [MLT^{-2} \times L] = ML^2T^{-2},\\
[\text{impulse}] &= [\text{force} \times \text{time}] = MLT^{-2} \times T = MLT^{-1},\\
[\text{momentum}] &= [\text{mass} \times \text{velocity}] = MLT^{-1},\\
[\text{work}] &= [\text{force} \times \text{distance}] = ML^2T^{-2},
\end{aligned}$$

$$[\text{kinetic energy}] = [\text{mass} \times (\text{velocity})^2] = ML^2T^{-2},$$
$$[\text{power}] = [\text{work/time}] = ML^2T^{-3}$$
$$[\text{pressure}] = [\text{force/unit area}] = ML^{-1}T^{-2}.$$

1.2 Exercises

Exercise 1.1 *Use the method of dimensions to check whether the following formula is dimensionally valid. Here t, m, l, F represent, time, mass, length, and tension, respectively.*

$$t = 2\pi\sqrt{\frac{ml}{F}}.$$

Exercise 1.2 *Use the method of dimensions to check whether the following formulae are dimensionally valid, where v_1 and v_2 are velocities, g is an acceleration, s is a distance, and t is a time.*

(a) $v_2^2 = v_1^2 + 2gs$

(b) $v_2 = v_1 - gt$

(c) $s = 0.5gt^2 + v_1t$

(d) $t = \sqrt{\frac{2s}{g}}$

Exercise 1.3 *When a spring is subjected to a tension force T, the resulting extension x is related to T according to* **Hooke's Law***:*

$$T = kx$$

where k denotes the spring constant *or* stiffness *of the spring. Determine the dimension of k and verify the dimensional correctness of the formula*

$$t = 2\pi\sqrt{\frac{m}{k}}$$

for the period t of oscillation of a mass m attached to a spring of stiffness k.

Exercise 1.4 *Suppose that a sphere of radius r moves at a slow speed v through a fluid with a coefficient of viscosity μ and density ρ. Find the way in which the resistive force acting on the sphere should depend on these three quantities in two different ways:*

(a) By omitting ρ from the list of variables.

(b) By omitting μ from the list of variables.

Ans: (a) $F \propto r^a v^b \mu^c$, (b) $F \propto \rho v^2 r^2$.

Exercise 1.5 *A wave of wavelength λ propagates along the surface of a fluid characterized by density ρ and surface tension σ. Taking into account the acceleration due to gravity g, determine possible expressions for the speed of the wave v as a function of λ by using the method of dimensional analysis in the following scenarios:*

(a) Exclude the density ρ from the set of variables and express v in terms of the remaining quantities.

(b) Exclude the gravitational acceleration g from the set of variables and express v in terms of the remaining quantities.

Ans: (a) v^2 *is proportional to* $g\lambda$, (b) v^2 *is proportional to* $\sigma/(\lambda\rho)$.

Exercise 1.6 *(1D heat conduction from a point source) Consider the temperature u in an "infinite" one-dimensional bar initially heated by a point source. We assume the temperature is determined by the following functional dependence:*

$$u = f(W_1, W_2, W_3, W_4, W_5, W_6),$$

where the variables are

$$\begin{aligned} W_1 &= x && \text{(distance along the bar from the point source)},\\ W_2 &= t && \text{(elapsed time after initial heating)},\\ W_3 &= \rho && \text{(mass density of the bar)},\\ W_4 &= c && \text{(specific heat of the bar)},\\ W_5 &= K && \text{(thermal conductivity of the bar)},\\ W_6 &= Q && \text{(total heat point at input (length units)}^2. \end{aligned}$$

Use The Pi Theorem to show there are two dimensionless quantities given by

$$\pi_1 = \frac{\rho c x^2}{Kt}, \qquad \pi_2 = \frac{u \rho c x}{Q}.$$

Exercise 1.7 *In the blast wave scenario discussed in Example 1.1.1 suppose that we consider a modified physical relation of the form*

$$f(t, R, \rho, E, P) = 0,$$

where P denotes the ambient atmospheric pressure. Determine two independent dimensionless quantities that can be constructed using the variables t, R, ρ, E, and P. Let these two dimensionless parameters be represented by π_1 and π_2. Assuming that the physical relation can be expressed in the equivalent form

$$\pi_1 = \phi(\pi_2),$$

investigate whether it still implies that the radius R depends on the time t according to a two-fifths power law.

Exercise 1.8 *(Height of a projectile) Consider the question of how high a baseball can be thrown straight upward. We wish to determine the factors influencing the maximum height h. It is reasonable to assume that h depends on the mass m of the ball, the acceleration due to gravity g, and the initial launch velocity v. For simplicity, we neglect air resistance, as it introduces additional complexity beyond the present scope. Using dimensional analysis, show that*

$$h = C\frac{v^2}{g},$$

for a dimensionless constant C.

Exercise 1.9 *Let us examine the problem of determining the speed of sound in a compressible fluid. The velocity c of sound in such a medium is influenced by the bulk modulus K of the material and its density ρ. Hence, the number of relevant physical quantities is $n = 3$. Note that the dimensions of bulk modulus $[K]$ are $ML^{-1}T^{-2}$, which correspond to pressure. Show that*

$$c = C\sqrt{\frac{K}{\rho}},$$

for a dimensionless constant C.

Exercise 1.10 *In this problem, we aim to determine the resonance frequency w of a freely suspended membrane. It is assumed that w depends on the air density ρ, the speed of sound c, the membrane tension T, the surface density σ of the membrane, and the diameter a of the membrane. Show that*

$$\pi_1 = \frac{wT}{\rho c^3}, \quad \pi_2 = \frac{w\sigma}{\rho c}, \quad \pi_3 = \frac{wa}{c}.$$

Exercise 1.11 *In this exercise, we find a formula for the period T of a simple pendulum. A simple pendulum has a period T that depends on the length l, the bob mass m, the gravitational acceleration g, and the initial angular amplitude θ_0 (dimensionless). The dimensions of the quantities involved are:*

$$[T] = T, \quad [l] = L, \quad [m] = M, \quad [g] = LT^{-2}, \quad \theta_0 \text{ is dimensionless}.$$

Show that

$$\pi_1 = \frac{T}{\sqrt{L/g}} = \sqrt{\frac{g}{l}}\,T, \qquad \pi_2 = \theta_0.$$

Conclude that

$$\pi_1 = f(\pi_2) \quad \text{implies} \quad T = \sqrt{\frac{L}{g}}\,f(\theta_0),$$

where f is an unknown dimensionless function. Note that the mass m does not appear in the relation: the period is independent of the bob's mass. For small amplitudes $\theta_0 \ll 1$ the dynamics give $f(\theta_0) \approx 2\pi$, hence the familiar small-angle formula

$$T \approx 2\pi\sqrt{\frac{l}{g}}.$$

Dimensional analysis cannot determine the numerical factor 2π or the explicit function $f(\theta_0)$; it only fixes the dependence on l, g, and on the dimensionless amplitude θ_0.

Exercise 1.12 *(Fluid mechanics) At low velocities (laminar flow), the volume flow Q through a small-bore tube is a function only of the tube radius R, the fluid viscosity μ, and the pressure drop per unit tube length $\frac{dp}{dx}$. Using the Pi Theorem, find an appropriate dimensionless relationship.*

Ans: $\pi = \dfrac{Q\mu}{\left(\dfrac{dp}{dx}\right)R^4}.$

Exercise 1.13 *Consider a projectile of mass m launched vertically upward from the surface of the Earth (radius R, mass M). Let $h(t)$ denote the height of the projectile above the Earth's surface at time t, and assume the projectile is given an initial upward speed V at $t = 0$. Neglect air resistance and any thrust after launch; the only force acting on the projectile is Earth's gravity, which varies with distance according to the inverse-square law. Using upward as the positive direction, Newton's second law gives*

$$m\frac{d^2h}{dt^2} = -\frac{GMm}{(R+h)^2},$$

where G is the gravitational constant. Canceling m yields the governing equation

$$\frac{d^2h}{dt^2} = -\frac{GM}{(R+h)^2},$$

with initial conditions

$$h(0) = 0, \qquad \frac{dh}{dt}(0) = V.$$

Show that there are three dimensionless quantities, namely,

$$\pi_1 = \frac{h}{R}, \qquad \pi_2 = t\frac{V}{R}, \qquad \pi_3 = \frac{VR}{\sqrt{GM}}.$$

Exercise 1.14 *(Relativistic kinetic energy scaling) Use the Buckingham Pi Theorem to determine the dimensionless parameter(s) that can be formed from the mass m, velocity v, and the speed of light c.*

(a) Determine the number of dimensionless π-groups.

(b) Construct the dimensionless combination explicitly.

(c) *Interpret the result and explain how it determines the onset of relativistic effects.*

Exercise 1.15 *(Dimensionless gravitational potential in relativity)*

Consider a massive spherical body of mass M and radius r. The relevant dimensional quantities governing the spacetime curvature near its surface are G, M, r, and c.

(a) *Apply the Buckingham Pi Theorem to determine the number of dimensionless π-groups and construct a unique such group.*

(b) *Show that the resulting dimensionless parameter $\pi_1 = GM/(rc^2)$ can be interpreted as the ratio of gravitational potential energy per unit mass to c^2, and hence measures the strength of the gravitational field in relativistic units.*

(c) *Compute π_1 for the following cases:*

- *Earth* ($M = 5.97 \times 10^{24}$ kg, $r = 6.37 \times 10^6$ m)
- *The Sun* ($M = 1.99 \times 10^{30}$ kg, $r = 6.96 \times 10^8$ m)
- *A Schwarzschild black hole of one solar mass (use the Schwarzschild radius for r).*

(d) *Comment on why relativistic corrections are negligible for Earth and the Sun, but crucial for a black hole.*

Hint: Compare your numerical results to unity and interpret physically.

Exercise 1.16 *Using the constants of nature G (gravitational constant), c (speed of light), and $\hbar$ (Planck's constant), form all possible dimensionless π-groups and characteristic scales.*

(a) *Determine the unique combinations of these constants that have the dimensions of length, mass, and time, respectively.*

(b) *Identify the resulting* Planck length l_P, Planck time t_P, *and* Planck mass m_P.

(c) *Explain briefly why these scales are believed to indicate the regime where quantum effects of gravity become significant.*

Answer check:

$$l_P = \sqrt{\frac{G\hbar}{c^3}}, \qquad t_P = \sqrt{\frac{G\hbar}{c^5}}, \qquad m_P = \sqrt{\frac{\hbar c}{G}}.$$

2

Scaling in Differential Equations

2.1 Introduction

Differential equations form the backbone of scientific and engineering modeling. However, for complex systems, the raw equations often become unwieldy due to the presence of numerous physical parameters and large disparities in variable scales. Scaling, therefore, emerges as a fundamental and indispensable technique. By rendering equations dimensionless, scaling reduces the number of independent parameters in a model, revealing the essential relationships among the remaining dimensionless quantities. This simplification not only facilitates analysis and interpretation but also clarifies the underlying physical structure of the system.

Scaling uses dimensionless parameters like the Reynolds and Prandtl numbers to show which physical effects dominate and how competing processes interact, giving valuable insight into the system's behavior. Moreover, scaling normalizes variables to be of order one, which is critical for numerical computation: it enhances stability, improves efficiency, and prevents numerical issues such as overflow or underflow.

At the heart of this process lies dimensionless analysis, a mathematical framework that eliminates units from equations, ensuring that relationships are universally valid, independent of any particular system of measurement.

2.2 Scaling in Differential Equations

Scaling, or nondimensionalization, is crucial for analyzing and simulating nonlinear differential equations by revealing the underlying physics, simplifying the model, and improving numerical stability. In complex nonlinear models, the wide range of physical parameters and variable scales can obscure which forces are dominant. By scaling variables to be of order one, the relative importance of different terms and

DOI: 10.1201/9781042009220-2

physical effects becomes apparent through the resulting dimensionless parameters, like the Reynolds or Péclet numbers. This process simplifies the model, making it easier to identify small or negligible nonlinear terms that can be dropped, which is often a necessary step for analysis and finding approximate solutions. Furthermore, scaling is vital for numerical simulations. Properly scaled equations have a reduced number of independent parameters, simplifying the setup and allowing for a single simulation to represent a wide range of physical cases. It also prevents computational issues like overflow or underflow by ensuring variables and equation residuals are on a stable numerical scale. This process not only makes complex nonlinear models more tractable but also provides greater physical insight into the system's behavior.

Intrinsic Time and Length Scales in Differential Equations.

In scaling and nondimensionalization of differential equations, the notions of *intrinsic time* and *intrinsic length* play a central role. These scales are not imposed externally but arise naturally from the dominant physical processes governing the system. Given a dimensional model, intrinsic scales are identified by balancing the leading terms in the equations, thereby revealing the characteristic magnitudes of time, length, velocity, or temperature over which significant evolution occurs. Once these intrinsic scales are used to nondimensionalize the variables, the resulting dimensionless parameters measure the relative importance of competing mechanisms and often expose small or large parameters responsible for asymptotic behavior.

Contrasting Physical Systems.

Intrinsic times can differ dramatically across physical systems because the underlying mechanisms operate at vastly different rates. A glacier, whose motion is governed by slow viscous flow and long-term mass balance, possesses an intrinsic time scale on the order of years to centuries. Scaling the governing equations for glacier dynamics reveals small dimensionless velocities and long diffusive or advective time scales, consistent with the observed gradual evolution of ice masses.

By contrast, the rise of temperature in a nuclear reactor is controlled by rapid heat generation and conduction, coupled with feedback mechanisms that can act on the order of seconds or less. The intrinsic time scale obtained by balancing heat capacity with thermal power is therefore extremely short. In this setting, nondimensionalization exposes large reaction or source terms, indicating that the system responds almost instantaneously compared to geological processes.

Insight from Scaling.

The comparison highlights that intrinsic time is not a property of "time itself," but of the physics encoded in the differential equations. Two moving or evolving systems may both be described by time-dependent models, yet their intrinsic times differ by many orders of magnitude. Scaling analysis makes this difference explicit, guiding model simplification, numerical simulation, and interpretation of solutions across diverse applications.

The concept of equilibrium solution or a fixed point in a system of differential equations plays an important role in the qualitative studies of solutions. We make the following definition.

Definition 2.1 *Consider the system of ordinary differential equations*

$$x' = f(t,x), \tag{2.1}$$

where $f \in C\big([0,\infty) \times D, \mathbb{R}^n\big)$ *and* $D \subset \mathbb{R}^n$ *is open. A vector* $x^* \in D$ *is called an equilibrium (or constant solution, or equilibrium solution) of* (2.1) *if*

$$f(t,x^*) = 0 \quad \text{for all } t \geq 0.$$

Equivalently, $x(t) \equiv x^*$ *is a solution of* (2.1).

2.2.1 Intrinsic (Maximum) Time Scale

In scaling and non-dimensionalization, it is often convenient to identify a natural or *intrinsic time scale* for the system. This time scale typically characterizes how fast the main process occurs. In some cases, it corresponds to the *maximum time* observed in the system, such as the time required to reach a steady state, the peak of a response, or the duration of a transient process.

Choosing this characteristic time, denoted by t_c, allows one to define the dimensionless time variable

$$\tilde{t} = \frac{t}{t_c},$$

which simplifies the governing equations and reveals the relative importance of different physical processes.

For example, the height y of a body projected vertically upward is given by

$$y = v_0 t - \tfrac{1}{2} g t^2,$$

where t denotes time, v_0 is the initial velocity at $t = 0$, and g is the acceleration due to gravity. We wish to scale this formula using two different choices of the characteristic time.

Time to reach the maximum height. The maximum height occurs when the vertical velocity becomes zero. That is

$$\frac{dy}{dt} = v_0 - gt = 0.$$

Hence, the time to reach the maximum height is

$$t_{\max} = \frac{v_0}{g}.$$

We choose this intrinsic time as the characteristic time,

$$t_c = \frac{v_0}{g}.$$

Let us define the dimensionless variables

$$\tilde{t} = \frac{t}{t_c}, \qquad \tilde{y} = \frac{y}{y_c}.$$

Then a natural choice for the characteristic height is the maximum. That is

$$y_{\max} = v_0\frac{v_0}{g} - \tfrac{1}{2}g\left(\frac{v_0}{g}\right)^2 = \frac{1}{2}\frac{v_0^2}{g}.$$

For the choice of

$$t_c = \frac{v_0}{g}, \quad y_c = \frac{v_0^2}{2g},$$

and by substituting

$$\tilde{t} = \frac{t}{t_c}, \qquad \tilde{y} = \frac{y}{y_c}$$

into the differential equations, one obtains

$$y_c\tilde{y} = v_0(t_c\tilde{t}) - \tfrac{1}{2}g(t_c\tilde{t})^2.$$

Dividing by y_c yields the dimensionless relation

$$\tilde{y} = 2\tilde{t} - \tilde{t}^2.$$

Time to return to the ground. The body returns to $y = 0$ at the time t_f satisfying

$$0 = v_0 t_f - \tfrac{1}{2}g t_f^2.$$

Solving for t_f gives

$$t_f = \frac{2v_0}{g}.$$

We now take this as the characteristic time,

$$t_c = \frac{2v_0}{g}.$$

Using the same height scale $y_c = \frac{v_0^2}{2g}$, we obtain

$$y_c\tilde{y} = v_0(t_c\tilde{t}) - \tfrac{1}{2}g(t_c\tilde{t})^2.$$

Dividing by y_c leads to

$$\tilde{y} = 4\tilde{t} - 4\tilde{t}^2 = 4\tilde{t}(1 - \tilde{t}).$$

Depending on the choice of the characteristic time, the scaled motion takes two equivalent but differently normalized forms,

$$\tilde{y} = 2\tilde{t} - \tilde{t}^2, \qquad \tilde{y} = 4\tilde{t}(1-\tilde{t}),$$

each corresponding to a distinct natural time scale: one emphasizing the ascent to maximum height, and the other describing the full flight duration.

To analyze physical models described by differential equations, it is often necessary to identify the intrinsic (or characteristic) time and distance scales that govern the system's behavior. These scales provide natural nondimensional variables that simplify the analysis.

Now, we consider the one-dimensional diffusion equation

$$\frac{\partial u}{\partial t} = D\frac{\partial^2 u}{\partial x^2}, \tag{2.2}$$

where $u(x,t)$ represents the concentration of a diffusing quantity, and $D > 0$ is the diffusion coefficient with dimensions

$$[D] = \frac{L^2}{T}.$$

Nondimensionalization and intrinsic time scale.

To determine the intrinsic time scale, introduce the nondimensional variables

$$x = Lx^*, \qquad t = Tt^*,$$

where L and T denote characteristic length and time scales.

Using the chain rule,

$$\frac{\partial}{\partial t} = \frac{1}{T}\frac{\partial}{\partial t^*}, \qquad \frac{\partial^2}{\partial x^2} = \frac{1}{L^2}\frac{\partial^2}{\partial x^{*2}}.$$

Substituting into (2.2) yields

$$\frac{1}{T}\frac{\partial u}{\partial t^*} = D\frac{1}{L^2}\frac{\partial^2 u}{\partial x^{*2}}.$$

Rewriting,

$$\frac{\partial u}{\partial t^*} = \frac{DT}{L^2}\frac{\partial^2 u}{\partial x^{*2}}.$$

For the nondimensional equation to be free of parameters, we choose T so that

$$\frac{DT}{L^2} = 1.$$

Thus, the intrinsic diffusion time scale is

$$T = \frac{L^2}{D}.$$

The characteristic time T represents the order of magnitude of the time required for diffusion over a distance L. Hence, if the diffusion coefficient D and a relevant spatial scale L are known, the intrinsic diffusion time follows directly from

$$T = \frac{L^2}{D}.$$

For example, for the heat diffusion in a metal rod with $D \approx 10^{-5}\,\text{m}^2/\text{s}$ and a rod length $L = 0.1\,\text{m}$, we obtain

$$T = \frac{(0.1)^2}{10^{-5}} = 10^3\,\text{s} \approx 17\,\text{minutes}.$$

2.3 Scaling in Nonlinear ODE

In many natural and social systems, populations or quantities evolve over time according to nonlinear laws of growth. While the exponential model provides a simple mathematical description, it fails to capture the effects of environmental limitations or resource constraints that inevitably slow down growth. To analyze such realistic scenarios, one often introduces nonlinear models such as the logistic equation. In this section, we study a general nonlinear growth model and demonstrate how to apply scaling techniques to express it in a dimensionless form, revealing the essential parameters that govern the system's behavior.

2.3.1 Logistic Growth Model

The logistic growth model describes how populations evolve over time in environments with limited resources. At the beginning, when individuals have plenty of space, food, and other essentials, the population increases quickly because every member can reproduce without much restriction. However, as the number of individuals grows, competition for resources intensifies, and food becomes scarcer, space becomes more crowded, and environmental pressures become stronger. This growing competition causes the rate of increase to slow down gradually. Eventually, the population reaches a point where the birth rate balances the death rate, leading to a stable equilibrium. This stable level, called the carrying capacity, represents the largest population that the environment can sustain indefinitely. The logistic growth model is widely used not only in biology to describe animal and human populations but also in fields like economics, sociology, and epidemiology to capture the idea that growth is naturally limited by available resources and environmental constraints.

We begin by considering the linear and simple model of the form

$$p'(t) = ap(t),$$

which describes populations that grow without limits. Such models are unrealistic in environments with limited resources, since real populations eventually experience constraints that slow down their growth. A more realistic representation is obtained by allowing the growth rate a to depend on the population size p. In this way, the effect of limited resources is incorporated through a simple nonlinear dependence. The simplest model of population growth is the *Malthus model*, which assumes that the rate of change of the population is proportional to the population itself. If $p = p(t)$ denotes the population at time t, then

$$\frac{dp}{dt} = rp,$$

where $r > 0$ is the constant per-capita growth rate. The solution of this differential equation is

$$p(t) = p_0 e^{rt},$$

where p_0 is the initial population. This model predicts unbounded exponential growth, which is unrealistic for large populations because it neglects limiting factors such as food supply, living space, and competition for resources. The model can then be written as

$$p' = a(p)p, \qquad p(0) = p_0. \tag{2.3}$$

If the population p cannot exceed a maximum value K, it is reasonable to require that the growth rate vanish at $p = K$, that is, $a(K) = 0$. A convenient and commonly used choice satisfying this condition is

$$a(p) = r\left(1 - \frac{p}{K}\right),$$

where r is a constant. When the initial population is small compared with the carrying capacity, that is $p_0/K \ll 1$, the model approximates the exponential growth equation $p' = rp$ at early times. Hence, r represents the initial exponential growth rate.

This particular form of $a(p)$ leads to the classical *logistic growth model*

$$p' = rp\left(1 - \frac{p}{K}\right), \qquad p(0) = p_0. \tag{2.4}$$

A slightly more general model can be defined by introducing an exponent $q > 0$ in the functional dependence of $a(p)$:

$$a(p) = r\left(1 - \frac{p}{K}\right)^q.$$

This choice still satisfies $a(K) = 0$ and approaches $a(p) \approx r$ for small p. We now proceed to nondimensionalize equation (2.3) with $a(p) = r(1-p/K)^q$. Since p varies between 0 and K, it is natural to choose K as the characteristic scale for p:

$$p_c = K, \qquad P = \frac{p}{K}, \qquad 0 < P \leq 1.$$

Let the dimensionless time variable be

$$\tau = \frac{t}{t_c},$$

where t_c is a characteristic time scale. Then,

$$\frac{dp}{dt} = \frac{d}{d\tau}\left(KP\right)\frac{d\tau}{dt} = \frac{K}{t_c}\frac{dP}{d\tau}.$$

Substituting the scaled variables into equation (2.3) gives

$$\frac{K}{t_c}\frac{dP}{d\tau} = a(KP)KP, \qquad P(0) = \frac{p_0}{K}. \tag{2.5}$$

Since $a(p) = r\left(1 - \frac{p}{K}\right)^q$, we have that

$$a(KP) = r\left(1 - \frac{KP}{K}\right)^q = r(1-P)^q.$$

Consequently, substituting into (2.5) gives, after simplification, that

$$\frac{dP}{d\tau} = t_c r(1-P)^q P, \qquad P(0) = \frac{p_0}{K}. \tag{2.6}$$

A natural choice for the time scale is

$$t_c = \frac{1}{r},$$

which makes the coefficient on the right-hand side of order one. This choice ensures that both sides of the equation are dimensionally consistent and of comparable magnitude.

When a scaling is chosen appropriately, the dimensionless variables, their derivatives, and the coefficients multiplying them are typically all of order one. Introducing the dimensionless parameter

$$\alpha = \frac{p_0}{K},$$

which measures the initial population as a fraction of the maximum possible value, we arrive at the dimensionless model

$$\frac{dP}{d\tau} = (1-P)^q P, \qquad P(0) = \alpha. \tag{2.7}$$

The model now depends on the dimensionless initial condition

$$\alpha = \frac{p_0}{K},$$

and on the exponent q. For the special case $q = 1$, the equation reduces to the standard logistic model, which involves only the single dimensionless parameter α. When $q = 1$, the solution of the dimensionless equation (2.7) is

$$P(\tau) = \frac{1}{1+Ce^{-\tau}}, \qquad C = \frac{1}{P(0)} - 1 = \frac{K-p_0}{p_0}.$$

Returning to dimensional variables, we obtain

$$p(t) = \frac{K}{1+\left(\frac{K-p_0}{p_0}\right)e^{-rt}}.$$

Notice that

$$\lim_{t\to\infty} p(t) = K, \quad \textit{which is the carrying capacity } K.$$

This nondimensionalization reveals that the qualitative behavior of the model depends only on the ratio of the initial population to the carrying capacity, p_0/K. For small populations ($p \ll K$), the model reduces to exponential growth, $p \approx p_0 e^{rt}$, consistent with the Malthus model. Note that the equilibrium solutions of (2.7) are $P = 0, \quad P = 1$.

Alternative scaling: Another possible scaling for p involves both scaling and translation:

$$P = \frac{p-p_0}{K}.$$

With this definition, we obtain

$$\frac{dP}{d\tau} = (1-\alpha-P)^q(P+\alpha), \qquad P(0) = 0. \tag{2.8}$$

In this formulation, P ranges from 0 to $1-\alpha$, whereas in the previous scaling (2.7), it ranged from α to 1. Both formulations involve the same nondimensional parameter α, although the dependent variable P is defined differently. An advantage of the first scaling (2.7) is that it allows easy reuse of computational tools developed for the unscaled model: the dimensionless form can be obtained simply by setting $p_0 = \alpha$, $K = 1$, and $r = 1$.

2.3.2 Vertical Launch from the Surface of the Earth

Consider a projectile of mass m launched vertically upward from the surface of the Earth (radius R, mass M). Let $h(t)$ denote the height of the projectile above the Earth's surface at time t, and assume the projectile is given an initial upward speed V at $t = 0$. Neglect air resistance and any thrust after launch; the only force acting on the projectile is Earth's gravity, which varies with distance according to the inverse-square law. Using upward as the positive direction, Newton's second law gives

$$m\frac{d^2h}{dt^2} = -\frac{GMm}{(R+h)^2},$$

where G is the gravitational constant. Canceling m yields the governing equation

$$\frac{d^2h}{dt^2} = -\frac{GM}{(R+h)^2}, \tag{2.9}$$

with initial conditions

$$h(0) = 0, \qquad \frac{dh}{dt}(0) = V.$$

Before we nondimensionalize, we compute the **escape velocity.** We begin by letting

$$v = \frac{dh}{dt}.$$

Using the chain rule, the second derivative of h with respect to time becomes

$$\frac{d^2h}{dt^2} = \frac{dv}{dt} = \frac{dv}{dh}\frac{dh}{dt} = v\frac{dv}{dh}.$$

Substituting this expression into the equation of motion (2.9) gives

$$v\frac{dv}{dh} = -\frac{GM}{(R+h)^2}.$$

To determine the escape velocity V_e, we integrate both sides from the moment of launch, when $h = 0$ and $v = V_e$, to the point where the projectile reaches infinity, $h \to \infty$ and $v \to 0$.

$$\int_{V_e}^{0} v\,dv = -GM\int_0^\infty \frac{dh}{(R+h)^2}.$$

The left-hand side evaluates to

$$\int_{V_e}^{0} v\,dv = \tfrac{1}{2}(0^2 - V_e^2) = -\tfrac{1}{2}V_e^2,$$

while the right-hand side becomes

$$-GM\int_0^\infty \frac{dh}{(R+h)^2} = -GM\left[-\frac{1}{R+h}\right]_0^\infty = -\frac{GM}{R}.$$

Equating the two expressions gives

$$-\tfrac{1}{2}V_e^2 = -\frac{GM}{R},$$

which simplifies to

$$V_e^2 = \frac{2GM}{R}.$$

Hence, the escape velocity is

$$V_e = \sqrt{\frac{2GM}{R}}.$$

To nondimensionalize, choose the length scale R and the time scale $t_c = R/V$ (so that the nondimensional initial velocity equals unity). Define the dimensionless height

$$x(\tau) = \frac{h(t)}{R}, \qquad \tau = \frac{t}{t_c} = \frac{tV}{R}.$$

Next, we compute the second derivative of h with respect to t. Starting from

$$h(t) = Rx(\tau), \qquad \tau = \frac{tV}{R},$$

we have, by the chain rule,

$$\begin{aligned} \frac{dh}{dt} &= \frac{dh}{d\tau}\frac{d\tau}{dt} \\ &= \frac{d}{d\tau}\Big(Rx(\tau)\Big)\frac{d\tau}{dt} \\ &= R\frac{dx}{d\tau}\frac{V}{R} = V\frac{dx}{d\tau}. \end{aligned} \tag{2.10}$$

Differentiating again with respect to t, we get by replacing h in (2.10) with $\frac{dh}{dt}$. That is

$$\frac{d^2h}{dt^2} = \frac{d}{d\tau}\left(\frac{dh}{dt}\right)\frac{d\tau}{dt} = \frac{d}{d\tau}\left(V\frac{dx}{d\tau}\right)\frac{V}{R} = \frac{V^2}{R}\frac{d^2x}{d\tau^2}.$$

Then, (2.9) becomes

$$\frac{V^2}{R}x''(\tau) = -\frac{GM}{R^2(1+x(\tau))^2}.$$

Introduce the dimensionless parameter

$$\lambda = \frac{GM}{RV^2}.$$

The nondimensional equation is therefore

$$x''(\tau) = -\frac{\lambda}{(1+x(\tau))^2}, \tag{2.11}$$

with nondimensional initial conditions

$$x(0) = 0, \qquad x'(0) = \frac{1}{V}h'(0) = 1.$$

Letting

$$\varepsilon = \frac{1}{\lambda} = \frac{RV^2}{GM},$$

then the equation becomes

$$\varepsilon x''(\tau) = -\frac{1}{(1+x(\tau))^2}. \tag{2.12}$$

If the launch speed V is much smaller than the escape velocity V_e, then

$$\varepsilon = \frac{RV^2}{GM} = 2\left(\frac{V}{V_e}\right)^2 \ll 1.$$

Hence, ε may be treated as a small parameter. Setting $\varepsilon = 0$ in (2.12) the equation reduces to

$$0 = \frac{-1}{(1+x(\tau))^2}.$$

The right-hand side, however, is always strictly negative for any finite value of $x(\tau)$, since

$$\frac{-1}{(1+x(\tau))^2} < 0 \quad \text{for all } x(\tau) \neq -1.$$

If $x(\tau) = -1$, the expression is undefined. Therefore, there is no value of $x(\tau)$ for which

$$0 = \frac{-1}{(1+x(\tau))^2}$$

can hold. In particular, evaluating at $\tau = 0$ and using $x(0) = 0$, we obtain

$$0 = \frac{-1}{(1+0)^2} = -1,$$

which is a contradiction. Hence, the reduced equation with $\varepsilon = 0$ admits *no solution* that satisfies the given initial conditions.

In the limit of small heights $x \ll 1$ compared with the Earth's radius, we can expand the right-hand side of (2.11) as

$$-\lambda \frac{1}{(1+x)^2} \approx -\lambda\,(1 - 2x + \cdots),$$

so that the leading constant term recovers the familiar nearly constant gravitational acceleration, now expressed in nondimensional form. This expansion justifies the usual approximation of uniform gravity for small heights above the Earth's surface.

Note that λ may be written in terms of the escape speed $V_e = \sqrt{2GM/R}$ as

$$\lambda = \frac{GM}{RV^2} = \frac{1}{2}\left(\frac{V_e}{V}\right)^2.$$

We have just seen that the previous scaling was inappropriate, since a term that appears small may, in fact, not be negligible. For example, consider the term

$$\varepsilon \frac{d^2x(\tau)}{d\tau^2}.$$

Although the parameter ε may be small, the quantity $\frac{d^2x(\tau)}{d\tau^2}$ could be large, and thus the product may not be negligible compared to other terms in the equation. Let's introduce a new scaling by introducing the dimensionless variables,

$$\tau = \frac{tGM}{VR^2}, \qquad x(\tau) = \frac{h(t)GM}{V^2R^2}.$$

Write the physical length and time scales as

$$h = \frac{V^2R^2}{GM}x, \qquad t = \frac{VR^2}{GM}\tau.$$

we have, by the chain rule,

$$\begin{aligned}
\frac{dh}{dt} &= \frac{dh}{d\tau}\frac{d\tau}{dt} \\
&= \frac{d}{d\tau}\left(\frac{V^2R^2}{GM}x(\tau)\right)\frac{d\tau}{dt} \\
&= \frac{V^2R^2}{GM}\frac{dx(\tau)}{d\tau}\frac{GM}{VR^2} \\
&= V\frac{dx}{d\tau}.
\end{aligned} \tag{2.13}$$

Differentiating again with respect to t, we get by replacing h in (2.13) with $\frac{dh}{dt}$. That is

$$\frac{d^2h}{dt^2} = \frac{d}{d\tau}\left(\frac{dh}{dt}\right)\frac{d\tau}{dt} = \frac{d}{d\tau}\left(V\frac{dx}{d\tau}\right)\frac{GM}{VR^2} = \frac{GM}{R^2}\frac{d^2x}{d\tau^2}.$$

Under these scalings, equation (2.9) becomes

$$\begin{aligned}
\frac{GM}{R^2}x''(\tau) &= -\frac{GM}{R^2(1+\frac{h}{R})^2} \\
&= -\frac{GM}{R^2(1+\frac{h}{R})^2} \\
&= -\frac{GM}{R^2(1+\frac{V^2R}{GM}x(\tau))^2}.
\end{aligned}$$

Simplifying both sides of the above equation and by setting the dimensionless parameter

$$\varepsilon = \frac{RV^2}{GM}$$

we obtain the nondimensional equation

$$x''(\tau) = -\frac{1}{(1+\varepsilon x(\tau))^2}, \tag{2.14}$$

with nondimensional initial conditions

$$x(0) = 0, \qquad x'(0) = \frac{1}{V}h'(0) = 1.$$

Remark 3

- *The scaling was chosen so the initial nondimensional speed is unity:* $x'(0) = 1$.
- *The parameter* ε *relates to the earlier parameter* $\lambda = GM/(RV^2)$ *by* $\varepsilon = 1/\lambda$. *In terms of the escape speed* $V_e = \sqrt{2GM/R}$ *one has*

$$\varepsilon = \frac{V^2 R}{GM} = 2\left(\frac{V}{V_e}\right)^2.$$

- *For small dimensionless height* $\beta x \ll 1$, *the right-hand side can be expanded:*

$$x''(\tau) = -1 + 2\varepsilon x(\tau) + O\big((\varepsilon x)^2\big),$$

so the leading term recovers the nearly constant downward acceleration -1 *in these units, and the first correction is proportional to* εx.

- *This scaling makes the initial speed exactly one (convenient for comparison of trajectories launched with different* V*), and places the dynamical dependence on the single parameter* ε. *If* $\varepsilon \ll 1$ *(very small launch speed relative to gravitational scaling), the deviation from constant gravity is small; if* $\varepsilon \gtrsim 1$, *the variable gravity term matters even for moderate* x.

Next, we solve the dimensionless equation (2.14) by neglecting ε. Thus, setting $\varepsilon = 0$ in (2.14), we arrive at the second order differential equation with initial conditions

$$x''(\tau) = -1, \quad x(0) = 0, \, x'(0) = 1.$$

This equation has the general solution

$$x(\tau) = -\frac{1}{2}\tau^2 + c_1\tau + c_2,$$

for constants c_1 and c_2. Applying the initial conditions, we arrive at

$$c_1 - 1, \quad c_2 = 0.$$

Consequently, the solution takes the form

$$x(\tau) = -\frac{1}{2}\tau^2 + \tau.$$

Therefore, the solution to our original equation (2.9) is

$$\frac{GM}{V^2R^2}h(t) = -\frac{1}{2}\left(\frac{GM}{VR^2}\right)^2 t^2 + \frac{GM}{VR^2}t.$$

Simplifying both sides of the above equation, we obtain

$$h(t) = -\frac{1}{2}\frac{GM}{R^2}t^2 + Vt. \tag{2.15}$$

When $h = 0$, that is, at the earth's surface, the gravitational force equals $-mg$ and hence

$$g = \frac{GM}{R^2}. \tag{2.16}$$

With this in mind, the solution given by (2.15) is equivalent to

$$h(t) = -\frac{1}{2}gt^2 + Vt.$$

The above solution is a fundamental kinematic equation that describes the vertical position of an object undergoing projectile motion near the Earth's surface, assuming negligible air resistance. This equation is a solution to the equations of motion under constant acceleration.

2.3.3 Bacterial Growth in a Chemostat

We consider a well-mixed chemostat of constant volume V. Let $C(t)$ denote the nutrient concentration (mass/volume) and $N(t)$ the bacterial density (cells/volume or biomass/volume) at time t. The volumetric flow rate in and out is F, and the nutrient concentration in the inflow (stock solution) is C_{in}. We assume perfect mixing and constant volume, so inflow equals outflow.

Let K be the reproduction coefficient. We assume a linear nutrient-limited growth law so that the specific growth rate (per-capita growth rate) is

$$\mu(C) = KC,$$

where K has units $(\text{time})^{-1}(\text{concentration})^{-1}$. (For greater realism, one could instead employ Monod kinetics $\mu(C) = \mu_{\max}C/(K_s + C)$.)

Let Y denote the yield coefficient, defined as the bacterial density produced per unit nutrient consumed. Thus, nutrient consumption occurs at rate $\frac{1}{Y}\mu(C)N$.

It is convenient to introduce the *dilution rate*

$$D = \frac{F}{V},$$

which has units of inverse time.

Bacterial balance. Growth minus washout gives

$$\frac{dN}{dt} = \mu(C)N - DN,$$

that is,

$$\frac{dN}{dt} = (KC - D)N.$$

The first term represents nutrient-driven growth, while the second represents dilution (washout) due to outflow.

Nutrient balance. Input from the feed, minus outflow and bacterial consumption, yields

$$\frac{dC}{dt} = D(C_{\text{in}} - C) - \frac{1}{Y}\mu(C)N,$$

or explicitly,

$$\frac{dC}{dt} = D(C_{\text{in}} - C) - \frac{1}{Y}KCN.$$

The first term accounts for nutrient inflow and outflow, while the second term represents consumption by bacterial growth.

This system provides a basic deterministic model of nutrient-limited microbial growth in a continuous-flow reactor. In particular, $KC > D$ is required for a positive equilibrium; otherwise, $N(t) \to 0$, a phenomenon known as *washout*.

To nondimensionalize the system, we choose a natural time scale equal to the residence time, $t_c = V/F$, and define the dimensionless time $\tau = t/t_c = (F/V)t$. Nutrient is scaled by the feed concentration, $c(\tau) = C(t)/C_{\text{in}}$, and bacterial density is scaled as $n(\tau) = N(t)/N_c$, where $N_c = C_{\text{in}}Y$. This choice simplifies the nutrient-consumption term in the nondimensional equations. Introduce the dimensionless growth–dilution parameter

$$\beta = KC_{\text{in}}t_c = KC_{\text{in}}\frac{V}{F} = \frac{KC_{\text{in}}}{D}.$$

The parameter β is defined as the growth rate at feed concentration multiplied by the residence time. It measures whether the growth per residence time is sufficient to overcome dilution. Consider the bacterial growth equation

$$\frac{dN}{dt} = KCN - DN.$$

Introduce dimensionless variables

$$N = N_c n, \quad C = C_{\text{in}} c, \quad t = t_c \tau,$$

so that

$$\frac{dN}{dt} = \frac{N_c}{t_c}\frac{dn}{d\tau}.$$

Substituting into the ODE gives

$$\frac{N_c}{t_c}\frac{dn}{d\tau} = KC_{\text{in}}c\,N_c n - DN_c n.$$

Dividing both sides by N_c/t_c yields the nondimensional form

$$\frac{dn}{d\tau} = KC_{\text{in}}t_c\,cn - Dt_c n.$$

Since the characteristic time scale

$$t_c = \frac{1}{D},$$

the equation simplifies to

$$\frac{dn}{d\tau} = KC_{\text{in}}t_c\,cn - n,$$

from which we obtain

$$\frac{dn}{d\tau} = \beta\,cn - n,$$

with

$$c(\tau) = \frac{C(t)}{C_{\text{in}}}, \quad n(\tau) = \frac{N(t)}{C_{\text{in}}Y}, \quad \tau = \frac{F}{V}t, \quad \beta = \frac{KC_{\text{in}}}{D},$$

which measures whether growth per residence time can overcome dilution. Here, β is dimensionless, and c, n, and τ are all nondimensional quantities. The nutrient concentration in the chemostat satisfies

$$\frac{dC}{dt} = \frac{F}{V}(C_{\text{in}} - C) - \frac{1}{Y}KCN.$$

Introducing dimensionless variables and dividing by C_{in}/t_c (noting that $\frac{dC}{dt} = \frac{C_{\text{in}}}{t_c}\frac{dc}{d\tau}$), we obtain

$$\frac{dc}{d\tau} = 1 - c - \left(\frac{t_c K}{Y}\right)cN.$$

Using $N/C_{\text{in}} = (N_c n)/C_{\text{in}} = Yn$, the coefficient simplifies:

$$\frac{t_c K}{Y}\cdot c\cdot Yn = t_c K\,cn.$$

Defining $\beta = t_c KC_{\text{in}}$, we have $t_c K = \beta/C_{\text{in}}$. Multiplying by C_{in} cancels, yielding the simple form

$$\frac{dc}{d\tau} = 1 - c - \beta cn.$$

In Example 8.13 we will consider the equilibrium points and study their behaviors and deduce some biological interpretation.

Comments and extensions

The linear form $\mu(C) = KC$ is suitable for low nutrient ranges. For saturating behavior, one may instead use Monod kinetics

$$\mu(C) = \mu_{\max}\frac{C}{K_s + C}.$$

Nondimensionalization proceeds similarly but produces slightly different parameter combinations, for example, the ratio K_s/C_{in} appears.

The yield Y determines the scaling of n; if one prefers a biomass-based N (mass per volume) rather than a cell density, a different N_c can be chosen.

Additional effects, such as death rates, maintenance consumption, multiple species, or product formation, can also be incorporated. The dimensionalization step follows the same principle: choose natural characteristic scales, such as the residence time for time and the feed concentration for the nutrient.

2.3.4 Continuously Stirred Tank Reactor (CSTR)

We develop a model for a single, isothermal, continuously stirred tank reactor (CSTR) with one inlet and one outlet stream. The reactor is assumed to satisfy the following conditions:

- Perfect mixing: concentration and temperature are spatially uniform and identical to outlet conditions.
- Constant temperature (isothermal operation).
- Constant reactor volume V and constant fluid density.
- Equal inlet and outlet volumetric flow rates.
- Single, irreversible reaction.

Under constant density and volume,

$$q_{\text{in}} = q_{\text{out}} = q.$$

As an example, consider the reaction

$$A \to B.$$

Next we discuss the general mole balance. The fundamental principle for reactor modeling is the species mole balance:

$$\text{Accumulation} = \text{Input} - \text{Output} + \text{Generation}.$$

For species A, this becomes

$$\frac{dN_A}{dt} = F_{A,\text{in}} - F_{A,\text{out}} + r_A V, \tag{2.17}$$

where

- N_A = moles of species A in the reactor,
- $F_{A,\text{in}}, F_{A,\text{out}}$ = molar flow rates,
- r_A = rate of formation of A per unit volume,
- V = reactor volume.

For a reactant, $r_A < 0$.

As for the conversion into concentration form, we use

$$N_A = C_A V, \qquad F_A = qC_A,$$

and assuming constant V and q, we obtain

$$\begin{aligned} \frac{d(C_A V)}{dt} &= qC_{A,\text{in}} - qC_A + r_A V, \\ V\frac{dC_A}{dt} &= q(C_{A,\text{in}} - C_A) + r_A V. \end{aligned}$$

Dividing by V gives

$$\frac{dC_A}{dt} = \frac{q}{V}(C_{A,\text{in}} - C_A) + r_A.$$

Define the *residence time*

$$\tau = \frac{V}{q},$$

which represents the average time a fluid element spends in the reactor. Then

$$\frac{dC_A}{dt} = \frac{1}{\tau}(C_{A,\text{in}} - C_A) + r_A.$$

For a first-order reaction,

$$r_A = -kC_A,$$

where k is the rate constant.

Substituting gives the dimensional model

$$\frac{dC_A}{dt} = \frac{1}{\tau}(C_{A,\text{in}} - C_A) - kC_A.$$

This is the governing equation for an isothermal, first-order CSTR. The steady-state solution is found by setting $dC_A/dt = 0$. This implies that

$$0 = \frac{1}{\tau}(C_{A,\text{in}} - C_A) - kC_A,$$

which yields

$$C_A^{(ss)} = \frac{C_{A,\text{in}}}{1 + k\tau}.$$

Now we are ready to dimensionalize the system by introducing the dimensionless variables

$$c = \frac{C_A}{C_{A,\text{in}}}, \qquad t^* = \frac{t}{\tau}.$$

Since

$$\frac{dC_A}{dt} = \frac{C_{A,\text{in}}}{\tau}\frac{dc}{dt^*},$$

substitution gives

$$\frac{dc}{dt^*} = (1-c) - k\tau c.$$

Define the *Damköhler number*

$$\text{Da} = k\tau,$$

which measures the ratio of reaction rate to flow rate.

The dimensionless model becomes

$$\frac{dc}{dt^*} = 1 - c - \text{Da}\,c.$$

This compact form shows that reactor performance is governed entirely by the Damköhler number:

- $\text{Da} \ll 1$: flow dominates, little reaction occurs.
- $\text{Da} \gg 1$: reaction dominates, strong conversion.

The model may be solved analytically or numerically to predict transient and steady-state behavior.

2.3.5 General Duffing Oscillator

The general Duffing oscillator is a differential equation that represents the motion of a weakly nonlinear oscillator, which is a system that behaves almost like a simple harmonic oscillator but includes small nonlinear effects. A weakly nonlinear Oscillator subject to an external forcing is governed by the nonlinear second–order differential equation

$$m\ddot{x} + c\dot{x} + kx + \alpha x^3 = F(t),$$
$$x(0) = x_0, \quad \dot{x}(0) = v_0,$$

where

- $m\ddot{x}$ is the inertial force, which represents the resistance of the mass m to acceleration, following Newton's second law $F = ma$,
- $c\dot{x}$ is the damping force that models energy loss due to friction or resistance (e.g., air drag, internal damping). The coefficient c determines how strongly the motion is damped,

- kx is the linear restoring force or Hooke's law term, describing the spring's tendency to restore the mass toward equilibrium. The constant k measures the spring stiffness.
- αx^3 is the nonlinear restoring force that introduces weak nonlinearity into the system. The sign of α determines whether the spring is
  - *Hardening* ($\alpha > 0$): restoring force grows faster than linearly with displacement,
  - *Softening* ($\alpha < 0$): restoring force grows more slowly than linearly.
- $F(t)$ is the external forcing that represents a time-dependent external influence, such as a periodic or impulsive driving force.

The initial conditions $x(0) = x_0$ and $\dot{x}(0) = v_0$ define the initial displacement and velocity of the oscillator. In general, the above system models a damped, driven oscillator with a small nonlinear stiffness. It describes systems such as vibrating beams, suspension springs, or electrical circuits where the restoring force deviates slightly from linearity. The nonlinear term can produce effects such as amplitude-dependent frequencies, bifurcations, and complex resonances not present in purely linear oscillator s. To remove the physical units, define appropriate characteristic scales for each variable. **Characteristic length scale (x_c):** A common choice is the initial position, x_0. If $x_0 = 0$, you might use the amplitude of the linear system's response. **Characteristic time scale (t_c):** A natural choice for an oscillator is the period of the unforced, undamped linear system. The frequency of this system is

$$\omega_0 = \sqrt{\frac{k}{m}},$$

so the characteristic time is

$$t_c = \omega_0^{-1} = \sqrt{\frac{m}{k}}.$$

Characteristic velocity scale (v_c): This can be derived from the length and time scales as

$$v_c = \frac{x_c}{t_c} = x_0\sqrt{\frac{k}{m}}.$$

Introduce new dimensionless variables by dividing each original variable by its characteristic scale.

$$\tilde{x} = \frac{x}{x_c} = \frac{x}{x_0} \quad \tilde{t} = \frac{t}{t_c}.$$

These substitutions allow the governing equation to be expressed in a dimensionless form, where the relative magnitudes of the dimensionless parameters reveal the physical importance of damping, forcing, and nonlinear effects. Rewrite the derivatives in terms of the new dimensionless time variable $\tilde{t}$ using the chain rule:

$$\dot{x} = \frac{dx}{dt} = \frac{d(\tilde{x}x_c)}{d(\tilde{t}t_c)} = \frac{x_c}{t_c}\frac{d\tilde{x}}{d\tilde{t}} = \frac{x_c}{t_c}\dot{\tilde{x}}.$$

Similarly,

$$\ddot{x} = \frac{d}{dt}\left(\frac{x_c}{t_c}\dot{\tilde{x}}\right) = \frac{1}{t_c}\frac{d}{d\tilde{t}}\left(\frac{x_c}{t_c}\dot{\tilde{x}}\right) = \frac{x_c}{t_c^2}\ddot{\tilde{x}}.$$

Now, substitute these expressions into the original IVP:

$$m\left(\frac{x_c}{t_c^2}\ddot{\tilde{x}}\right) + c\left(\frac{x_c}{t_c}\dot{\tilde{x}}\right) + k(x_c\tilde{x}) + \alpha(x_c\tilde{x})^3 = F(t_c\tilde{t}).$$

Divide the entire equation by the coefficient of the highest-order term, mx_c/t_c^2 gives

$$\ddot{\tilde{x}} + \frac{ct_c}{m}\dot{\tilde{x}} + \frac{kt_c^2}{m}\tilde{x} + \frac{\alpha x_c^2 t_c^2}{m}\tilde{x}^3 = \frac{t_c^2}{mx_c}F(t_c\tilde{t}).$$

The terms $\frac{kt_c^2}{m}$ and $\frac{ct_c}{m}$ are important. By our choice of $t_c = \sqrt{m/k} = \omega_0^{-1}$, we have

$$\frac{kt_c^2}{m} = \frac{k(m/k)}{m} = 1.$$

This choice ensures that the linear stiffness term becomes unity, simplifying the equation significantly.

Thus, the equation becomes

$$\ddot{\tilde{x}} + \left(\frac{c}{\sqrt{mk}}\right)\dot{\tilde{x}} + \tilde{x} + \left(\frac{\alpha x_0^2}{k}\right)\tilde{x}^3 = \frac{F_0}{\omega_0^2 m x_0}\tilde{F}(\tilde{t}),$$

where we assume a characteristic forcing amplitude F_0 and a dimensionless forcing function

$$\tilde{F}(\tilde{t}) = \frac{F(t_c\tilde{t})}{F_0}.$$

Since $\omega_0^2 = k/m$, we have

$$\frac{F_0}{\omega_0^2 m x_0} = \frac{F_0}{kx_0} = f_0.$$

To simplify notation, drop the tildes ($\sim$) and define the dimensionless parameters:

$$\mu = \frac{c}{\sqrt{mk}}, \quad \beta = \frac{\alpha x_0^2}{k}, \quad f_0 = \frac{F_0}{kx_0}, \quad \Omega = \frac{\omega}{\omega_0}.$$

To simplify notation, drop the tildes ($\sim$) and define the dimensionless parameters:

$$\mu = \frac{c}{\sqrt{mk}}, \quad \beta = \frac{\alpha x_0^2}{k}, \quad f_0 = \frac{F_0}{kx_0}, \quad \Omega = \frac{\omega}{\omega_0}$$

If we assume the dimensional forcing term is for the form

$$F_0\cos(\omega t) = F_0\cos(\omega t_c\tilde{t}),$$

and since

$$t_c = \omega_0^{-1}$$

we obtain

$$\cos(\omega t) = \cos\left(\omega\,\omega_0^{-1}\,\tilde{t}\right) = \cos\left(\frac{\omega}{\omega_0}\tilde{t}\right) = \cos(\Omega\tilde{t}),$$

where

$$\Omega = \frac{\omega}{\omega_0}.$$

We then rename the dimensionless time variable (standard practice):

$$\tilde{t} \mapsto t.$$

Thus the forcing term becomes

$$F_0 \cos(\Omega t).$$

Consequently, the final dimensionless equation becomes:

$$\ddot{x} + \mu\dot{x} + x + \beta x^3 = f_0 \cos(\Omega t),$$

assuming $F(t) = F_0 \cos(\omega t)$.

The weakly nonlinear assumption means that $\beta \ll 1$. Thus, the cubic restoring force remains small compared with the linear restoring force for oscillations of typical amplitude. As for the initial conditions, using the same characteristic scales, we arrive at $x(0) = x_0$ from which we have

$$\tilde{x}(0) = \frac{x_0}{x_c} = \frac{x_0}{x_0} = 1.$$

Similarly, since $\dot{x}(0) = v_0$, implies that

$$\dot{\tilde{x}}(0) = \frac{v_0 t_c}{x_c} = \frac{v_0}{x_0\omega_0}.$$

Thus, the final dimensionless initial value problem is

$$\begin{aligned} &\ddot{x} + \mu\dot{x} + x + \beta x^3 = f_0 \cos(\Omega t), \\ &x(0) = 1, \quad \dot{x}(0) = \frac{v_0}{x_0\omega_0}. \end{aligned} \tag{2.18}$$

2.3.6 Nonlinear Vertical Motion with Air Resistance

Consider the vertical motion of a body moving through a fluid under the influence of gravity, buoyancy, and quadratic air resistance. Let $v(t)$ denote the velocity of the body, where positive velocity is taken to be upward. Then $v(t)$ satisfies

$$(m+\mu)\frac{dv}{dt} = -\frac{1}{2}C_D\rho A|v|v - (m - \rho V)g,$$

where m is the mass of the body, μ is the added mass due to the surrounding fluid, C_D is the drag coefficient, ρ is the density of the fluid, A is the cross-sectional area perpendicular to the motion, g is the gravitational acceleration, and V is the volume of the body. The initial condition is

$$v(0) = v_0.$$

We assume throughout this subsection that

$$m > \rho V.$$

Thus, the body is heavier than the fluid it displaces, and the effective weight $(m - \rho V)g$ acts downward.

The terminal speed v_T is found by balancing quadratic drag with the effective weight:

$$\frac{1}{2} C_D \rho A v_T^2 = (m - \rho V)g.$$

Thus

$$v_T = \sqrt{\frac{2(m-\rho V)g}{C_D \rho A}}.$$

We introduce the dimensionless variables

$$v = v_T \bar{v}, \qquad t = t_c \bar{t},$$

where t_c is a characteristic time scale to be determined.

Substituting these scalings into the differential equation gives

$$(m+\mu)\frac{v_T}{t_c}\frac{d\bar{v}}{d\bar{t}} = -\frac{1}{2} C_D \rho A v_T^2 |\bar{v}|\bar{v} - (m - \rho V)g.$$

Using the definition of v_T, this becomes

$$(m+\mu)\frac{v_T}{t_c}\frac{d\bar{v}}{d\bar{t}} = -(m - \rho V)g\big(|\bar{v}|\bar{v} + 1\big).$$

Dividing by $(m+\mu)v_T/t_c$, we obtain

$$\frac{d\bar{v}}{d\bar{t}} = -\frac{(m-\rho V)g t_c}{(m+\mu)v_T}\big(|\bar{v}|\bar{v} + 1\big).$$

We now choose the intrinsic time scale

$$t_c = \frac{(m+\mu)v_T}{(m-\rho V)g}.$$

Equivalently, using the formula for v_T,

$$t_c = \frac{m+\mu}{(m-\rho V)g}\sqrt{\frac{2(m-\rho V)g}{C_D \rho A}}.$$

With this choice, the dimensionless equation becomes

$$\frac{d\bar{v}}{d\bar{t}} = -\left(|\bar{v}|\bar{v} + 1\right),$$

with a dimensionless initial condition

$$\bar{v}(0) = \frac{v_0}{v_T}.$$

Thus, all physical parameters have been absorbed into the characteristic velocity v_T and the characteristic time t_c. The resulting parameter-free equation describes the scaled vertical motion of a body subject to quadratic drag, gravity, and buoyancy.

2.3.7 Predator–Prey Model (Lotka–Volterra Equations)

The interaction between two biological species, a prey population $Y(t)$ and a predator population $X(t)$, can be modeled by the classical Lotka–Volterra system

$$\begin{cases} \dfrac{dY}{dt} = aY - bXY, \\ \dfrac{dX}{dt} = -cX + dXY, \end{cases} \tag{2.19}$$

where

- $Y(t)$ denotes the prey population at time t,
- $X(t)$ denotes the predator population at time t,
- $a > 0$ is the natural growth rate of the prey in the absence of predators,
- $b > 0$ is the predation rate coefficient,
- $c > 0$ is the natural death rate of the predators in the absence of prey,
- $d > 0$ measures the efficiency with which consumed prey are converted into new predators.

The first equation states that the prey population grows exponentially at a rate a when predators are absent, but decreases through encounters with predators at a rate proportional to the product XY. The second equation shows that the predator population decreases exponentially at a rate c in the absence of prey, but increases through successful predation.

We now introduce dimensionless variables

$$\bar{t} = \frac{t}{t_c}, \qquad \bar{Y} = \frac{Y}{Y_c}, \qquad \bar{X} = \frac{X}{X_c},$$

where t_c, Y_c, and X_c are characteristic scales to be determined. Substituting

$$Y = Y_c\bar{Y}, \qquad X = X_c\bar{X}, \qquad t = t_c\bar{t},$$

into system (2.19) gives

$$\frac{Y_c}{t_c}\frac{d\bar{Y}}{d\bar{t}} = aY_c\bar{Y} - bX_cY_c\bar{X}\bar{Y},$$

$$\frac{X_c}{t_c}\frac{d\bar{X}}{d\bar{t}} = -cX_c\bar{X} + dX_cY_c\bar{X}\bar{Y}.$$

Dividing the first equation by Y_c/t_c and the second by X_c/t_c, we obtain

$$\frac{d\bar{Y}}{d\bar{t}} = at_c\bar{Y} - bX_ct_c\bar{X}\bar{Y},$$

$$\frac{d\bar{X}}{d\bar{t}} = -ct_c\bar{X} + dY_ct_c\bar{X}\bar{Y}.$$

To determine appropriate scales, we examine the nontrivial equilibrium solution of system (2.19). Setting

$$\frac{dY}{dt} = \frac{dX}{dt} = 0,$$

yields

$$(a - bX)Y = 0, \qquad X(-c + dY) = 0.$$

Hence, the equilibrium solutions are

$$(0,0), \qquad \left(\frac{a}{b}, \frac{c}{d}\right).$$

This suggests the characteristic population scales

$$X_c = \frac{a}{b}, \qquad Y_c = \frac{c}{d}.$$

Since a^{-1} represents the characteristic growth time of the prey, we choose

$$t_c = \frac{1}{a}.$$

With these choices, the dimensionless system becomes

$$\frac{d\bar{Y}}{d\bar{t}} = \bar{Y}(1 - \bar{X}),$$

$$\frac{d\bar{X}}{d\bar{t}} = \lambda\bar{X}(\bar{Y} - 1),$$

where

$$\lambda = \frac{c}{a}$$

is a dimensionless parameter representing the ratio of the predator death rate to the prey growth rate.

The dimensionless initial conditions are

$$\bar{Y}(0) = \frac{Y_0}{Y_c} = \frac{dY_0}{c}, \qquad \bar{X}(0) = \frac{X_0}{X_c} = \frac{bX_0}{a}.$$

Thus, the original four-parameter system has been reduced to a simpler dimensionless form involving only one essential parameter, namely the ratio $\lambda = c/a$. This reduction reveals the intrinsic structure of the predator–prey interaction and facilitates the qualitative analysis of the system.

Remark 4 *The dimensionless system possesses the equilibrium solutions*

$$(0,0) \qquad and \qquad (1,1),$$

where the equilibrium point $(1,1)$ corresponds to the nontrivial equilibrium

$$\left(\frac{a}{b}, \frac{c}{d}\right)$$

of the original system. The stability and qualitative behavior of these equilibria will be studied in Examples 8.12 and 8.27.

2.4 Exercises

Exercise 2.1 *Consider the heat exchange between a body with temperature $T(t)$ and the surroundings maintained at a constant temperature T_s. This process can be described by Newton's law of cooling:*

$$\frac{dT}{dt} = -k(T - T_s), \qquad T(0) = T_0,$$

where k denotes the heat transfer coefficient. From a physical perspective, the temperature starts at T_0 and gradually approaches the surrounding temperature T_s. Therefore, we anticipate that $T(t)$ remains between T_0 and T_s for all $t > 0$. This behavior is also evident from the exact solution. To define an appropriate dimensionless temperature variable, we translate and scale T as follows:

$$\bar{T} = \frac{T - T_0}{T_s - T_0}.$$

This transformation normalizes the temperature such that initially $\bar{T}(0) = 0$ and as $t \to \infty$, $\bar{T} \to 1$, providing a convenient nondimensional representation of the cooling process. With this scaling, it follows that $0 \leq \bar{T} \leq 1$. Next, we introduce a dimensionless time variable by defining

$$\bar{t} = \frac{t}{t_c}.$$

Show that under the proposed scaling, the original differential equation is transformed into the dimensionless one

$$\frac{d\bar{T}}{d\bar{t}} = -(\bar{T} - 1), \qquad \bar{T}(0) = 0,$$

where

$$t_c k = 1.$$

Moreover, find the exact solution to the original problem and

$$\lim_{t\to\infty} T(t).$$

Exercise 2.2 *(SIR Model for the Spread of Diseases) The study of epidemiology often employs systems of ordinary differential equations (ODEs) to model the spread of infectious diseases—such as smallpox, measles, plague, influenza, swine flu, and HIV—through a population. Various models capture different aspects of disease transmission, typically characterized by dimensionless parameters. Most of these effects can be modeled through exponential growth or decay of the dependent variables. The most basic framework divides the population into three groups: susceptible individuals $S(t)$, who can contract the disease, infectious individuals $I(t)$ who are currently infected and can transmit the disease, and recovered individuals $R(t)$ who have gained immunity after infection.*

Let $S(t)$, $I(t)$, and $R(t)$ denote the number of people in each respective category. The model, known as the SIR model, is represented by the following system of ODEs:

$$\frac{dS}{dt} = -\beta SI,$$

$$\frac{dI}{dt} = \beta SI - \nu I,$$

$$\frac{dR}{dt} = \nu I,$$

where β is the transmission rate, and ν is the recovery rate. Here, β and ν are empirical parameters. The average recovery time from the disease is given by ν^{-1}, while the transmission rate β is generally more difficult to estimate. It is therefore advantageous to formulate a scaled model in which β is absorbed into dimensionless parameters. A natural way to scale the variables S, I, and R is by using the initial susceptible population $S(0)$:

$$\bar{S} = \frac{S}{S(0)}, \qquad \bar{I} = \frac{I}{S(0)}, \qquad \bar{R} = \frac{R}{S(0)}.$$

Introducing a dimensionless time variable

$$\bar{t} = \frac{t}{t_c}, \quad \text{where} \quad t_c = \frac{1}{\nu}.$$

The choice of t_c corresponds to the average recovery time.

(a) For $R(0) = 0$, show that the scaled form of the system is

$$\frac{d\bar{S}}{d\bar{t}} = -R_0\bar{S}\bar{I},$$
$$\frac{d\bar{I}}{d\bar{t}} = R_0\bar{S}\bar{I} - \bar{I},$$
$$\frac{d\bar{R}}{d\bar{t}} = \bar{I},$$

with initial conditions

$$\bar{S}(0) = 1, \qquad \bar{I}(0) = \alpha, \qquad \bar{R}(0) = 0,$$

and the dimensionless quantity

$$R_0 = \frac{S(0)\beta}{\nu},$$

is known as the reproduction number.

(b) Show that for the infection to spread, we must have

$$R_0 > 1.$$

Hence, R_0 characterizes the ability of the disease to spread through the population and is known as the basic reproduction number. It represents the expected number of secondary infections produced by a single infected individual during the course of the disease.

Exercise 2.3 *(Vertical motion with air resistance)*

A small sphere of mass m falls under gravity and air resistance proportional to the square of its velocity. Its velocity $v(t)$ satisfies

$$m\frac{dv}{dt} = -mg - cv|v|,$$

where g is the acceleration due to gravity and c is a drag coefficient.

(a) Determine the physical dimensions (units) of the drag coefficient c.

(b) Define the terminal velocity v_t by balancing drag and gravity (write the defining equation for v_t).

(c) Introduce the nondimensional variables

$$v = v_t\,\bar{v}, \qquad t = t_c\bar{t},$$

where t_c is a characteristic time scale. Substitute these into the dimensional ODE and write the scaled equation (leave any dimensionless grouping parameters symbolically).

(d) *Identify the natural choice for the time scale t_c (give it symbolically in terms of the problem parameters).*

Exercise 2.4 *(Logistic population growth with harvesting)*

A population $N(t)$ follows the differential equation

$$\frac{dN}{dt} = rN\left(1 - \frac{N}{K}\right) - H,$$

where r is the intrinsic growth rate, K is the carrying capacity, and H is a constant harvesting rate.

(a) *Determine the physical dimensions (units) of the parameters r, K, and H.*

(b) *Introduce the nondimensional variables*

$$\bar{N} = \frac{N}{K}, \qquad \bar{t} = rt.$$

Substitute these variables into the dimensional equation and write the nondimensionalized form of the logistic model.

(c) *Identify the dimensionless group that represents the effect of harvesting.*

Exercise 2.5 *Set $\beta = 0$ in (2.18) and then discuss the solution of the differential equation.*

Exercise 2.6 *In the predator-prey model, assume the scalings:*

$$\bar{t} = at, \quad \bar{Y} = \frac{Yd}{c}, \quad \bar{X} = \frac{Xb}{a}$$

(a) *Identify t_c, Y_c, and X_c.*

(b) *Let α, β, γ, and δ be given as in the discussion of the predator-prey model. Show that under the given scalings, the new dimensionless system takes the form*

$$\frac{d\bar{Y}}{d\bar{t}} = \bar{Y}(1 - \bar{X}),$$

$$\frac{d\bar{X}}{d\bar{t}} = \mu\bar{X}(\bar{Y} - 1) = \frac{\gamma}{\alpha}\bar{X}(\bar{Y} - 1),$$

$$\bar{Y}(0) = \eta = \frac{\beta}{\gamma} = \frac{a}{c}Y_0 d, \qquad \bar{X}(0) = \omega = \frac{\delta}{\alpha} = \frac{X_0 b}{a}.$$

$$\text{where} \quad \mu = \frac{c}{a}, \qquad \eta = \frac{Y_0 d}{c}, \qquad \omega = \frac{X_0 b}{a}.$$

(c) *Argue why we must have $\eta \neq 1$ and $\omega \neq 1$.*

Exercise 2.7 *Derive equation (2.16).*

Exercise 2.8 *In the projectile problem, we define the* escape velocity *as*

$$V_e = \sqrt{\frac{2GM}{R}}.$$

Using this velocity, we introduce the following dimensionless variables:

$$x(\tau) = \frac{h(t)}{R}, \qquad \tau = \frac{tV_e}{R}.$$

(a) Using these scalings, derive the corresponding nondimensional differential equation governing the projectile motion.

(b) Solve the resulting dimensionless equation, if possible, and interpret the physical meaning of the solution.

Exercise 2.9 *(Escape velocity's alternative derivation) In the projectile problem, starting from the energy principle, show that a projectile of mass m launched from the Earth's surface with initial speed V will escape to infinity if and only if*

$$\tfrac{1}{2}mV^2 \geq G\frac{Mm}{R}.$$

Use this to recover the escape speed $V_e = \sqrt{2GM/R}$.
Hint: *equate initial kinetic plus potential energy at* $h = 0$ *to the limit of total energy as* $h \to \infty$.

Exercise 2.10 *(Energy integral and time to turn-around). Starting from the dimensional equation* (2.9) *use the substitution* $v = dh/dt$ *and separate variables to obtain the energy integral*

$$\tfrac{1}{2}v^2 - \frac{GM}{R+h} = \textit{constant}.$$

(a) Determine the constant using initial conditions $h(0) = 0$, $v(0) = V$ *and write the implicit relation between v and h.*

(b) Using that relation, derive an integral expression for the time $T_{\max}$ *it takes the projectile (with* $V < V_e$*) to reach its maximum height* $h_{\max}$. *That is, express*

$$T_{\max} = \int_0^{h_{\max}} \frac{dh}{\sqrt{V^2 - 2GM\left(\frac{1}{R} - \frac{1}{R+h}\right)}}.$$

(c) Show that in the limit $h \ll R$, *this integral reduces to the familiar constant-g formula* $T_{\max} \approx V/g$.

Hint: *solve for* $h_{\max}$ *from* $v(h_{\max}) = 0$ *using the energy relation; then use* $dt = dh/v(h)$ *to form the time integral.*

Exercise 2.11 *(Quadratic air resistance, nondimensional groups, terminal velocity) Consider the same vertical launch as in the main text, but now include a quadratic air resistance force. Let A be the projectile cross-sectional area, ρ the air density, and C_D the drag coefficient. The drag force (opposing the motion) is modeled by*

$$F_{\text{drag}} = -\tfrac{1}{2} C_D \rho A\, v|v|,$$

where $v = dh/dt$ and upward is positive. The gravitational force still follows the inverse-square law. The dimensional equation of motion is therefore

$$m\frac{d^2h}{dt^2} = -\frac{GMm}{(R+h)^2} - \tfrac{1}{2} C_D \rho A\, v|v|.$$

(a) Introduce the drag constant

$$k := \tfrac{1}{2} C_D \rho A,$$

and write the governing ODE in the form

$$\frac{d^2h}{dt^2} = -\frac{GM}{(R+h)^2} - \frac{k}{m} v|v|.$$

(b) Nondimensionalize using the length scale R and the velocity scale V (the initial launch speed). Define

$$x = \frac{h}{R}, \quad \tau = \frac{tV}{R}, \quad v = \frac{dh}{dt} = V x'(\tau).$$

Derive the nondimensional ODE and identify the two (independent) dimensionless groups that appear. Write the equation in the form

$$x''(\tau) = -\beta \frac{1}{(1+x)^2} - \Gamma x'|x'|,$$

and give β and Γ in terms of the physical parameters (G, M, R, V, k, m) and, if helpful, in terms of the escape velocity $V_e = \sqrt{2GM/R}$ or $g = GM/R^2$.

(c) Using the constant-gravity approximation (i.e., set $R/(R+h) \approx 1$) and neglecting the curvature of gravity, find the dimensional terminal speed $v_t > 0$ for a falling object (downward motion) by balancing weight and drag. Express v_t in terms of m, g, k. Then nondimensionalize this result to give v_t/V in terms of β and Γ, or directly in terms of V, m, and k.

(d) Discuss qualitatively how the drag parameter Γ affects (i) the time to reach maximum height, (ii) the maximum height, and (iii) whether escape is possible for a given V. What happens in the limits $\Gamma \to 0$ and $\Gamma \to \infty$?

(e) Propose a simple numerical strategy (ODE solver and nondimensional form to use) to compute $h(t)$ and $h_{\max}$ for given physical parameters. Suggest nondimensional parameter values for a small projectile (e.g., a baseball) and for a dense compact projectile (e.g., a small metal sphere), then explain how you would compare results to the no-drag case.

2.5 O (Big Oh); o (little oh)

The symbols Big O and little o play a fundamental role in perturbation theory, where the behavior of solutions is analyzed as a small parameter approaches zero or infinity. Big-O notation provides a convenient way to express the order of magnitude of terms, helping to estimate the size of neglected quantities and justify approximations. In contrast, little-o notation captures terms that become negligible compared to leading terms, indicating higher-order corrections that vanish faster in the limit. Together, these notations offer a precise mathematical framework for quantifying errors and simplifying complex problems in asymptotic and perturbation analyses.

Let $\mathbb{R}_{\geq 0}$ denote the set of nonnegative real numbers. Let $D \subseteq \mathbb{R}_{\geq 0}$ be defined in some neighborhood (or punctured neighborhood) of $\varepsilon = 0$. Below, we give a formal definition of

$$f(\varepsilon) = O(g(\varepsilon)), \quad \text{as } \varepsilon \to 0.$$

Definition 2.2 *Let $f(\varepsilon)$ and $g(\varepsilon)$ be functions defined on the set D. We say that*

$$f(\varepsilon) = O(g(\varepsilon)), \quad 0 < \varepsilon \leq \varepsilon_0,$$

if there exists a positive constant M such that

$$|f(\varepsilon)| \leq M|g(\varepsilon)|, \quad \textit{for all} \quad 0 < \varepsilon \leq \varepsilon_0.$$

Equivalently, $f(\varepsilon) = O(g(\varepsilon))$ means that the ratio

$$\left|\frac{f(\varepsilon)}{g(\varepsilon)}\right|$$

remains bounded for $0 \leq \varepsilon \leq \varepsilon_0$. In simpler terms, $f = O(g)$ indicates that the growth rate of f does not exceed that of g.

Remark 5 *Roughly speaking,*

$$f(\varepsilon) = O(g(\varepsilon))$$

means that f grows no faster than g.

Example 2.1 *Let $f(\varepsilon) = \frac{5}{\varepsilon} + 6$. We claim that $f(\varepsilon) = O(\frac{1}{\varepsilon})$. In order to show this, all we need to do is exhibit two constants $M > 0, \varepsilon_0 \geq 0$ such that for all $0 < \varepsilon \leq \varepsilon_0$, we have $\frac{5}{\varepsilon} + 6 \leq M\frac{1}{\varepsilon}$. If we set $\varepsilon_0 = \frac{1}{6}$, then we get for all $\varepsilon \leq \frac{1}{6}$, that $\frac{5}{\varepsilon} + 6 \leq \frac{5}{\varepsilon} + \frac{1}{\varepsilon} \leq 6\frac{1}{\varepsilon}$. Thus, we have shown that $\frac{5}{\varepsilon} + 6 = O(\frac{1}{\varepsilon})$ with $M = 6$.* □

Example 2.2 *Let* $f(\varepsilon) = \frac{1}{\varepsilon^2}\log\frac{1}{\varepsilon} + \frac{1}{\varepsilon^3}$ *and* $g(\varepsilon) = \frac{1}{\varepsilon^3}$, *for* $0 < \varepsilon \leq \varepsilon_0$. *We are interested in showing that* $f(\varepsilon) = O(g(\varepsilon))$. *Let*

$$h(\varepsilon) = \frac{f(\varepsilon)}{g(\varepsilon)} = \varepsilon\log\frac{1}{\varepsilon} + 1, \quad 0 < \varepsilon \leq \varepsilon_0.$$

Using L'Hôpital's rule, we find that

$$\lim_{\varepsilon\to 0^+} h(\varepsilon) = 1 + \lim_{\varepsilon\to 0^+} \varepsilon\log\frac{1}{\varepsilon} = 1.$$

Moreover, the function $h(\varepsilon)$ *attains its maximum at* $\varepsilon = \frac{1}{e}$. *To see this, we try to maximize* $h(\varepsilon)$. *Thus,*

$$h'(\varepsilon) = \log\frac{1}{\varepsilon} - 1.$$

Setting $h'(\varepsilon) = 0$ *and solving for* ε, *we obtain* $\varepsilon = \frac{1}{e}$. *Additionally,*

$$h''(\varepsilon) = -\frac{1}{\varepsilon} < 0 \quad \textit{for all } 0 < \varepsilon \leq \varepsilon_0 = \frac{1}{e}.$$

Thus, we conclude

$$\frac{f(\varepsilon)}{g(\varepsilon)} \leq h\left(\frac{1}{e}\right) = 1 + \frac{1}{e}, \quad \textit{for all } 0 < \varepsilon \leq \frac{1}{e}.$$

Hence, we have shown that $f(\varepsilon) = O\left(\frac{1}{\varepsilon^3}\right)$ *with* $M = 1 + \frac{1}{e}$ *and* $\varepsilon_0 = \frac{1}{e}$. □

Example 2.3 *Let* $f(\varepsilon) = \frac{\varepsilon}{1+\varepsilon}$ *as* $\varepsilon \to 0$. *As* $\varepsilon \to 0$, *the denominator* $1 + \varepsilon \to 1$. *To prove this, we need to find a constant M such that*

$$\frac{\varepsilon}{1+\varepsilon} \leq M\varepsilon$$

for small ε. *We can choose*

$$M = 1,$$

since

$$\frac{f(\varepsilon)}{\varepsilon} = \frac{1}{1+\varepsilon} < 1 \quad \textit{for } \varepsilon > 0.$$

□

We have the following important remark.

Remark 6 *The Big-O notation* $f(\varepsilon) = O(g(\varepsilon))$ *is not symmetric, meaning* $f(\varepsilon) = O(g(\varepsilon))$ *does not imply* $g(\varepsilon) = O(f(\varepsilon))$.

Here is an example

Example 2.4 *Let* $f(\varepsilon)=\varepsilon$ *and* $g(\varepsilon)=\varepsilon^2$ *as* $\varepsilon\to 0^+$. *We observe that*

$$\frac{f(\varepsilon)}{g(\varepsilon)}=\frac{\varepsilon}{\varepsilon^2}=\frac{1}{\varepsilon}\to\infty \quad \text{as } \varepsilon\to 0^+.$$

Hence, $f(\varepsilon)$ *is* not $O(g(\varepsilon))$. *However,*

$$\frac{g(\varepsilon)}{f(\varepsilon)}=\frac{\varepsilon^2}{\varepsilon}=\varepsilon\to 0 \quad \text{as } \varepsilon\to 0^+,$$

so we have $g(\varepsilon)=O(f(\varepsilon))$. *Therefore, the Big-O relation is* ***not symmetric****:* $g(\varepsilon)=O(f(\varepsilon))$ *does not imply* $f(\varepsilon)=O(g(\varepsilon))$. □

However, the Big-O relation is *transitive as the next corollary shows.*

Corollary 1 *Assume we have three functions,* $f(\varepsilon)$, $g(\varepsilon)$, *and* $h(\varepsilon)$.

$$f(\varepsilon)=O(g(\varepsilon)) \quad \text{as } \varepsilon\to 0, \quad g(\varepsilon)=O(h(\varepsilon)) \quad \text{as } \varepsilon\to 0.$$

Then, $f(\varepsilon)=O(h(\varepsilon))$ *as* $\varepsilon\to 0$.

Proof *From the definition of Big-O notation as* $\varepsilon\to 0$, *we have that*

since $f(\varepsilon)=O(g(\varepsilon))$, *there are constants* $C_1>0$ *and* $\varepsilon_1>0$ *such that*
$|f(\varepsilon)|\le C_1|g(\varepsilon)|$ *for all* $0<\varepsilon\le\varepsilon_1$.
Similarly, $g(\varepsilon)=O(h(\varepsilon))$, *there are constants* $C_2>0$ *and* $\varepsilon_2>0$ *such that*
$|g(\varepsilon)|\le C_2|h(\varepsilon)|$ *for all* $0<\varepsilon\le\varepsilon_2$.

By substituting the second inequality into the first, we arrive at

$$|f(\varepsilon)|\le C_1|g(\varepsilon)|\le C_1(C_2|h(\varepsilon)|)=(C_1C_2)|h(\varepsilon)|.$$

This inequality holds when both initial conditions are satisfied, that is, for

$$0<\varepsilon<\min(\varepsilon_1,\varepsilon_2).$$

Letting

$$M=C_1C_2, \qquad \varepsilon_0=\min(\varepsilon_1,\varepsilon_2),$$

we have for all ε *such that* $0<\varepsilon\le\varepsilon_0$,

$$|f(\varepsilon)|\le M|h(\varepsilon)|, \quad \text{for all } 0<\varepsilon\le\varepsilon_0.$$

We have the following example.

Example 2.5 *Consider the following relationships as* $\varepsilon\to 0$:

$$f(\varepsilon)=5\varepsilon^3+2\varepsilon^4, \qquad g(\varepsilon)=\varepsilon^2, \qquad h(\varepsilon)=\varepsilon.$$

We observe:

$$f(\varepsilon) = O(g(\varepsilon)),$$

since for small ε,

$$5\varepsilon^3 + 2\varepsilon^4 \leq 5\varepsilon^2 + 2\varepsilon^2 \leq 7\varepsilon^2 \quad \textit{for } 0 < \varepsilon < 1.$$

Also,

$$g(\varepsilon) = O(h(\varepsilon)),$$

because $\varepsilon^2 \leq \varepsilon$ *when* $0 < \varepsilon < 1$. *By transitivity, we conclude*

$$f(\varepsilon) = O(h(\varepsilon)).$$

Indeed, $5\varepsilon^3 + 2\varepsilon^4$ *is bounded by a constant multiple of* ε *as varepsilon* $\to 0$. □

Definition 2.3 *We say that* $f(\varepsilon) \in \Theta(g(\varepsilon))$ *if there exists constants* $\varepsilon_0 \geq 0$ *and* $M_1, M_2 > 0$ *such that for all* $0 < \varepsilon \leq \varepsilon_0$,

$$M_1|g(\varepsilon)| \leq |f(\varepsilon)| \leq M_2|g(\varepsilon)|.$$

We provide the following example.

Example 2.6 *Let*

$$f(\varepsilon) = 5\varepsilon^2 + 2\varepsilon^3 \quad \textit{and} \quad g(\varepsilon) = \varepsilon^2.$$

We claim that $f(\varepsilon) \in \Theta(g(\varepsilon))$ *as* $\varepsilon \to 0$. *For* $0 < \varepsilon \leq 1$, *we have*

$$|f(\varepsilon)| = |5\varepsilon^2 + 2\varepsilon^3| \leq 5\varepsilon^2 + 2\varepsilon^3 \leq 7\varepsilon^2 = 7|g(\varepsilon)|.$$

Also,

$$|f(\varepsilon)| = |5\varepsilon^2 + 2\varepsilon^3| \geq 5\varepsilon^2 - 2\varepsilon^3 \geq 3\varepsilon^2 = 3|g(\varepsilon)|.$$

Hence, we can choose $M_1 = 3$, $M_2 = 7$, *and* $\varepsilon_0 = 1$, *which shows that*

$$M_1|g(\varepsilon)| \leq |f(\varepsilon)| \leq M_2|g(\varepsilon)| \quad \textit{for all } 0 < \varepsilon \leq \varepsilon_0.$$

Therefore, $f(\varepsilon) \in \Theta(g(\varepsilon))$ *as* $\varepsilon \to 0$. □

Definition 2.4 *Let* $f(\varepsilon)$ *and* $g(\varepsilon)$ *be functions defined on the set D. If*

$$\lim_{\varepsilon \to 0} \frac{f(\varepsilon)}{g(\varepsilon)} = 0,$$

then we say that

$$f(\varepsilon) = o(g(\varepsilon)), \quad \varepsilon \to 0.$$

The "little-oh" notation, denoted by $o(\varepsilon)$, is often used to signify "very small errors."

Remark 7 *If*

$$f(\varepsilon) = o(g(\varepsilon)),$$

then automatically

$$f(\varepsilon) = O(g(\varepsilon)).$$

However, the converse is generally false. Additionally,

$$f(\varepsilon) = O(g(\varepsilon))$$

means that f grows no faster than g, while

$$f(\varepsilon) = o(g(\varepsilon))$$

means that f is negligible compared to g.

Example 2.7 *Consider the following examples illustrating the use of the little-oh notation $o(\varepsilon)$, where $f(\varepsilon) = o(g(\varepsilon))$ as $\varepsilon \to 0$ means that*

$$\lim_{\varepsilon \to 0} \frac{f(\varepsilon)}{g(\varepsilon)} = 0.$$

1. ***Polynomial comparison.***

   Let

   $$f(\varepsilon) = \varepsilon^2, \qquad g(\varepsilon) = \varepsilon.$$

   Then

   $$\frac{f(\varepsilon)}{g(\varepsilon)} = \varepsilon \to 0,$$

   so

   $$\varepsilon^2 = o(\varepsilon).$$

2. ***Exponential vs. polynomial.***

   Let

   $$f(\varepsilon) = e^{-1/\varepsilon}, \qquad g(\varepsilon) = \varepsilon^n, \quad (n > 0).$$

   Then

   $$\frac{f(\varepsilon)}{g(\varepsilon)} = \frac{e^{-1/\varepsilon}}{\varepsilon^n} \to 0,$$

   hence

   $$e^{-1/\varepsilon} = o(\varepsilon^n).$$

   (Exponentials decay faster than any power.)

3. ***Logarithmic correction.***

   Let

   $$f(\varepsilon) = \varepsilon \ln \varepsilon, \qquad g(\varepsilon) = \varepsilon.$$

   Then

   $$\frac{f(\varepsilon)}{g(\varepsilon)} = \ln \varepsilon \to -\infty \qquad \text{as } \varepsilon \to 0^+,$$

so $f(\varepsilon)$ *is not* $o(g(\varepsilon))$.

Although $f(\varepsilon)$ *is not* $o(\varepsilon)$, *we still have*

$$f(\varepsilon) = o(1),$$

since

$$\varepsilon \ln \varepsilon \to 0 \qquad \text{as } \varepsilon \to 0^+.$$

4. ***Trigonometric example.***

Let

$$f(\varepsilon) = \sin \varepsilon - \varepsilon, \qquad g(\varepsilon) = \varepsilon.$$

Using the Taylor expansion $\sin \varepsilon = \varepsilon - \frac{\varepsilon^3}{6} + O(\varepsilon^5)$,

$$f(\varepsilon) = -\frac{\varepsilon^3}{6} + O(\varepsilon^5) = o(\varepsilon).$$

□

Example 2.8 *Let*

$$f(\varepsilon) = \log(\varepsilon + 1), \quad g(\varepsilon) = \varepsilon.$$

As $\varepsilon \to 0$, *we have*

$$\lim_{\varepsilon \to 0} \frac{\log(\varepsilon + 1)}{\varepsilon} = 1.$$

Since the limit is a nonzero constant, $f(\varepsilon)$ *is not* $o(g(\varepsilon))$. *However, if we take* $g(\varepsilon) = \sqrt{\varepsilon}$, *the limit becomes*

$$\lim_{\varepsilon \to 0} \frac{\log(\varepsilon + 1)}{\sqrt{\varepsilon}} = \lim_{\varepsilon \to 0} \left(\frac{\log(\varepsilon + 1)}{\varepsilon} \cdot \sqrt{\varepsilon} \right) = 1 \cdot 0 = 0.$$

Therefore,

$$\ln(\varepsilon + 1) = o(\sqrt{\varepsilon}) \quad \text{as } \varepsilon \to 0.$$

□

2.5.1 Exercises

Exercise 2.12 *Show that* $\frac{10}{\varepsilon} + 4 = O(\frac{1}{\varepsilon})$.

Exercise 2.13 *Let*

$$f(\varepsilon) = \frac{2}{\varepsilon^4} + \frac{1}{\varepsilon^2} \log \frac{1}{\varepsilon}, \qquad g(\varepsilon) = \frac{1}{\varepsilon^4}.$$

Then $f(\varepsilon) = O\left(\frac{1}{\varepsilon^4}\right)$ *as* $\varepsilon \to 0^+$.

Exercise 2.14 *Let*

$$f(\varepsilon) = \frac{\sin(\varepsilon)}{\varepsilon^2} + \frac{3}{\varepsilon}, \qquad g(\varepsilon) = \frac{1}{\varepsilon^2}.$$

Then as $\varepsilon \to 0^+$, $f(\varepsilon) = O\left(\frac{1}{\varepsilon^2}\right)$.

Exercise 2.15 *Let*

$$f(\varepsilon) = \frac{1}{\varepsilon}\log^2\frac{1}{\varepsilon} + \frac{5}{\sqrt{\varepsilon}}, \qquad g(\varepsilon) = \frac{1}{\varepsilon}.$$

Then $f(\varepsilon) = O\left(\frac{1}{\varepsilon}\log^2\frac{1}{\varepsilon}\right)$, *as* $\varepsilon \to 0^+$

Exercise 2.16 *Let*

$$f(\varepsilon) = e^{1/\varepsilon} + \frac{1}{\varepsilon^3}, \qquad g(\varepsilon) = e^{1/\varepsilon}.$$

Then $f(\varepsilon) = O(e^{1/\varepsilon})$ *as* $\varepsilon \to 0^+$.

Exercise 2.17 *Below all functions take nonnegative values.*

(a) If $f(\varepsilon) = \Theta(g(\varepsilon))$ *and* $g(\varepsilon) = \Theta(h(\varepsilon))$, *then* $f(\varepsilon) = \Theta(h(\varepsilon))$.

(b) If $f(\varepsilon) = O(g_1(\varepsilon))$ *and* $h(\varepsilon) = O(g_2(\varepsilon))$, *then* $f(\varepsilon) + h(\varepsilon) = O(g_1(\varepsilon) + g_2(\varepsilon))$.

(c) If $f(\varepsilon) = O(g_1(\varepsilon))$ *and* $h(\varepsilon) = O(g_2(\varepsilon))$, *then* $f(\varepsilon) \cdot h(\varepsilon) = O(g_1(\varepsilon) \cdot g_2(\varepsilon))$.

(d) Let $f(\varepsilon) = O(g(\varepsilon))$ *and let* $h(\varepsilon) \to 0$ *as* $\varepsilon \to 0$.
Then

$$(f \circ h)(\varepsilon) = O\big((g \circ h)(\varepsilon)\big).$$

Display an example of such functions.

Exercise 2.18 *Let* $f(\varepsilon) = 4\varepsilon^3 + 2\varepsilon^4$ *and* $g(\varepsilon) = \varepsilon^3$. *Show that* $f(\varepsilon) = \Theta(g(\varepsilon))$ *as* $\varepsilon \to 0$.

Exercise 2.19 *Let* $f(\varepsilon) = \frac{1}{\varepsilon^2} + \frac{3}{\varepsilon}$ *and* $g(\varepsilon) = \frac{1}{\varepsilon^2}$. *Determine whether* $f(\varepsilon) = \Theta(g(\varepsilon))$ *as* $\varepsilon \to 0$.

Exercise 2.20 *Let* $f(\varepsilon) = \varepsilon\log(1/\varepsilon) + \varepsilon^2$ *and* $g(\varepsilon) = \varepsilon\log(1/\varepsilon)$. *Show that* $f(\varepsilon) = \Theta(g(\varepsilon))$ *as* $\varepsilon \to 0$.

Exercise 2.21 *Let*

$$f(\varepsilon) = \frac{\varepsilon^3 + \varepsilon^4\log(1/\varepsilon)}{\varepsilon^2 + \varepsilon(2 + \sin(1/\varepsilon))} \quad \textit{and} \quad g(\varepsilon) = \varepsilon.$$

Determine whether $f(\varepsilon) = \Theta(g(\varepsilon))$ *as* $\varepsilon \to 0$.

Exercise 2.22 *Let*

$$f(\varepsilon) = \varepsilon e^{\varepsilon^2} + \varepsilon^2 \log(1+\varepsilon) \quad \text{and} \quad g(\varepsilon) = \varepsilon.$$

Show that $f(\varepsilon) = O(g(\varepsilon))$ *as* $\varepsilon \to 0$.

Exercise 2.23 *Let*

$$f(\varepsilon) = \frac{\log(1+\varepsilon^2)}{\sqrt{\varepsilon^2+\varepsilon^4}} + \varepsilon \sin(1/\varepsilon) \quad \text{and} \quad g(\varepsilon) = 1.$$

Decide whether $f(\varepsilon) = \Theta(g(\varepsilon))$ *as* $\varepsilon \to 0$.

Exercise 2.24 *Let*

$$f(\varepsilon) = \frac{\varepsilon^3 + \varepsilon^4 \log(1/\varepsilon)}{\varepsilon^2 + \varepsilon \sin(1/\varepsilon)}, \quad g(\varepsilon) = \varepsilon.$$

Determine whether $f(\varepsilon) \in O(g(\varepsilon))$ *as* $\varepsilon \to 0$.

Exercise 2.25 *Let*

$$f(\varepsilon) = \varepsilon e^{\varepsilon^2} + \varepsilon^2 \log(1+\varepsilon), \quad g(\varepsilon) = \varepsilon.$$

Show that $f(\varepsilon) \in \Theta(g(\varepsilon))$ *as* $\varepsilon \to 0$.

Exercise 2.26 *Let*

$$f(\varepsilon) = \frac{\log(1+\varepsilon^2)}{\sqrt{\varepsilon^2+\varepsilon^4}} + \varepsilon \sin(1/\varepsilon), \quad g(\varepsilon) = 1.$$

Decide whether $f(\varepsilon) \in \Theta(g(\varepsilon))$ *as* $\varepsilon \to 0$.

Exercise 2.27 *(a) Prove that if* $f(\varepsilon) = o(g(\varepsilon))$ *and* $h(\varepsilon)$ *is bounded as* $\varepsilon \to 0$, *then* $h(\varepsilon)f(\varepsilon) = o(g(\varepsilon))$.

(b) Let $f(\varepsilon) = \varepsilon \ln(1+\varepsilon)$ *and* $g(\varepsilon) = \varepsilon^2$. *Determine whether* $f(\varepsilon) = o(g(\varepsilon))$ *as* $\varepsilon \to 0$.

(c) Let $f(\varepsilon) = e^{\sqrt{\varepsilon}} - 1 - \sqrt{\varepsilon}$. *Find the smallest power p such that* $f(\varepsilon) = o(\varepsilon^p)$ *as* $\varepsilon \to 0$.

(d) Let $f(\varepsilon) = \sin(\varepsilon) - \varepsilon + \frac{\varepsilon^3}{6}$. *Show that* $f(\varepsilon) = o(\varepsilon^3)$ *as* $\varepsilon \to 0$.

(e) Prove that if $f(\varepsilon) = o(g(\varepsilon))$ *and* $g(\varepsilon) = o(h(\varepsilon))$, *then* $f(\varepsilon) = o(h(\varepsilon))$.

Exercise 2.28 *Evaluate whether the following statements are true or false (justify each answer):*

(a) $\ln(1+\varepsilon) = o(\varepsilon)$

(b) $\varepsilon^2 = o(\varepsilon \ln(1/\varepsilon))$

(c) $1 - \cos\varepsilon = o(\varepsilon)$

(d) $\tan\varepsilon - \varepsilon = o(\varepsilon^2)$

Exercise 2.29 *Let* $f(\varepsilon) = e^{\varepsilon} - 1 - \varepsilon - \dfrac{\varepsilon^2}{2}$. *Find the smallest integer n such that* $f(\varepsilon) = o(\varepsilon^n)$.

Exercise 2.30 *Assume* $f(\varepsilon) = o(\varepsilon^2)$ *and* $g(\varepsilon) = o(\varepsilon)$. *Show that* $f(\varepsilon)+g(\varepsilon) = o(\varepsilon)$.

Exercise 2.31 *Let* $f(\varepsilon) = \sin\varepsilon - \varepsilon + \dfrac{\varepsilon^3}{6}$.

(a) Show that

$$f(\varepsilon) = o(\varepsilon^3) \quad \text{as } \varepsilon \to 0.$$

In other words, prove that

$$\lim_{\varepsilon\to 0} \frac{\sin\varepsilon - \varepsilon + \varepsilon^3/6}{\varepsilon^3} = 0.$$

(b) Deduce from part (a) that

$$\sin\varepsilon = \varepsilon - \frac{\varepsilon^3}{6} + o(\varepsilon^3) \quad \text{as } \varepsilon \to 0.$$

(c) Using only the definition of small-o notation (and not any pre-stated Taylor expansions), establish the stronger estimate

$$\sin\varepsilon = \varepsilon - \frac{\varepsilon^3}{6} + o(\varepsilon^4) \quad \text{as } \varepsilon \to 0.$$

Hint: *Use the identity*

$$\sin\varepsilon - \varepsilon + \frac{\varepsilon^3}{6} = \varepsilon^3\left(\frac{\sin\varepsilon}{\varepsilon^3} - \frac{1}{\varepsilon^2} + \frac{1}{6}\right),$$

together with the fact that

$$\frac{\sin\varepsilon}{\varepsilon} \to 1, \qquad \frac{1-\cos\varepsilon}{\varepsilon^2} \to \frac{1}{2},$$

and suitable algebraic manipulations.

Exercise 2.32 *Define, for* $\varepsilon \neq 0$,

$$f(\varepsilon) = \frac{e^{\varepsilon} - 1 - \varepsilon}{\varepsilon^2}, \qquad g(\varepsilon) = \frac{\ln(1+\varepsilon) - \varepsilon + \dfrac{\varepsilon^2}{2}}{\varepsilon^2}.$$

(a) Show that

$$e^{\varepsilon} - 1 - \varepsilon = \frac{\varepsilon^2}{2} + o(\varepsilon^2) \quad \text{as } \varepsilon \to 0,$$

by proving directly from the definition that

$$\lim_{\varepsilon \to 0} f(\varepsilon) = \frac{1}{2}.$$

Hint: *You may use the inequality* $e^{\varepsilon} \geq 1 + \varepsilon$ *and the fact that*

$$\frac{e^{\varepsilon} - 1}{\varepsilon} \to 1 \quad \text{as } \varepsilon \to 0.$$

(b) Prove that

$$\ln(1+\varepsilon) - \varepsilon + \frac{\varepsilon^2}{2} = o(\varepsilon^2) \quad \text{as } \varepsilon \to 0$$

by establishing that

$$\lim_{\varepsilon \to 0} g(\varepsilon) = 0.$$

Hint: *Start from the basic limit*

$$\lim_{\varepsilon \to 0} \frac{\ln(1+\varepsilon)}{\varepsilon} = 1$$

and use suitable rearrangements and estimates to bound $g(\varepsilon)$ *in absolute value by a function that tends to* 0.

(c) Combine parts (a) and (b) to show that

$$e^{\varepsilon} \ln(1+\varepsilon) = \varepsilon + \frac{\varepsilon^2}{2} + o(\varepsilon^2) \quad \text{as } \varepsilon \to 0.$$

Hint: *Write both* e^{ε} *and* $\ln(1+\varepsilon)$ *in the form*

$$1 + \varepsilon + \frac{\varepsilon^2}{2} + (\text{small-}o \text{ term}),$$

carefully expand the product, and keep track of which terms are $o(\varepsilon^2)$.

2.6 Asymptotic Expansions

Asymptotic expansions provide a fundamental framework for approximating the behavior of functions under extreme conditions, such as when a variable becomes very large or very small. In applied mathematics, physics, and engineering, exact formulas are often unavailable or too complicated to be useful in such regimes. While Taylor and Maclaurin series are powerful tools for local approximation near a fixed point, they rely on convergence and frequently fail to describe the behavior of functions far from that point. Asymptotic expansions overcome this limitation by providing approximations whose accuracy improves as the variable approaches a limiting value, even though the resulting series may not converge. This makes asymptotic methods indispensable in the analysis of large-parameter problems, singular perturbations, special functions, and many applications throughout applied mathematics. Asymptotic expansions play a fundamental role in the analysis of integrals containing large parameters. When a parameter $\lambda \to \infty$, exact evaluation of integrals of the form

$$I(\lambda) = \int_a^b f(t)\, e^{\lambda \phi(t)}\, dt$$

is often difficult or impossible in closed form. In such cases, asymptotic methods provide accurate approximations by identifying the dominant contributions to the integral as the parameter becomes large.

To illustrate the basic idea of asymptotic approximation, consider the function

$$(4t^2 + t)^{1/3}, \qquad t \to \infty.$$

Since $4t^2$ dominates t for large t, we expect

$$(4t^2 + t)^{1/3} \approx (4t^2)^{1/3} = 4^{1/3} t^{2/3}.$$

The relative error in this approximation is

$$\frac{(4t^2 + t)^{1/3} - 4^{1/3} t^{2/3}}{4^{1/3} t^{2/3}}.$$

Observe that

$$(4t^2 + t)^{1/3} = 4^{1/3} t^{2/3} \left(1 + \frac{1}{4t}\right)^{1/3}.$$

Hence

$$\frac{(4t^2 + t)^{1/3} - 4^{1/3} t^{2/3}}{4^{1/3} t^{2/3}} = \left(1 + \frac{1}{4t}\right)^{1/3} - 1.$$

By the Mean Value Theorem,

$$(1 + x)^{1/3} = 1 + \frac{1}{3}(1 + c)^{-2/3} x, \qquad 0 < c < x.$$

Therefore,

$$\left(1 + \frac{1}{4t}\right)^{1/3} - 1 = \frac{1}{3}(1 + c)^{-2/3} \frac{1}{4t}, \qquad 0 < c < \frac{1}{4t}.$$

Since $\frac{1}{4t} \to 0$ as $t \to \infty$, the relative error tends to zero. Equivalently,

$$\lim_{t \to \infty} \frac{(4t^2 + t)^{1/3}}{4^{1/3} t^{2/3}} = 1.$$

Definition 2.5 (Asymptotic expansion near a finite point) *A function $f(x)$ is said to have an asymptotic expansion in a sequence of functions $\{\phi_n(x)\}$ as $x \to x_0$ if the following condition holds for every N:*

$$f(x) - \sum_{n=0}^{N} a_n \phi_n(x) = o(\phi_N(x)) \quad \text{as } x \to x_0. \tag{2.20}$$

Here, a_n are constant coefficients, and the notation $g(x) = o(h(x))$ means

$$\lim_{x \to x_0} \frac{g(x)}{h(x)} = 0,$$

meaning $g(x)$ goes to zero faster than $h(x)$. The sequence $\{\phi_n(x)\}$ is called an asymptotic sequence, *typically ordered such that*

$$\phi_{n+1}(x) = o(\phi_n(x)).$$

The expansion is formally written as

$$f(x) \sim \sum_{n=0}^{\infty} a_n \phi_n(x) \quad \text{as} \quad x \to x_0.$$

The crucial difference from convergent series is that the error is small relative to the *last term included* for a given large x, rather than the entire series converging to the true value as the number of terms goes to infinity. The successive coefficients a_n can be obtained from a given function $f(x)$ and a given asymptotic sequence by the successive application of the following limits:

$$a_0 = \lim_{x \to x_0} \frac{f(x)}{\phi_0(x)},$$

$$a_1 = \lim_{x \to x_0} \frac{f(x) - a_0\phi_0(x)}{\phi_1(x)},$$

$$\vdots$$

$$a_k = \lim_{x \to x_0} \frac{f(x) - \sum_{n=1}^{k-1} a_n \phi_n(x)}{\phi_k(x)}.$$

Thus, the expansion of a given function $f(x)$ in terms of a given asymptotic sequence is unique. A different asymptotic expansion can result in terms of a different sequence ϕ_n. Various linear operations and multiplications can be performed with asymptotic expansions. For certain purposes, a slightly more general asymptotic expansion can be defined.

Definition 2.6 (General asymptotic expansion) *A sequence of functions $\psi_n(x)$ forms an asymptotic expansion of $f(x)$ in terms of an asymptotic sequence of gauge functions $\phi_n(x)$ if*

$$\lim_{x \to x_0} \frac{f(x) - \sum_{n=1}^{N} \psi_n(x)}{\phi_N(x)} = 0.$$

Example 2.9 *(Asymptotic expansion near a singularity) When a function has a singularity at the point of interest (e.g., $x = 0$), a convergent power series (like a Taylor series) cannot be formed. Asymptotic expansion, which can use a variety of gauge functions beyond simple powers of x (such as $\log x$ or $1/x$), provides a powerful way to approximate the function's behavior near that singularity. Consider the function*

$$f(x) = \frac{1}{1 - e^{-x}}.$$

As $x \to 0^+$, $e^{-x} \to 1$, so the denominator approaches zero, and $f(x)$ approaches infinity (a singularity). A Taylor series around $x = 0$ is not possible for $f(x)$ itself.

To derive an asymptotic expansion for small x, we use the Taylor expansion of e^{-x} in the denominator:

$$e^{-x} = 1 - x + \frac{x^2}{2!} - \frac{x^3}{3!} + \dots.$$

Substitute this into $f(x)$:

$$f(x) = \frac{1}{1 - (1 - x + \frac{x^2}{2!} - \frac{x^3}{3!} + \dots)} = \frac{1}{x - \frac{x^2}{2!} + \frac{x^3}{3!} - \dots}.$$

Factor out an x:

$$f(x) = \frac{1}{x\left(1 - \frac{x}{2!} + \frac{x^2}{3!} - \dots\right)} = \frac{1}{x} \frac{1}{1 - \left(\frac{x}{2!} - \frac{x^2}{3!} + \dots\right)}.$$

Let

$$\varepsilon(x) = \frac{x}{2!} - \frac{x^2}{3!} + \dots.$$

Since $\varepsilon(x) \to 0$ as $x \to 0$, we apply

$$\frac{1}{1 - \varepsilon} = 1 + \varepsilon + \varepsilon^2 + \dots.$$

Thus,

$$f(x) \sim \frac{1}{x}\left(1 + \left(\frac{x}{2!} - \frac{x^2}{3!}\right) + \left(\frac{x}{2!} - \frac{x^2}{3!}\right)^2 + \dots\right).$$

Expanding and collecting powers of x yields:

$$f(x) \sim \frac{1}{x} + \frac{1}{2} + \frac{x}{12} - \frac{x^2}{24} + \dots, \qquad x \to 0^+.$$

This expansion, which uses the asymptotic sequence $\{x^{-1}, x^0, x^1, x^2, \ldots\}$*, provides an effective approximation for* $f(x)$ *near its singularity at* $x = 0$*, something a standard Taylor series cannot do.* □

In many problems arising in analysis and applications, one is interested in understanding the behavior of a function for very large values of the independent variable. In such regimes, exact formulas are often either unavailable or too complicated to be useful. Instead, it is common to seek approximate representations that capture the dominant behavior of the function and provide systematic corrections. Asymptotic expansions offer a powerful framework for this purpose. Unlike Taylor series, which rely on local analyticity and convergence, asymptotic expansions are defined through an error estimate and do not require the underlying series to converge. Their significance lies in the fact that truncating the series after finitely many terms yields increasingly accurate approximations as the variable tends to the limiting value, here infinity. For functions defined on a semi-infinite interval, expansions in inverse powers of the variable naturally arise and play a central role in areas such as perturbation theory, special functions, and applied mathematics. The following definition formalizes the notion of an asymptotic expansion of a function as the independent variable tends to infinity.

Definition 2.7 (Asymptotic expansion at infinity) *Let* $f(x)$ *be a function defined for all sufficiently large* x*. A formal series*

$$c_0 + \frac{c_1}{x} + \frac{c_2}{x^2} + \frac{c_3}{x^3} + \cdots$$

(where convergence is not *required) is called an* asymptotic expansion *of* $f(x)$ *as* $x \to \infty$ *if, for every fixed integer* $N \geq 0$*,*

$$f(x) - \sum_{k=0}^{N} \frac{c_k}{x^k} = o\left(\frac{1}{x^N}\right) \quad \text{as } x \to \infty.$$

This is often written as

$$f(x) \sim c_0 + \frac{c_1}{x} + \frac{c_2}{x^2} + \frac{c_3}{x^3} + \cdots, \quad x \to \infty.$$

The meaning of the definition is the following:

- The expression
$$c_0 + \frac{c_1}{x} + \frac{c_2}{x^2} + \cdots$$
is a *formal series*. It is not required to converge for any value of x.
- The condition
$$f(x) - \sum_{k=0}^{N} \frac{c_k}{x^k} = o\left(\frac{1}{x^N}\right) \tag{2.21}$$
means that truncating the series after N terms gives an approximation whose error is much smaller than x^{-N} as $x \to \infty$.

- Equivalently, multiplying both sides by x^N yields

$$\left[f(x) - \sum_{k=0}^{N} \frac{c_k}{x^k}\right] x^N \longrightarrow 0 \quad \text{as } x \to \infty, \tag{2.22}$$

which matches the structure of the definition you wrote (after correcting powers and notation).

- Each additional term improves the approximation for large x, even if the infinite series does not converge.

- Big-O Variant: Sometimes, one finds the weaker statement

$$f(x) - \sum_{k=0}^{N} \frac{c_k}{x^k} = O\left(\frac{1}{x^{N+1}}\right).$$

This still implies an asymptotic expansion, but it gives less precise information than the little-o form.

- By making use of (2.21) we can easily determine that 6

$$c_0 = \lim_{x\to\infty} f(x),$$

$$c_1 = \lim_{x\to\infty} x\big(f(x) - c_0\big),$$

$$c_2 = \lim_{x\to\infty} x^2\left(f(x) - c_0 - \frac{c_1}{x}\right),$$

and so on. This shows that the coefficients c_k are uniquely determined (if the expansion exists).

Remark 8 • *An asymptotic series may diverge, yet still give excellent approximations when truncated.*

- *Convergence of the series is* not *the goal; accuracy of finite truncations for large x is.*

- *Asymptotic expansions are local descriptions (here, near $x = \infty$), not global representations.*

Example 2.10 *Consider*

$$f(x) = \sqrt{x^2+1}, \qquad x \to \infty.$$

We write

$$\sqrt{x^2+1} = x\sqrt{1+\frac{1}{x^2}}.$$

Using the binomial expansion,

$$\sqrt{1+u} = 1 + \frac{1}{2}u - \frac{1}{8}u^2 + \frac{1}{16}u^3 + O(u^4), \quad u \to 0,$$

with $u = \frac{1}{x^2}$, *we obtain*

$$\sqrt{x^2+1} = x\left(1 + \frac{1}{2x^2} - \frac{1}{8x^4} + \frac{1}{16x^6} + \cdots\right).$$

Hence,

$$\sqrt{x^2+1} \sim x + \frac{1}{2x} - \frac{1}{8x^3} + \frac{1}{16x^5} + \cdots, \qquad x \to \infty.$$

Truncating after the first two terms gives

$$\sqrt{x^2+1} = x + \frac{1}{2x} + o\left(\frac{1}{x}\right),$$

and including more terms systematically improves the approximation for large x, even though the full series does not converge. □

The study of integral equations is important since they arise from nonlinear differential equations, as the next example shows.

Example 2.11 *Consider the differential equation*

$$y'(t) - y(t) = -\frac{1}{t}, \qquad t > 0.$$

Multiply the equation by the integrating factor e^{-t}, *we obtain*

$$e^{-t}y'(t) - e^{-t}y(t) = -\frac{e^{-t}}{t}.$$

The left-hand side is an exact derivative, and hence it can be written in the form

$$\frac{d}{dt}\left(e^{-t}y(t)\right) = -\frac{e^{-t}}{t}.$$

Integrate both sides from t to ∞ *to have*

$$\int_t^\infty \frac{d}{ds}\left(e^{-s}y(s)\right)ds = -\int_t^\infty \frac{e^{-s}}{s}ds.$$

This gives

$$-e^{-t}y(t) = -\int_t^\infty \frac{e^{-s}}{s}ds,$$

where we have assumed

$$\lim_{s\to\infty} e^{-s}y(s) = 0.$$

Multiplying by $-e^t$ *yields*

$$y(t) = e^t \int_t^\infty \frac{e^{-s}}{s}ds.$$

Thus, the integral equation

$$y(t) = e^t \int_t^\infty \frac{e^{-s}}{s}\, ds$$

arises naturally as the solution of the inhomogeneous differential equation

$$y'(t) - y(t) + \frac{1}{t} = 0, \qquad t > 0,$$

under the decay condition $e^{-t} y(t) \to 0$ *as* $t \to \infty$. □

Commonly Used Taylor Series

$$e^x = \sum_{n=0}^{\infty} \frac{x^n}{n!} = 1 + x + \frac{x^2}{2!} + \frac{x^3}{3!} + \cdots$$

$$\sin(x) = \sum_{n=0}^{\infty} (-1)^n \frac{x^{2n+1}}{(2n+1)!} = x - \frac{x^3}{3!} + \frac{x^5}{5!} - \frac{x^7}{7!} + \cdots$$

$$\cos(x) = \sum_{n=0}^{\infty} (-1)^n \frac{x^{2n}}{(2n)!} = 1 - \frac{x^2}{2!} + \frac{x^4}{4!} - \frac{x^6}{6!} + \cdots$$

$$\ln(1+x) = x - \frac{x^2}{2} + \frac{x^3}{3} - \frac{x^4}{4} + \frac{x^5}{5} - \cdots = \sum_{n=1}^{\infty} (-1)^{n+1} \frac{x^n}{n}, \quad |x| < 1.$$

$$\frac{1}{1-x} = 1 + x + x^2 + \cdots = \sum_{n=0}^{\infty} x^n, \quad |x| < 1.$$

$$(1+x)^\alpha = 1 + \alpha x + \frac{\alpha(\alpha-1)}{2!} x^2 + \frac{\alpha(\alpha-1)(\alpha-2)}{3!} x^3 + \cdots = \sum_{n=0}^{\infty} \binom{\alpha}{n} x^n, \qquad |x| < 1,$$

where α is any real number and

$$\binom{\alpha}{n} = \frac{\alpha(\alpha-1)\cdots(\alpha-n+1)}{n!}, \qquad n \geq 1,$$

and

$$\binom{\alpha}{0} = 1.$$

2.6.1 Exercises

Exercise 2.33 *Consider the function*

$$f(t) = (9t^2 + 5t - 1)^{1/3}.$$

(a) Find the leading-order asymptotic approximation of $f(t)$ as $t \to \infty$.

(b) Compute the relative error of this approximation.

(c) Show explicitly that the relative error tends to zero as $t \to \infty$.

Exercise 2.34 *Let*

$$f(t) = (t^3 + 2t^2)^{1/2}.$$

(a) Factor out the dominant power of t and rewrite $f(t)$ in the form

$$t^{3/2}(1 + g(t))^{1/2},$$

where $g(t) \to 0$ as $t \to \infty$.

(b) Use the binomial expansion to obtain the first two terms of an asymptotic expansion of $f(t)$.

(c) Verify that the neglected remainder is $o(t^{1/2})$.

Exercise 2.35 *Let*

$$f(x) = \frac{\ln(1+x)}{x}.$$

(a) Determine an asymptotic expansion of $f(x)$ as $x \to 0^+$ up to and including terms of order x^2.

(b) Identify the asymptotic sequence used.

(c) Explain why the expansion remains useful even if the resulting series is truncated.

Exercise 2.36 *Consider the function*

$$f(x) = \frac{1}{\sqrt{1 - e^{-x}}}.$$

(a) Show that $f(x)$ has a singularity at $x = 0$.

(b) Using the Taylor expansion of e^{-x}, derive the first two terms in an asymptotic expansion of $f(x)$ as $x \to 0^+$.

(c) Verify that the error after truncation is small relative to the last retained term.

Exercise 2.37 *Let $f(x)$ admit an asymptotic expansion*

$$f(x) \sim a_0 + \frac{a_1}{x} + \frac{a_2}{x^2} + \cdots, \qquad x \to \infty.$$

(a) Explain why the series may diverge for fixed x.

(b) *Show that, for any fixed N,*

$$f(x) - \sum_{n=0}^{N} \frac{a_n}{x^n} = o(x^{-N}), \quad x \to \infty.$$

(c) *Interpret this result in terms of the practical accuracy of asymptotic approximations.*

Exercise 2.38 *Consider the differential equation*

$$y'(t) - 2y(t) + \frac{1}{t} = 0, \qquad t > 0.$$

(a) *Rewrite the equation in standard linear form and determine the integrating factor.*

(b) *Multiply the equation by the integrating factor and show that the left-hand side becomes an exact derivative.*

(c) *By integrating from t to ∞ and assuming $\lim_{s\to\infty} e^{-2s}y(s) = 0$, derive an integral representation for the solution $y(t)$ analogous to*

$$y(t) = e^t \int_t^\infty \frac{e^{-s}}{s}\, ds.$$

Exercise 2.39 *Consider the differential equation*

$$y'(t) - y(t) + \frac{\ln t}{t} = 0, \qquad t > 1.$$

(a) *Rewrite the equation as*

$$y'(t) - y(t) = -\frac{\ln t}{t},$$

and find the integrating factor.

(b) *Show that after multiplying by the integrating factor, the equation can be written as*

$$\frac{d}{dt}\left(e^{-t}y(t)\right) = -\frac{e^{-t}\ln t}{t}.$$

(c) *Assuming $\lim_{s\to\infty} e^{-s}y(s) = 0$, integrate from t to ∞ to obtain an integral formula*

$$y(t) = e^t \int_t^\infty \frac{e^{-s}\ln s}{s}\, ds.$$

(d) *Verify directly (by differentiation) that this integral expression satisfies the original differential equation.*

3

Perturbation in Differential Equations

3.1 An Introduction to Perturbation Methods

In the vast landscape of mathematical modeling, many physical phenomena are described by differential equations that are too complex to solve exactly. From the intricate orbital mechanics of celestial bodies to the tumultuous dynamics of turbulent fluids, the exact, closed-form solutions for these equations are often out of reach. When confronted with such intractable problems, mathematicians and scientists turn to approximation techniques. Among the most powerful and versatile of these is *perturbation theory*, a systematic collection of methods for finding approximate solutions by leveraging the existence of a small parameter. The fundamental idea of perturbation theory is deceptively simple: if a problem that cannot be solved exactly is "close" to one that can, then its solution should be "close" to the known, exact solution. The proximity of these problems is defined by a small, dimensionless parameter, often denoted by ε. When $\varepsilon = 0$, the problem simplifies to an *unperturbed* state that is solvable. By assuming the full solution can be expressed as an expansion in powers of ε, we can systematically calculate a series of corrections to the unperturbed solution. In differential equations, singular perturbation and boundary layers emerge when a small parameter (ε) multiplies the highest-order derivative term. This causes a fundamental change in the character of the equation as ε approaches zero, a phenomenon known as *singular perturbation* . In this limiting case, the order of the differential equation is reduced, meaning that a lower-order differential equation must approximate the solution. However, the lower-order approximation cannot satisfy all the original boundary conditions, which leads to the formation of a narrow region of rapid variation, known as the *boundary layer*.

Inside the boundary layer, the small parameter term becomes significant, and a different asymptotic expansion is needed to capture the rapid changes. A *matched asymptotic expansion* is used to resolve this issue by constructing an *inner solution* for the boundary layer and an *outer solution* for the rest of the domain. These two

DOI: 10.1201/9781042009220-3

approximations are then matched to create a single, uniformly valid approximation for the entire domain.

This technique is essential for modeling phenomena where dynamics occur on multiple scales, such as in fluid mechanics with small viscosity, where the fluid's velocity changes rapidly within a narrow layer near a surface.

This chapter begins with the most straightforward approach: *regular perturbation theory*. A problem is classified as a regular perturbation if its solution depends smoothly on the small parameter ε. In this case, setting $\varepsilon = 0$ provides a valid leading-order approximation that captures the essential behavior of the system. The solution can then be expressed as a simple power series, such as

$$y(t) = y_0(t) + \varepsilon y_1(t) + \varepsilon^2 y_2(t) + \cdots,$$

where $y_0(t)$ is the *unperturbed solution*, or the *leading term*. By substituting this series into the governing differential equation and collecting terms by powers of ε, we can construct a hierarchy of linear, more manageable equations that are solved sequentially. This iterative process allows us to build increasingly accurate approximations.

The true value of perturbation methods is realized when an exact analytical solution is impossible to find. Many important differential equations in physics and engineering contain nonlinear terms or have complicated forcing functions, such as those with exponential or logarithmic dependencies, that make them analytically intractable. Perturbation theory provides a crucial pathway forward, delivering approximate analytical solutions that are often more insightful than purely numerical results. These solutions can reveal the underlying physical principles, highlight parameter sensitivities, and inform our understanding of complex systems.

Perturbation methods are particularly indispensable for solving nonlinear differential equations. Unlike linear problems, for which superposition holds, nonlinear systems often exhibit complex behaviors, such as bifurcations, chaos, and limit cycles. Exact analytical solutions are often difficult or impossible to obtain. By treating nonlinear terms as small perturbations, we can linearize the problem around a known solution or equilibrium point. This allows for the systematic analysis of stability and the prediction of how a system's behavior changes as parameters vary. When small nonlinearities are incorporated into a system, they can lead to rich dynamics that a simple linear analysis would entirely miss.

This chapter lays the groundwork for understanding the core concepts of perturbation theory, from the simple and intuitive regular expansions to the more advanced techniques necessary for handling complex nonlinearities. While seemingly a tool for solving "small" problems, perturbation theory offers deep analytical insights into some of the most challenging problems in applied mathematics and theoretical science.

Applications of Perturbation Methods in Nonlinear Differential Equations

Perturbation methods arise naturally in many areas of applied mathematics, including fluid mechanics, nonlinear oscillations, quantum mechanics, wave propagation, and celestial mechanics. In these settings, small parameters often govern weak nonlinearities, slow variations, or multiscale behavior, making perturbation techniques indispensable analytical tools. Here are some real-world examples where perturbation methods solved nonlinear differential equations:

Celestial Mechanics and Orbital Dynamics

Discovery of Neptune.

The most famous success of perturbation theory was the prediction of Neptune's existence and location. Astronomers observed irregularities in the orbit of Uranus that could not be explained by the known gravitational forces of the Sun and other planets. These deviations were treated as small perturbations on Uranus's orbit, caused by an unknown planet. By applying perturbation methods to the nonlinear differential equations of motion for the solar system, mathematicians Urbain Le Verrier and John Couch Adams independently calculated the new planet's position.

Lunar Theory.

Predicting the Moon's orbit is a famous "three-body problem," involving the gravitational pull of the Earth and the Sun. Isaac Newton acknowledged that solving this exactly was beyond his capabilities. Early perturbation theory, developed by mathematicians like Pierre–Simon Laplace and Siméon Denis Poisson, provided highly accurate approximations for the Moon's motion by treating the Sun's gravity as a small perturbation.

Satellite Trajectories.

Modern space exploration and satellite navigation rely on perturbation methods to accurately predict and maintain satellite orbits. A satellite's motion is primarily governed by the Earth's gravity, but atmospheric drag, the gravitational pull of the Sun and Moon, and the Earth's non-uniform mass distribution all act as small perturbations. Engineers use perturbation theory to account for these nonlinear effects and compute precise trajectories.

Fluid Dynamics

Aerofoil Design.

In aerodynamics, perturbation methods are used to model the complex flow of air over an aircraft's wing (aerofoil). The basic, linear approximation assumes a simple potential flow. Perturbation theory is then used to account for nonlinear effects, such as a wing's thickness and angle of attack. This provides a more accurate prediction of lift and drag.

Water Waves.

The propagation of waves in water is governed by nonlinear equations. Perturbation methods, such as the *method of multiple scales*, are used to analyze wave phenomena like solitons (stable, localized waves) and the breakdown of wave trains. This is critical for understanding ocean dynamics and coastal engineering.

boundary layers.

A boundary layer is a thin layer of fluid near a surface where viscous effects are dominant. When the Reynolds number is large, the full fluid equations are singular. Perturbation methods, especially the *method of matched asymptotic expansions*, are used to create separate solutions for the inner boundary layer and the outer inviscid flow, which are then matched together.

Nonlinear Vibrations

Engine Mechanics.

In mechanical engineering, many vibrating systems, such as a spring–mass system, are not perfectly linear. The restoring force may include nonlinear terms. The *Duffing equation*, which models a mass–spring system with a nonlinear spring, is commonly solved using perturbation techniques. This helps predict the system's response to external forces and understand phenomena like resonance and damping in real-world machinery.

Structural Analysis.

Engineers use perturbation methods to analyze the nonlinear vibrations of elastic structures like beams and plates, especially at large displacements. The theory helps determine how finite displacements affect the frequency, period, and dynamic stresses of a structure, which is crucial for safety and design.

Quantum Mechanics

Atomic Structure.

The Schrödinger equation is fundamental to quantum mechanics. While it can be solved exactly for simple systems like the hydrogen atom, the presence of an external electric field (Stark effect) or magnetic field (Zeeman effect) adds nonlinear terms. Perturbation theory is used to find approximate solutions for the resulting energy shifts, which are in excellent agreement with experimental observations.

Electronic Structure.

In computational chemistry, methods like *Møller–Plesset perturbation theory* are used to find approximate solutions for the many-electron Schrödinger equation. By treating electron correlation as a small perturbation to a simpler model, these methods allow for the calculation of molecular properties.

These applications highlight the wide-ranging impact of perturbation methods across fluid mechanics, structural dynamics, and quantum systems. Through systematic approximations, these techniques transform complex nonlinear problems into tractable forms that yield profound physical insight.

Simple Pendulum with Damping

A classic and concrete example of identifying a small parameter ε is found in the equation of a simple pendulum with a small amount of damping.

The nonlinear differential equation for a simple pendulum is:

$$\frac{d^2\theta}{dt^2} + \frac{g}{L}\sin\theta = 0$$

where:

- θ is the angular displacement,
- g is the acceleration due to gravity,
- L is the length of the pendulum.

If we add a small amount of linear air resistance or friction, the equation becomes:

$$\frac{d^2\theta}{dt^2} + \gamma\frac{d\theta}{dt} + \frac{g}{L}\sin\theta = 0$$

where γ is the damping coefficient.

Next, we identify the small parameter. The equation as written still has dimensions (units). To nondimensionalize the equation, we define dimensionless time

$$\tau = \omega_0 t,$$

where $\omega_0 = \sqrt{\frac{g}{L}}$ is the natural frequency of the undamped pendulum. This transforms the equation into:

$$\frac{d^2\theta}{d\tau^2} + \varepsilon \frac{d\theta}{d\tau} + \sin\theta = 0$$

where the dimensionless parameter ε is identified as:

$$\varepsilon = \frac{\gamma}{\omega_0} = \gamma\sqrt{\frac{L}{g}}.$$

The parameter ε is the ratio of the damping frequency (γ) to the natural frequency (ω_0) of the system. If the physical damping is very small (meaning γ is a small number), then ε will also be a small, dimensionless number.

In this case, the unperturbed problem ($\varepsilon = 0$) is the familiar, simple harmonic motion for the undamped pendulum, which is exactly solvable. The small damping effect (the ε term) perturbs the system away from this solvable case, and we can use perturbation methods to find an approximate solution. This solution captures the gradual decay in the pendulum's amplitude over time.

3.2 Regular Perturbation

Regular perturbation theory is a mathematical technique used to obtain approximate solutions to problems involving a small parameter, typically denoted by ε. When an exact solution is unavailable or difficult to compute, one seeks an approximate solution in the form of a power series expansion in ε. Substituting this expansion into the governing equation and equating coefficients of like powers of ε produces a sequence of simpler problems that can be solved successively. This approach provides valuable insight into how small changes in the parameter influence the behavior of the system. Throughout this book, the notation

$$\varepsilon \ll 1$$

indicates that ε is a small positive parameter satisfying

$$0 < \varepsilon \ll 1.$$

A typical perturbation expansion has the form

$$x = x_0 + \varepsilon x_1 + \varepsilon^2 x_2 + O(\varepsilon^3),$$

where x_0 is called the *leading-order* or *unperturbed* solution, obtained by setting $\varepsilon = 0$. The terms

$$\varepsilon x_1, \qquad \varepsilon^2 x_2, \qquad \dots$$

represent successive corrections generated by the perturbation parameter ε. The coefficients $x_1, x_2, \dots$ are determined by solving a hierarchy of equations obtained from the perturbation procedure.

The term

$$O(\varepsilon^3)$$

denotes the remainder term and indicates that the truncation error is of order ε^3 as $\varepsilon \to 0$.

Perturbation expansions are closely related to Taylor expansions. If $x = x(\varepsilon)$ is sufficiently smooth near $\varepsilon = 0$, then its Taylor expansion about $\varepsilon = 0$ is

$$x(\varepsilon) = x(0) + x'(0)\varepsilon + \frac{x''(0)}{2!}\varepsilon^2 + O(\varepsilon^3).$$

Comparing this with the perturbation expansion shows that

$$x_0 = x(0), \qquad x_1 = x'(0), \qquad x_2 = \frac{x''(0)}{2!}.$$

Thus, regular perturbation expansions may be viewed as asymptotic Taylor expansions in the small parameter ε.

3.2.1 Algebraic Equations

Regular perturbation methods for algebraic equations are best illustrated through examples. We begin with a simple quadratic equation for which the perturbation expansion reproduces the Taylor expansion of the exact solution.

Example 3.1 *Consider the quadratic equation with parameter* ε

$$x^2 - \varepsilon x - 1 = 0, \quad \varepsilon \ll 1. \tag{3.1}$$

Assume that the solution x *of* (3.1) *can be expanded in a Taylor series in* ε. *That is,*

$$x = x_0 + \varepsilon x_1 + \varepsilon^2 x_2 + \cdots + x_n = \sum_{k=0}^{n} \varepsilon^k x_k.$$

By considering only the first three terms, we adopt the notation

$$x = x_0 + \varepsilon x_1 + \varepsilon^2 x_2 + O(\varepsilon^3). \tag{3.2}$$

The term x_0 *is called the* leading-order *term. Substituting* (3.2) *into* (3.1), *we obtain*

$$\left(x_0 + \varepsilon x_1 + \varepsilon^2 x_2 + O(\varepsilon^3)\right)^2 - \varepsilon\left(x_0 + \varepsilon x_1 + \varepsilon^2 x_2 + O(\varepsilon^3)\right) - 1 = 0.$$

By expanding the factors and collecting terms in powers of ε we arrive at

$$x_0^2 - 1 + \varepsilon\left(2x_0x_1 - x_0\right) + \varepsilon^2\left(x_1^2 + 2x_0x_2 - x_1\right) + O(\varepsilon^3) = 0.$$

Equating coefficients of like powers of ε gives

$$\begin{aligned} O(\varepsilon^0): &\quad x_0^2 - 1 = 0, \\ O(\varepsilon^1): &\quad 2x_0x_1 - x_0 = 0, \\ O(\varepsilon^2): &\quad x_1^2 + 2x_0x_2 - x_1 = 0, \\ O(\varepsilon^3): &\quad \cdots \end{aligned} \tag{3.3}$$

Next, we successively solve the equations in (3.3). *Since the two roots of the first equation $x_0^2 - 1 = 0$ are $x_0 = \pm 1$, we obtain*

$$\begin{aligned} x_0 &= 1, \quad x_1 = \frac{1}{2}, \quad x_2 = \frac{1}{8}, \\ x_0 &= -1, \quad x_1 = \frac{1}{2}, \quad x_2 = -\frac{1}{8}. \end{aligned}$$

Substituting the above values into the perturbed solution in (3.2) *we obtain the perturbed solution*

$$\begin{aligned} x_1 &= 1 + \frac{\varepsilon}{2} + \frac{\varepsilon^2}{8} + O(\varepsilon^3), \\ x_2 &= -1 + \frac{\varepsilon}{2} - \frac{\varepsilon^2}{8} + O(\varepsilon^3). \end{aligned} \tag{3.4}$$

Next, we find the exact (or true) solution of (3.1) *and compare it with the approximate solutions obtained in* (3.4). *Since* (3.1) *is quadratic in x, its roots are given by*

$$\begin{aligned} x_1^* &= \frac{\varepsilon}{2} + \sqrt{1 + \frac{\varepsilon^2}{4}}, \\ x_2^* &= \frac{\varepsilon}{2} - \sqrt{1 + \frac{\varepsilon^2}{4}}. \end{aligned}$$

Using the fifth entry of (Commonly Used Taylor Series), for small ε, these roots are approximated by the first few terms of their Taylor series expansions

$$\begin{aligned} x_1^* &= 1 + \frac{\varepsilon}{2} + \frac{\varepsilon^2}{8} + O(\varepsilon^3), \\ x_2^* &= -1 + \frac{\varepsilon}{2} - \frac{\varepsilon^2}{8} + O(\varepsilon^3), \end{aligned}$$

which agree with the perturbation expansions obtained in (3.4). □

While Example 3.1 may not have clearly demonstrated the benefits of regular perturbation, the next example illustrates its power. We will consider a nonlinear algebraic equation that is impossible to solve for its exact solution, yet regular perturbation provides an effective method for approximating its solution when $\varepsilon \ll 1$.

Example 3.2 *Consider the nonlinear algebraic equation*

$$x^2 - \varepsilon e^x - 4 = 0, \quad \varepsilon \ll 1. \tag{3.5}$$

As before, we assume a perturbed solution of the form

$$x = x_0 + \varepsilon x_1 + \varepsilon^2 x_2 + O(\varepsilon^3).$$

Substituting into the main (3.5)*, we obtain*

$$\left(x_0 + \varepsilon x_1 + \varepsilon^2 x_2 + O(\varepsilon^3)\right)^2 - \varepsilon e^{\left(x_0 + \varepsilon x_1 + \varepsilon^2 x_2 + O(\varepsilon^3)\right)} - 4 = 0. \tag{3.6}$$

Using

$$e^z = 1 + z + \frac{z^2}{2!} + O(z^3),$$

the second term in (3.6) *reduces to*

$$\begin{aligned} \varepsilon e^{x_0 + \varepsilon x_1 + \varepsilon^2 x_2 + O(\varepsilon^3)} &= \varepsilon e^{x_0} e^{\varepsilon x_1 + \varepsilon^2 x_2 + O(\varepsilon^3)} \\ &= \varepsilon e^{x_0} \Big[1 + (\varepsilon x_1 + \varepsilon^2 x_2 + O(\varepsilon^3)) \\ &+ \tfrac{1}{2}(\varepsilon x_1 + \varepsilon^2 x_2 + O(\varepsilon^3))^2 + O(\varepsilon^3)\Big] \\ &= \varepsilon e^{x_0} + \varepsilon^2 x_1 e^{x_0} + O(\varepsilon^3). \end{aligned}$$

where we took

$$z = \varepsilon x_1 + \varepsilon^2 x_2 + O(\varepsilon^3).$$

Substituting into (3.6)*, and by expanding the factors and collecting terms in powers of ε, we arrive at*

$$x_0^2 - 4 + \varepsilon\left(2x_0 x_1 - e^{x_0}\right) + \varepsilon^2\left(x_1^2 + 2x_0 x_2 - x_1 e^{x_0}\right) + O(\varepsilon^3) = 0.$$

Equating to zero the successive terms of the expression on the left-hand side provides us with the equations:

$$\begin{aligned} O(\varepsilon^0): &\quad x_0^2 - 4 = 0, \\ O(\varepsilon^1): &\quad 2x_0 x_1 - e^{x_0} = 0, \\ O(\varepsilon^2): &\quad x_1^2 + 2x_0 x_2 - x_1 e^{x_0} = 0. \\ O(\varepsilon^3): &\quad \cdots \end{aligned} \tag{3.7}$$

Next, we successively solve the equations in (3.7). Since the two roots of the first equation $x_0^2 - 4 = 0$ are $x_0 = \pm 2$, we obtain

$$x_0 = 2, \quad x_1 = \frac{e^2}{4}, \quad x_2 = \frac{3e^4}{64},$$
$$x_0 = -2, \quad x_1 = -\frac{e^{-2}}{4}, \quad x_2 = \frac{5e^{-4}}{64}$$

Substituting these coefficients into the perturbation expansion yields the perturbed solution

$$x_1^* = 2 + \varepsilon\frac{e^2}{4} + \varepsilon^2\frac{3e^4}{64} + O(\varepsilon^3),$$

$$x_2^* = -2 - \varepsilon\frac{e^{-2}}{4} + \varepsilon^2\frac{5e^{-4}}{64} + O(\varepsilon^3).$$

□

Regular perturbation fails when setting the small parameter (ε) to zero, fundamentally changes the nature of the problem. These are called **singular perturbation** problems and are commonly characterized by a change in the number of solutions, or solutions that are not uniformly valid as $\varepsilon \to 0$. We provide the following example to enforce the idea.

Example 3.3 *(Singular perturbation; failure of regular perturbation) Consider the quadratic equation*

$$\varepsilon x^2 + 2x + 1 = 0, \quad \varepsilon \ll 1.$$

Assume a regular perturbation series of the form $x = x_0 + \varepsilon x_1 + O(\varepsilon^2)$. We begin by solving the unperturbed equation by setting $\varepsilon = 0$. Setting $\varepsilon = 0$ results in the linear equation

$$2x + 1 = 0 \quad \text{that has the solution} \quad x_0 = -\frac{1}{2}.$$

This gives one unperturbed root, which is a clear sign that regular perturbation will fail, since the full quadratic equation has two roots. We have lost a root by setting $\varepsilon = 0$. Using regular perturbation with

$$x = x_0 + \varepsilon x_1 + \varepsilon^2 x_2 + O(\varepsilon^3)$$

we have from the main equation that

$$\varepsilon(x_0 + \varepsilon x_1 + \dots)^2 + 2(x_0 + \varepsilon x_1 + \dots) + 1 = 0.$$

After expanding the factors, we have

$$\varepsilon(x_0^2 + 2\varepsilon x_0 x_1 + \dots) + (2x_0 + 2\varepsilon x_1 + \dots) + 1 = 0.$$

Group by powers of ε *to obtain*

$$(2x_0+1)+\varepsilon(x_0^2+2x_1)+O(\varepsilon^2)=0.$$

Equating coefficients to zero yields:

$$\begin{aligned} O(\varepsilon^0): &\quad 2x_0+1=0 \\ O(\varepsilon^1): &\quad x_0^2+2x_1=0. \end{aligned}$$

Solving the above equations, we get the values

$$x_0=-\frac{1}{2}, \quad x_1=-\frac{1}{8}.$$

Consequently, the regular perturbation solution is

$$x=-\frac{1}{2}-\frac{1}{8}\varepsilon+O(\varepsilon^2).$$

This approximation only finds one of the two roots. The failure occurs because the equation changes from a quadratic to a linear equation when ε *is zero, and a linear equation cannot capture the full solution structure of the quadratic equation, which has two solutions. The lost root is a singularly perturbed root, and it is not of the same order as the regular root. To find the second root, we must rescale the variables. We look for a dominant balance where terms that were dropped in the regular perturbation are now relevant. Consider the full equation:*

$$\varepsilon x^2+2x+1=0.$$

We will try to rescale our terms to capture the effect of every term in the equation. When x *is large, the term* $2x+1$ *can be ignored. A balance must exist between the first and last two terms,* εx^2+2x*, or between* εx^2 *and* 1*. Let's balance* εx^2 *and* $2x$*. This suggests* x *is on the order of* $1/\varepsilon$*. Let's set* $x=y/\varepsilon$*, where* y *is an* $O(1)$ *variable. Substitute* $x=y/\varepsilon$ *into the equation*

$$\begin{aligned} \varepsilon\left(\frac{y}{\varepsilon}\right)^2+2\left(\frac{y}{\varepsilon}\right)+1=0 \\ \frac{y^2}{\varepsilon}+\frac{2y}{\varepsilon}+1=0 \end{aligned}$$

Multiplying by ε *we obtain*

$$y^2+2y+\varepsilon=0$$

Next, we perform a regular perturbation on the new equation by assuming a solution of the form

$$y=y_0+\varepsilon y_1+O(\varepsilon^2).$$

Substituting this perturbed solution into the equation yields,

$$(y_0+\varepsilon y_1)^2+2(y_0+\varepsilon y_1)+\varepsilon=0.$$

Or,

$$(y_0^2 + 2\varepsilon y_0 y_1 + \varepsilon^2 y_1^2) + (2y_0 + 2\varepsilon y_1) + \varepsilon = 0.$$

$$O(\varepsilon^0): \quad y_0^2 + 2y_0 = 0, \qquad \text{which implies that} \qquad y_0(y_0+2) = 0.$$

$$\text{Hence } y_0 = 0 \text{ or } y_0 = -2.$$

$$O(\varepsilon^1): \quad 2y_0y_1 + 2y_1 + 1 = 0.$$

If $y_0 = 0$, *then from* $2y_0y_1 + 2y_1 + 1 = 0$, *we obtain* $y_1 = -\frac{1}{2}$. *Consequently,*

$$x = y/\varepsilon = \frac{y_0 + \varepsilon y_1}{\varepsilon} = \frac{0 - \varepsilon\frac{1}{2}}{\varepsilon} = -\frac{1}{2},$$

which is the same regular root already found. On the other hand, if $y_0 = -2$, *then* $2y_0y_1 + 2y_1 + 1 = 0$ *gives the solution* $y_1 = \frac{1}{2}$. *Consequently, we have the perturbed solution for y,*

$$y = y_0 + \varepsilon y_1 + O(\varepsilon^2) = -2 + \frac{1}{2}\varepsilon + O(\varepsilon^2).$$

Then the relation $x = y/\varepsilon$ *gives*

$$x = \frac{-2 + \frac{1}{2}\varepsilon + O(\varepsilon^2)}{\varepsilon} = -\frac{2}{\varepsilon} + \frac{1}{2} + O(\varepsilon).$$

This is the second root, which is large and could not be captured by regular perturbation. As a matter of fact, As a matter of fact, Example 3.3 is a singular perturbation problem because setting $\varepsilon = 0$ *changes the structure of the equation and causes one of the solution branches to be lost. The missing root exists on a different asymptotic scale and can only be recovered through an appropriate rescaling of variables. In summary, regular perturbation fails for this problem because:*

- ***The number of roots changes.*** *The unperturbed equation (*$\varepsilon = 0$*) has one root, while the full equation has two.*
- ***The solution scaling is different.*** *One root is* $O(1)$ *and easily found by regular perturbation, but the second root is* $O(1/\varepsilon)$ *and requires a change of variables (rescaling) to find.*

□

3.2.2 Finding the Proper Scaling

In Example 3.3, we observed that the regular perturbation method failed, and a rescaling was necessary to capture the full behavior of the equation in relation to its roots. Setting $\varepsilon \to 0^+$ altered the equation and yielded only one root, even though the original equation had two. To illustrate the idea of rescaling, consider the equation

$$\varepsilon^3 x^3 + x^2 + 2x + \varepsilon = 0, \qquad \varepsilon \ll 1. \tag{3.8}$$

The reader may verify that applying a regular perturbation directly to (3.8) does not fully capture the nature of the equation or its roots. Our goal is to shift ε to appear in front of the term with the smallest power of x. Hence, we introduce a new scaling of the form

$$x = \delta(\varepsilon)X,$$

where $\delta(\varepsilon)$ is a function of ε, and X is independent of ε. In other words, as

$$\varepsilon \to 0, \quad X \text{ is neither small nor large.}$$

Substituting into the original equation (3.8) gives

$$\varepsilon^3(\delta(\varepsilon))^3X^3 + (\delta(\varepsilon))^2X^2 + 2\delta(\varepsilon)X + \varepsilon = 0. \tag{3.9}$$

We now seek to create a balance between the terms.

(a) Suppose

$$\varepsilon^3(\delta(\varepsilon))^3X^3 \approx 2\delta(\varepsilon)X.$$

Ignoring constant multiplicative factors, dominant balance requires

$$\varepsilon^3\delta(\varepsilon)^3 \sim \delta(\varepsilon).$$

$$\varepsilon^3(\delta(\varepsilon))^2 = 1 \quad ,$$

which implies that

$$\delta(\varepsilon) = \varepsilon^{-3/2}.$$

Substituting this value of $\delta(\varepsilon)$ into (3.9) and simplifying yields the scaled equation

$$X^3 + \varepsilon^{-3/2}X^2 + 2X + \varepsilon^{5/2} = 0.$$

Since the coefficient of X^2 becomes unbounded as $\varepsilon \to 0$, this scaling does not produce a useful asymptotic balance. which is a singular perturbation problem.

(b) Suppose instead that

$$\varepsilon^3(\delta(\varepsilon))^3X^3 \approx (\delta(\varepsilon))^2X^2.$$

Equating the coefficients of X^3 and X^2 gives

$$\varepsilon^3\delta(\varepsilon) = 1,$$

so that

$$\delta(\varepsilon) = \varepsilon^{-3}.$$

Substituting this value of $\delta(\varepsilon)$ into (3.9) and simplifying gives

$$\varepsilon^{-6}X^3 + \varepsilon^{-6}X^2 + 2\varepsilon^{-3}X + \varepsilon = 0.$$

Multiplying both sides by ε^6 leads to the scaled equation

$$X^3 + X^2 + 2\varepsilon^3 X + \varepsilon^7 = 0. \tag{3.10}$$

In the scaled equation (3.10), the dominant terms remain balanced as $\varepsilon \to 0$, while the remaining terms appear as small perturbations. This produces a meaningful asymptotic structure in which regular perturbation methods may now be applied successfully to obtain an approximate solution for X. Once the perturbed solution for X is determined, the corresponding approximation for the original variable x follows from the scaling relation

$$x = \varepsilon^{-3} X.$$

Other dominant balances may also be examined, although not all lead to useful asymptotic scalings. The effectiveness of a scaling is determined by whether it preserves a balance among the dominant terms while relegating the remaining terms to higher-order perturbations.

3.2.3 Exercises

Exercise 3.1 *(a) Find a three-term perturbation approximation of each root to the real solution of*

$$x^2 + \varepsilon x - 1 = 0.$$

(b) Expand the exact solution in powers of ε using a Taylor series, and compare the result with your approximation in part (a).

Exercise 3.2 *(a) Find a three-term perturbation approximation of each root to the real solution of*

$$x^2 - (3 + 2\varepsilon)x + 2 + \varepsilon = 0.$$

(b) Expand the exact solution in powers of ε using a Taylor series, and compare the result with your approximation in part (a).

Ans: (a)

$$x = \begin{cases} 1 - \varepsilon + 3\varepsilon^2 + O(\varepsilon^3) \\ 2 + 3\varepsilon - 3\varepsilon^2 + O(\varepsilon^3) \end{cases}$$

Exercise 3.3 *(a) Find a three-term perturbation approximation of each root to the real solution of*

$$(x-1)(x-\beta) + \varepsilon x = 0, \quad \beta \neq 1.$$

(b) Expand the exact solution in powers of ε using a Taylor series, and compare the result with your approximation in part (a).

Ans: (a)

$$x = \begin{cases} 1 - \frac{\varepsilon}{1-\beta} - \frac{\varepsilon^2\beta}{(1-\beta)^3} + O(\varepsilon^3) \\ \beta + \frac{\varepsilon\beta}{1-\beta} + \frac{\varepsilon^2\beta}{(1-\beta)^3} + O(\varepsilon^3) \end{cases}$$

Exercise 3.4 *Find a three-term perturbation approximation of each root to the real solution*

$$x^2 - 3 = 2\varepsilon x.$$

Exercise 3.5 *Find a three-term perturbation approximation of each root to the real solution*

$$x^2 + 2x + \varepsilon = 0.$$

Exercise 3.6 *Consider the perturbed equation*

$$(1-\varepsilon)x^2 - 2x + 1 = 0.$$

(a) Try a regular perturbation expansion and explain what goes wrong.

(b) Instead, seek an expansion of the form

$$x = x_0 + \varepsilon^{1/2}x_1 + \varepsilon x_2 + O(\varepsilon^{3/2}).$$

Determine x_0, x_1, *and* x_2.

(c) Compare the perturbation expansion with the exact roots.

This is a singular perturbation problem because $x = 1$ *is a repeated root of the unperturbed equation when* $\varepsilon = 0$.

Exercise 3.7 *Find a three-term perturbation approximation of each root to the real solution*

$$x^3 - x + \varepsilon = 0.$$

Exercise 3.8 *Repeat Example 3.3 for the perturbed algebraic equation*

$$\varepsilon x^3 \quad x + 1 - 0.$$

Exercise 3.9 *Find a three-term perturbation approximation of each root to the real solution*

$$x^2 - 1 = \varepsilon e^{2x}.$$

Exercise 3.10 *Find the first three terms of an expansion for each root of the following equation*

$$x^3 - (2-\varepsilon)x^2 - x + 2 + \varepsilon = 0.$$

Exercise 3.11 *Find a two-term solution of* (3.10).

Exercise 3.12 *Rescale the singular problem*

$$\varepsilon x^2 + x - 1 = 0,$$

and then find a two-term approximation of each root of the equation.

Exercise 3.13 *Consider the singularly perturbed equation*

$$\varepsilon x^5 - 16x + 1 = 0, \qquad 0 < \varepsilon \ll 1.$$

(a) Use regular perturbation to determine the $O(1)$ *root.*

(b) Determine an appropriate rescaling that captures the remaining large roots.

(c) Find a two-term asymptotic approximation for the large roots.

3.3 Regular Perturbation in Differential Equations

When introducing the main concept of regular perturbation in differential equations (second-order), it is advantageous to use a general notation such as

$$F(t, y, y', y'', \varepsilon) = 0, \quad t \in I, \tag{3.11}$$

with initial conditions

$$y(t_0) = \alpha, \qquad y'(t_0) = \beta, \tag{3.12}$$

because it captures the essential structure of a wide class of problems without being restricted to a specific example. This general form highlights the dependence of the equation on the small parameter ε and shows how the solution $y(t)$ and its derivatives interact with it. By avoiding unnecessary details tied to particular functions, one can focus on the systematic procedure of the regular perturbation method, namely expanding the solution in powers of ε, substituting into the governing equation, and equating terms of equal order. Such abstraction helps students and readers grasp the underlying logic and methodology of perturbation analysis before applying it to concrete examples.

In (3.11) the variable t is the independent variable, I is the time interval, and y is the dependent variable. Next, we give a quick guidelines on the procedure. Consider systems (3.11) and (3.12). Assume a solution $y(t)$ that can be represented as a Taylor series in the small parameter ε:

$$y(t) = y_0(t) + \varepsilon y_1(t) + \varepsilon^2 y_2(t) + \cdots \tag{3.13}$$

where $y_0(t)$ is the solution of the unperturbed problem (corresponding to $\varepsilon = 0$), commonly referred to as the *leading term*, and $y_1(t), y_2(t), \ldots$ represent successive higher–order correction terms. Next, we substitute the series expansion for $y(t)$ given by (3.13) and its derivatives into the original differential equation. The derivatives are also expanded as a power series:

$$y'(t) = y_0'(t) + \varepsilon y_1'(t) + \varepsilon^2 y_2'(t) + \cdots$$

$$y''(t) = y_0''(t) + \varepsilon y_1''(t) + \varepsilon^2 y_2''(t) + \cdots$$

Substituting these into the equation $F(t,y,y',y'',\varepsilon) = 0$ results in a complicated expression involving various powers of ε. Expand the resulting equation in powers of ε and group the terms by the power of ε. Since this equation must hold for all small values of ε, the coefficient of each power of ε must be zero. This yields a sequence of simpler differential equations for $y_0(t), y_1(t), y_2(t), \ldots$ The initial conditions $y(t_0) = \alpha$ and $y'(t_0) = \beta$ are also expanded according to the perturbation series:

$$y(t_0) = y_0(t_0) + \varepsilon y_1(t_0) + \varepsilon^2 y_2(t_0) + \cdots = \alpha,$$
$$y'(t_0) = y_0'(t_0) + \varepsilon y_1'(t_0) + \varepsilon^2 y_2'(t_0) + \cdots = \beta.$$

Equating the coefficients of like powers of ε gives a set of initial conditions for each order of the problem:

$$\begin{aligned}
O(1): &\quad y_0(t_0) = \alpha, \quad y_0'(t_0) = \beta,\\
O(\varepsilon): &\quad y_1(t_0) = 0, \quad y_1'(t_0) = 0,\\
O(\varepsilon^2): &\quad y_2(t_0) = 0, \quad y_2'(t_0) = 0,
\end{aligned}$$

and so on for higher orders. The final step is to solve the sequence of initial value problems starting with the $O(1)$ problem, which is the unperturbed equation. The solution to each subsequent problem $y_1(t), y_2(t), \ldots$ depends on the solution of the previous order. This sequential solving process provides the terms of the perturbation series.

For example, the $O(1)$ problem is:

$$F(t, y_0, y_0', y_0'', 0) = 0, \quad y_0(t_0) = \alpha, \quad y_0'(t_0) = \beta.$$

The solutions for $y_0(t), y_1(t), \ldots$ are then assembled to form the approximate solution $y(t)$. The accuracy of the approximation improves as more terms are included in the series. The first few terms of such a series yield an approximate representation of the solution, referred to as a **perturbation solution** or **perturbation approximation**. In most applications, only the first two or three terms are required to obtain satisfactory accuracy. The method is generally valid provided that the approximation is **uniform**; that is, the difference between the approximate and exact solutions tends to zero at a well-defined rate as $\varepsilon \to 0$, uniformly on the interval I. A rigorous formulation of these concepts will be presented later in the chapter.
A solution is a *uniform approximation* if the difference between the exact solution $y_{\text{exact}}(t)$ and the approximate solution $y(t)$ remains small for all t within the interval I.

For the given approximate solution

$$y(t) = y_0(t) + \varepsilon y_1(t) + \varepsilon^2 y_2(t) + \cdots,$$

the condition for a uniform approximation of order N is that the error term

$$\left| y_{\text{exact}}(t) - \sum_{n=0}^{N} \varepsilon^n y_n(t) \right|$$

must be of the order of the first neglected term, i.e.,

$$O(\varepsilon^{N+1}),$$

for all $t \in I$ and for small ε.

To determine if the approximation is uniform, one must check if the terms in the series expansion remain bounded and do not grow with t. An approximation becomes non-uniform when one or more of the perturbation terms, such as $y_1(t)$, grow as t increases. This leads to a breakdown of the approximation, as the term $\varepsilon y_1(t)$ might no longer be small compared to the leading-order term $y_0(t)$, even for small ε.

Non-uniformity typically arises from *secular terms*, which are terms that grow with t (e.g., $t\cos(t)$ or t). These secular terms can appear in the higher-order terms of the series solution (e.g., in $y_1(t)$ or $y_2(t)$). If such a term is present, the approximation is only valid for a limited time interval, not for all $t \in I$. Special techniques, such as the *Poincaré–Lindstedt method*, or *method of multiple scales* or the *averaging method*, are required to find a uniformly valid solution.

Example 3.4 *Let*

$$y(t,\varepsilon) = (1+t\varepsilon)^{-1/2}, \qquad t \in [0,\infty), \quad \varepsilon \ll 1.$$

We use the first three terms of the Taylor expansion in powers of ε to obtain the approximation

$$y_a(t,\varepsilon) = 1 - \frac{1}{2}t\varepsilon + \frac{3}{8}t^2\varepsilon^2.$$

Then the corresponding error is

$$E(t,\varepsilon) = y(t,\varepsilon) - y_a(t,\varepsilon).$$

Since

$$(1+z)^{-1/2} = 1 - \frac{1}{2}z + \frac{3}{8}z^2 - \frac{5}{16}z^3 + \frac{35}{128}z^4 - \frac{63}{256}z^5 + \cdots,$$

with $z = t\varepsilon$, we have

$$E(t,\varepsilon) = -\frac{5}{16}t^3\varepsilon^3 + \frac{35}{128}t^4\varepsilon^4 - \frac{63}{256}t^5\varepsilon^5 + \cdots.$$

For each fixed t, the error satisfies

$$E(t,\varepsilon) = O(\varepsilon^3), \qquad \varepsilon \to 0.$$

However, if ε is fixed, no matter how small, t may be chosen large enough so that the approximation becomes invalid. Hence, the approximation is not uniform on $I = [0,\infty)$.

For instance, choosing $t = 1/\varepsilon$ gives

$$E(1/\varepsilon,\varepsilon) = -\frac{5}{16} + \frac{35}{128} - \frac{63}{256} + \cdots,$$

which is not small. Therefore, we cannot write

$$E(t,\varepsilon) = O(\varepsilon^3)$$

uniformly for all $t \in [0,\infty)$. □

In most cases, we do not have access to the exact solution y_{exact} to test for uniform approximation. The following definition clarifies how such a test can be performed.

Definition 3.1 *Let* $y_a(t,\varepsilon)$ *be the perturbed solution of*

$$F(t,y,y',y'',\varepsilon) = 0, \quad t \in I. \tag{3.14}$$

If

$$\lim_{\varepsilon\to 0} F\left(t, y_a(t,\varepsilon), y_a'(t,\varepsilon), y_a''(t,\varepsilon)\right) = 0 \text{ uniformly for all } t \in I, \tag{3.15}$$

then we say $y_a(t,\varepsilon)$ *is a uniform approximation of* (3.14).

3.3.1 Vertical Motion of an Object

For an object falling vertically under gravity, Newton's second law gives

$$m\frac{dv}{d\tau} = mg - av - bv^2, \quad \tau \geq 0,$$

where $v(\tau)$ is the velocity (taken positive downward), $m > 0$ is the mass, g is the gravitational acceleration, and $a, b \geq 0$ are the linear and quadratic air-resistance coefficients, respectively. A typical initial condition is $v(0) = 0$.

- $m\frac{dv}{d\tau}$: the net force on the object (Newton's second law).
- mg: the gravitational force pulling the object downward.
- $-av$: a linear drag term, which models resistance proportional to velocity (dominant at low speeds, e.g., in viscous fluids).
- $-bv^2$: a quadratic drag term, which models resistance proportional to the square of the velocity (dominant at higher speeds, e.g., in air).

Typically, the object starts from rest ($v(0) = 0$) with acceleration downward due to gravity. As v increases, drag forces ($av + bv^2$) increase in magnitude and oppose motion. We try to rescale the problem. We begin with time τ. Eventually, the forces balance when

$$mg = av + bv^2,$$

at which point

$$\frac{dv}{d\tau} = 0,$$

and the object reaches a terminal velocity v_τ that is a solution of

$$mg = av_\tau + bv_\tau^2.$$

If we neglect the quadratic term (bv^2) initially, the terminal velocity is $v_\tau = \frac{mg}{a}$. By introducing the new variable u and rescaling by setting

$$u = \frac{v}{v_\tau} = \frac{av}{mg},$$

we obtain

$$v = \frac{mg}{a}u.$$

Next, we try to rescale time τ. A natural time scale is the characteristic time, τ_c, associated with the linear damping. For the linear drag model, the velocity approaches terminal velocity exponentially on the timescale m/a. Let's choose $\tau_c = \frac{m}{a}$. By setting a new time scale

$$t = \frac{\tau}{\tau_c} = \frac{a\tau}{m},$$

we obtain

$$\tau = \frac{m}{a}t.$$

Now, we rescale the differential equation with respect to the new variable u and t.

We use the chain rule to express the derivative with respect to t in terms of the new variables we have

$$\frac{dv}{d\tau} = \frac{d}{d\tau}\left(\frac{mg}{a}u\right) = \frac{mg}{a}\frac{du}{d\tau} = \frac{mg}{a}\frac{du}{dt}\frac{dt}{d\tau} = \frac{mg}{a}\frac{du}{dt}\left(\frac{a}{m}\right) = g\frac{du}{dt}.$$

Substitute the rescaled velocity and derivative into the original equation, $m\frac{dv}{d\tau} = mg - av - bv^2$, we arrive at

$$m\left(g\frac{du}{dt}\right) = mg - a\left(\frac{mg}{a}u\right) - b\left(\frac{mg}{a}u\right)^2,$$

or

$$mg\frac{du}{dt} = mg - mgu - b\frac{m^2g^2}{a^2}u^2.$$

Dividing all terms by mg we get the dimensionless form

$$\frac{du}{dt} = 1 - u - \frac{bmg}{a^2}u^2.$$

The coefficient of the quadratic term, $\frac{bmg}{a^2}$, can be identified as our dimensionless parameter. Let's call it ε: that is

$$\varepsilon = \frac{bmg}{a^2}.$$

The dimensionless equation becomes

$$\frac{du}{dt} = 1 - u - \varepsilon u^2, \quad \varepsilon \ll 1, \quad t \geq 0 \tag{3.16}$$

with initial velocity

$$u(0) = 0.$$

Physical interpretation of ε

The parameter ε measures the relative strength of the quadratic drag compared to the linear and gravitational effects in the system. It is defined by

$$\varepsilon = \frac{bmg}{a^2}.$$

Its magnitude determines the behavior of the system:

- If $\varepsilon \ll 1$: the quadratic drag is weak. The dynamics are dominated by the terms $1 - u$, and the nonlinear contribution $-\varepsilon u^2$ provides only a small correction.
- If $\varepsilon \sim 1$: the quadratic drag is comparable to the linear effects, and nonlinear resistance plays an important role in the motion.
- If $\varepsilon \gg 1$: the quadratic drag dominates the dynamics, strongly limiting the growth of the velocity.

Thus, ε determines whether the system behaves approximately linearly or exhibits strongly nonlinear behavior. Returning to equation (3.16), we note that it is nonlinear, and an exact solution is not readily available. In a perturbation framework, we assume that the quadratic term is small compared to the linear and gravitational terms, which corresponds to taking $\varepsilon \ll 1$. Physically, this regime occurs when the linear drag coefficient a is large compared to the quadratic drag coefficient b, the mass m, and the gravitational acceleration g (that is, when a^2 is large). In this case, the problem can be viewed as a small perturbation of the exactly solvable linear damping equation, making it amenable to perturbation methods.

We assume a solution of the form:

$$u(t) = u_0(t) + \varepsilon u_1(t) + O(\varepsilon^2). \tag{3.17}$$

Differentiating with respect to t we arrive at

$$u'(t) = u_0'(t) + \varepsilon u_1'(t) + O(\varepsilon^2),$$

Substitute these series expansions into the original equation (3.16) we obtain

$$\Big(u_0'(t) + \varepsilon u_1'(t) + O(\varepsilon^2)\Big) - 1 \quad + \quad \Big(u_0(t) + \varepsilon u_1(t) + O(\varepsilon^2)\Big)$$
$$+ \quad \varepsilon\Big(u_0(t) + \varepsilon u_1(t) + O(\varepsilon^2)\Big)^2 = 0.$$

By expanding and collecting powers of ε, we obtain

$$\left(u_0'(t) - 1 + u_0(t)\right) + \varepsilon\left(u_1'(t) + u_1(t) + u_0^2(t)\right) + O(\varepsilon^2) = 0. \tag{3.18}$$

Similarly, expanding the initial condition gives

$$u(0) = u_0(0) + \varepsilon u_1(0) + O(\varepsilon^2) = 0. \tag{3.19}$$

Equating the coefficients of each power of ε to zero in (3.18) and (3.19) leads to the following set of initial value problems:

$$\frac{du_0}{dt} + u_0(t) - 1 = 0, \qquad u_0(0) = 0,$$

$$\frac{du_1}{dt} + u_1(t) + u_0^2(t) = 0, \qquad u_1(0) = 0.$$

The first equation is separable and can be written as

$$\frac{du_0}{1 - u_0} = dt.$$

Integrating both sides gives

$$\ln|1 - u_0| = -t + C_1.$$

Exponentiating both sides and absorbing the constant into a single term, we obtain

$$u_0(t) = 1 - Ke^{-t}.$$

Applying the initial condition $u_0(0) = 0$. Hence, $0 = 1 - K$, and the final solution is

$$u_0(t) = 1 - e^{-t}. \tag{3.20}$$

Next, using (3.20), we compute $u_1(t)$. Since

$$u_0(t) = 1 - e^{-t},$$

we have

$$u_0^2(t) = \left(1 - e^{-t}\right)^2 = 1 - 2e^{-t} + e^{-2t}.$$

Thus, the equation for u_1 becomes

$$u_1'(t) + u_1(t) = -u_0^2(t) = -1 + 2e^{-t} - e^{-2t}, \qquad u_1(0) = 0.$$

We solve this linear equation using the integrating factor e^t:

$$e^t u_1'(t) + e^t u_1(t) = -e^t + 2 - e^{-t},$$

so that

$$\frac{d}{dt}\left(e^{t}u_1(t)\right) = -e^{t} + 2 - e^{-t}.$$

Integrating both sides gives

$$e^{t}u_1(t) = -e^{t} + 2t + e^{-t} + C.$$

Hence,

$$u_1(t) = -1 + 2te^{-t} + e^{-2t} + Ce^{-t}.$$

Applying the initial condition $u_1(0) = 0$, we obtain

$$0 = -1 + 0 + 1 + C,$$

so $C = 0$. Therefore,

$$u_1(t) = -1 + 2te^{-t} + e^{-2t}. \tag{3.21}$$

Substituting (3.20) and (3.21) into (3.17), we obtain the approximation

$$u_a(t) = 1 - e^{-t} + \varepsilon\left(-1 + 2te^{-t} + e^{-2t}\right) + O(\varepsilon^2).$$

Having constructed the perturbation approximation, we now examine whether it remains valid uniformly for large times. In our setting, we have

$$F(t,u,u',\varepsilon) = \frac{du}{dt} + u + \varepsilon u^2 - 1,$$

and hence,

$$F(t,u_a,u_a',\varepsilon) = \frac{du_a}{dt} + u_a + \varepsilon u_a^2 - 1 = O(\varepsilon^2).$$

More precisely, after simplification, we obtain

$$F(t,u_a,u_a',\varepsilon) = \varepsilon^2 g_2(t) + O(\varepsilon^3),$$

where

$$g_2(t) = 2\Big((1-2t)e^{t} + (2t+1)e^{2t} - e^{3t} - 1\Big)e^{-3t}.$$

The function $g_2(t)$ is continuous and bounded for $t \geq 0$; indeed,

$$\lim_{t\to\infty} g_2(t) = -2,$$

so there exists a constant $C > 0$ such that

$$|g_2(t)| \leq C \quad \text{for all } t \geq 0.$$

Finally, solving for $\cos^3(\tau)$ we obtain the formula

$$\cos^3\tau = \tfrac{1}{4}\Big(3\cos(\tau)+\cos(3\tau)\Big). \tag{3.24}$$

Back to our problem. Using (3.24) in (3.23) we arrive at

$$\theta_2''+\theta_2 = \frac{1}{8}\cos(\tau)+\frac{1}{24}\cos(3\tau), \quad \theta_2(0)=0, \quad \theta_2'(0)=0. \tag{3.25}$$

The above equation has a homogeneous solution

$$\theta_{2h} = A\cos(\tau)+B\sin(\tau),$$

for constants A and B to be determined. Due to the existence of $\cos(\tau)$ in the homogeneous solution, we must assume a particular solution $\theta_{2p}(\tau)$ of the form

$$\theta_{2p}(\tau) = c_1\tau\cos(\tau)+c_2\tau\sin(\tau)+c_3\cos(3\tau)+c_4\sin(3\tau).$$

The term $\cos\tau$ is resonant with the homogeneous solution and produces a secular term. Substituting the particular solution into (3.25) leads to, after some calculations and then simplification to

$$-2c_1\sin(\tau)+2c_2\cos(\tau)-8c_3\cos(3\tau)-8c_4\sin(3\tau) = \frac{1}{8}\cos(\tau)+\frac{1}{24}\cos(3\tau).$$

Equating both sides leads to

$$c_1=0, \quad c_2=\frac{1}{16}, \quad c_3=-\frac{1}{192}, \quad c_4=0.$$

Consequently, the particular solution is

$$\theta_{2p}(\tau) = \frac{1}{16}\tau\sin(\tau)-\frac{1}{192}\cos(3\tau).$$

Hence, the solution of

$$\theta_2(\tau) = \theta_{2h}(\tau)+\theta_{2p}(\tau) = A\cos(\tau)+B\sin(\tau)+\frac{1}{16}\tau\sin(\tau)-\frac{1}{192}\cos(3\tau).$$

Applying the initial condition $\theta_2(0)=0$, gives $A=\frac{1}{192}$, $B=0$. Finally, the solution of (3.25) is

$$\theta_2(\tau) = \frac{1}{192}\cos\tau+\frac{1}{16}\tau\sin\tau-\frac{1}{192}\cos 3\tau.$$

Hence the two-term approximation becomes

$$\theta(\tau) = \cos\tau+\varepsilon^2\left(\frac{1}{192}\cos\tau+\frac{1}{16}\tau\sin\tau-\frac{1}{192}\cos 3\tau\right)+O(\varepsilon^3).$$

The secular term or the correction term $\frac{\varepsilon^2}{16}\tau\sin\tau$ grows without bound as τ increases, and is therefore *secular*. The regular perturbation approximation is valid only for

$$\tau \ll \varepsilon^{-2},$$

beyond which the approximation breaks down. To obtain a uniformly valid approximation for all time, a frequency correction (via the Poincaré–Lindstedt or multiple scales method) must be introduced. Next, we try to find the exact period of oscillation. Consider the scaled pendulum equation

$$\frac{d^2\theta}{d\tau^2} + \frac{\sin(\varepsilon\theta)}{\varepsilon} = 0.$$

Multiply the differential equation by θ' to obtain

$$\theta''\,\theta' + \frac{\sin(\varepsilon\theta)}{\varepsilon}\,\theta' = 0.$$

Recognizing total derivatives,

$$\frac{d}{d\tau}\left(\tfrac{1}{2}\theta'^2\right) - \frac{1}{\varepsilon^2}\frac{d}{d\tau}\big(\cos(\varepsilon\theta)\big) = 0,$$

because

$$\frac{d}{d\tau}\big(\cos(\varepsilon\theta)\big) = -\varepsilon\sin(\varepsilon\theta)\,\theta'.$$

Integrating with respect to τ gives

$$\frac{1}{2}\theta'^2 = \frac{1}{\varepsilon^2}\cos(\varepsilon\theta) + C.$$

Applying the initial conditions

$$\theta(0) = 1, \qquad \theta'(0) = 0,$$

gives

$$\frac{1}{2}\theta'^2(0) = \frac{1}{\varepsilon^2}\cos(\varepsilon\theta(0)) + C.$$

Hence

$$C = -\frac{1}{\varepsilon^2}\cos(\varepsilon),$$

and consequently, we arrive at the *energy integral*

$$\frac{1}{2}\theta'^2 = \frac{1}{\varepsilon^2}\big(\cos(\varepsilon\theta) - \cos\varepsilon\big).$$

Solving for θ' (taking the positive or negative root depending on the direction of motion),

$$\frac{d\theta}{d\tau} = \pm\frac{\sqrt{2}}{\varepsilon}\sqrt{\cos(\varepsilon\theta) - \cos\varepsilon}.$$

A quarter-period corresponds to the motion from $\theta = 1$ (maximum displacement) to $\theta = 0$. Hence, the full period P is four times the quarter-period. That is

$$P = 4\int_0^1 \frac{d\tau}{d\theta}\,d\theta.$$

Using the relation derived above,

$$\frac{d\tau}{d\theta} = \frac{\varepsilon}{\sqrt{2}}\frac{1}{\sqrt{\cos(\varepsilon\theta) - \cos\varepsilon}},$$

we obtain

$$P = 4\int_0^1 \frac{\varepsilon}{\sqrt{2}}\frac{d\theta}{\sqrt{\cos(\varepsilon\theta) - \cos\varepsilon}}.$$

Renaming the integration variable $\theta \mapsto x$, we arrive at the exact expression for the period

$$P = \frac{4\varepsilon}{\sqrt{2}}\int_0^1 \frac{dx}{\sqrt{\cos(\varepsilon x) - \cos\varepsilon}}.$$

This integral gives the exact period of oscillation for the scaled pendulum.

3.3.3 Exercises

Exercise 3.14 *Let*

$$y(t,\varepsilon) = e^{-\varepsilon t}, \qquad t \geq 0, \quad 0 < \varepsilon \ll 1,$$

and let $y_a(t,\varepsilon)$ denote the approximation obtained by retaining the first four terms of the Taylor series of $e^{-\varepsilon t}$ about $\varepsilon = 0$. Determine whether the error

$$E(t,\varepsilon) = y(t,\varepsilon) - y_a(t,\varepsilon)$$

tends to zero uniformly on $[0,\infty)$ as $\varepsilon \to 0^+$.

Exercise 3.15 *(a) Find a three-term approximate solution of the first-order differential equation*

$$y'(t) + y(t) - \varepsilon y^2 = 0, \quad y(0) = 2,\ t \geq 0.$$

(b) Use part (a) and write down a closed form of the solution $y(t)$.

(c) Find the exact solution by directly solving the differential equation.

(d) Obtain the exact solution in part (c) using the result from part (b).

(e) Is the perturbed solution obtained in part (a) a uniform approximation of the exact solution?

Ans: (b) $y(t) = \sum\limits_{n=0}^{\infty} \varepsilon^n 2^{n+1} e^{-t}(1-e^{-t})^n.$

Exercise 3.16 *Next, let us consider a differential equation describing the motion of a body of mass m. Assume the body starts with an initial velocity V_0 and moves in a straight line through a medium that exerts a resistive force of magnitude $av - bv^2$, where $v = v(\tau)$ denotes the velocity of the object as a function of time τ. Here, a and b are positive constants, with $b \ll a$, indicating that the nonlinear correction is much smaller than the linear resistance. Thus, the nonlinear term in the force is considered a small correction to the dominant linear term. The constants a and b have dimensions of force per unit velocity and force per unit velocity squared, respectively. According to Newton's second law, the corresponding equation of motion is*

$$m\frac{dv}{d\tau} = -av + bv^2, \qquad v(0) = V_0.$$

(a) *Use the independent variable t for time and the dependent variable y for velocity v to nondimensionalize the problem and obtain*

$$\frac{dy}{dt} + y(t) - \varepsilon y^2(t) = 0, \quad \varepsilon \ll 1, \quad t \geq 0 \tag{3.26}$$

where $\varepsilon = \frac{bV_0}{a}$.

(b) *Find the exact solution of* (3.26).

(c) *Use regular perturbation to obtain a three–term perturbation solution of* (3.26).

(d) *Show that perturbed solution obtained in part (c) is a uniform approximation.*

Exercise 3.17 *Find a two–term perturbation solution of*

$$y'(t) + y(t) - \frac{1}{1+\varepsilon y(t)} = 0, \quad y(0) = 0, \ \varepsilon \ll 1.$$

Exercise 3.18 *Consider the second–order nonlinear initial value problem*

$$y'' + y + \varepsilon y^3 = 0, \quad t \geq 0,$$

with initial conditions $y(0) - 1$ *and* $y'(0) = 0$.

(a) *Use regular perturbation to obtain a two–term perturbation solution.*

(b) *Is the perturbed solution obtained in part (a) a uniform approximation? Explain.*

Exercise 3.19 *Try regular perturbation expansion to obtain the first three terms of the following differential equation*

$$y'' + 2\varepsilon y' + (1+\varepsilon^2)y = 1, \quad y(0) = 0, \quad y\left(\frac{\pi}{2}\right) = 0.$$

Exercise 3.20 *Consider the first-order nonlinear initial value problem*

$$y' + y = \varepsilon y^3, \qquad t \geq 0,$$

with initial condition $y(0) = A$.

(a) *Use regular perturbation to obtain a two-term perturbation solution.*

(b) *Is the perturbation solution obtained in part (a) a uniform approximation on* $[0,\infty)$*? Explain.*

Exercise 3.21 *Consider the first-order initial value problem*

$$y' + y = \varepsilon g(t), \qquad y(0) = A,$$

where $g(t)$ *is continuous and satisfies*

$$\sup_{t\geq 0} |g(t)| \leq M < \infty.$$

Seek an expansion of the form

$$y(t) = y_0(t) + \varepsilon y_1(t) + O(\varepsilon^2).$$

Show that

$$y_0(t) = Ae^{-t}$$

and

$$y_1(t) = \int_0^t e^{-(t-s)} g(s)\, ds.$$

Conclude that the first-order approximation is

$$y_a(t) = Ae^{-t} + \varepsilon \int_0^t e^{-(t-s)} g(s)\, ds,$$

and show that this approximation is uniformly valid on $[0,\infty)$.

Exercise 3.22 *Consider the first-order initial value problem*

$$y' + P(t)y = \varepsilon g(t), \qquad y(0) = 1,$$

where $g(t)$ *and* $P(t)$ *are continuous for all* $t \geq 0$.

(a) *Use regular perturbation to obtain the first two terms* $y_0(t)$ *and* $y_1(t)$.

(b) *Give sufficient conditions on* $g(t)$ *and* $P(t)$ *that guarantee* $y_0(t)$ *and* $y_1(t)$ *are bounded on* $[0,\infty)$*. That is, find conditions under which there exist positive constants* M *and* Q *such that*

$$|y_0(t)| \leq M, \qquad |y_1(t)| \leq Q, \quad t \geq 0.$$

Exercise 3.23 *Consider the nondimensional equation*

$$x''(\tau) = -\frac{1}{(1+\varepsilon x(\tau))^2}, \qquad x(0) = 0, \quad x'(0) = 1,$$

where $0 < \varepsilon \ll 1$*. Seek a regular expansion*

$$x(\tau) = x_0(\tau) + \varepsilon x_1(\tau) + O(\varepsilon^2).$$

(a) *Derive the differential equations satisfied by x_0 and x_1, together with their initial conditions.*

(b) *Solve for $x_0(\tau)$ and obtain an explicit expression for $x_1(\tau)$.*

(c) *Use the expansion to estimate, to first order in ε, the time at which the object reaches its maximum height.*

3.4 The Poincaré–Lindstedt Method

The **Poincaré–Lindstedt method** is a perturbation technique used to obtain approximate periodic solutions of weakly nonlinear oscillators when the regular perturbation method fails due to the appearance of *secular terms*. These terms typically grow linearly in time, producing expressions such as

$$t\sin(\omega t), \qquad t\cos(\omega t),$$

and consequently destroy the periodicity and uniform validity of the perturbation expansion for large t. The origin of secular terms is that nonlinear effects slightly modify the oscillation frequency, whereas a regular perturbation expansion assumes the same time scale as the corresponding linear problem. As a result, resonant forcing terms appear in the perturbation equations and generate unbounded growth. To eliminate this difficulty, the Poincaré–Lindstedt method introduces a stretched time variable

$$t = \Omega(\varepsilon)\tau,$$

where the corrected frequency is expanded in powers of the small parameter:

$$\Omega(\varepsilon) = \omega_0 + \varepsilon\omega_1 + \varepsilon^2\omega_2 + \cdots.$$

At the same time, the solution is written in the form

$$y(t) = y_0(t) + \varepsilon y_1(t) + \varepsilon^2 y_2(t) + \cdots.$$

The constants $\omega_1, \omega_2, \ldots$ are then chosen so that secular terms are removed at each order of approximation. Consequently, the resulting perturbation expansion remains bounded and uniformly valid for long times. This method is particularly useful in nonlinear oscillatory systems such as the *Duffing oscillator*

$$\ddot{x} + x + \varepsilon x^3 = 0,$$

and the *van der Pol oscillator*, where nonlinear effects alter the oscillation frequency Using the stretched variable $t = \Omega(\varepsilon)\tau$, the derivatives transform according to

$$\frac{dy}{d\tau} = \Omega\frac{dy}{dt}, \qquad \frac{d^2y}{d\tau^2} = \Omega^2\frac{d^2y}{dt^2}.$$

We begin by presenting an example illustrating how the regular perturbation method breaks down due to the presence of a secular term. We then revisit the pendulum problem from Example 3.3.2 and apply the Poincaré–Lindstedt method to construct a uniformly valid approximation.

Example 3.5 *Consider the second–order differential equation*

$$y''(t)+\varepsilon y'(t)+y(t)=0, \quad y(0)=1, \quad y'(0)=0.$$

Applying the regular perturbation method, we assume a solution in the form of a power series in ε*:*

$$y(t)=y_0(t)+\varepsilon y_1(t)+\varepsilon^2 y_2(t)+\cdots.$$

Substituting this expansion into the differential equation and equating coefficients of powers of ε *yields a sequence of problems.*

$$y_0''+y_0=0, \qquad y_0(0)=1, \qquad y_0'(0)=0.$$

The solution to this simple harmonic oscillator is

$$y_0(t)=\cos(t).$$

In addition, we have the differential equation

$$y_1''+y_1=-y_0', \qquad y_1(0)=0, \qquad y_1'(0)=0.$$

Substituting $y_0(t)=\cos(t)$ *gives*

$$y_1''+y_1=\sin(t).$$

This is a driven harmonic oscillator problem. We seek a particular solution in the form

$$y_p=c_1 t\sin t+c_2 t\cos t,$$

and determine the constants by the method of undetermined coefficients. Substituting into the differential equation gives

$$c_1=0, \qquad c_2=-\frac{1}{2}.$$

Thus, the particular solution contains a secular term, and the general solution is

$$y_1(t)=A\cos(t)+B\sin(t)-\tfrac{1}{2}t\cos(t).$$

Applying the initial conditions $y_1(0)=0$ *and* $y_1'(0)=0$ *gives* $A=0$ *and* $B=0$*. Hence,*

$$y_1(t)=-\tfrac{1}{2}t\cos(t).$$

Breakdown of the approximation. *The complete two-term approximate solution is*

$$y(t)=\cos(t)-\tfrac{\varepsilon}{2}t\cos(t).$$

For this approximation to be valid, the correction term $\varepsilon y_1(t)$ must remain much smaller than the leading-order term $y_0(t)$:

$$|\varepsilon y_1(t)| \ll |y_0(t)|,$$

that is,

$$\left|-\tfrac{\varepsilon}{2} t\cos(t)\right| \ll |\cos(t)|.$$

This inequality holds for small t. However, as time increases, the linear growth of the secular term $t\cos(t)$ becomes significant. When $t \sim 1/\varepsilon$, the correction term $\varepsilon y_1(t)$ becomes comparable in magnitude to $y_0(t)$. Thus, the perturbation expansion breaks down, and the approximation is no longer uniform for all t. The exact solution of the initial value problem is given by

$$y(t) = e^{-\varepsilon t/2}\left[\cos\left(t\sqrt{1-\tfrac{\varepsilon^2}{4}}\right) + \frac{\varepsilon}{2\sqrt{1-\tfrac{\varepsilon^2}{4}}}\sin\left(t\sqrt{1-\tfrac{\varepsilon^2}{4}}\right)\right].$$

This expression represents a decaying oscillation, *with an exponentially decreasing amplitude and a slightly perturbed frequency. In contrast, the regular perturbation approximation predicts an oscillatory solution with a linearly growing amplitude due to the presence of a secular term. This qualitative discrepancy indicates that the approximation is not uniformly valid for large times. This behavior illustrates the failure of the regular perturbation method and reveals the singular nature of the problem.* □

3.4.1 Pendulum Problem–Revisited

To remove the secular growth found in the regular expansion of the pendulum problem of Section 3.3.2 we apply the Poincaré–Lindstedt method. To do so, we introduce a stretched time

$$t = \Omega(\varepsilon)\,\tau$$

with the frequency expansion

$$\Omega(\varepsilon) = 1 + \alpha_1\varepsilon + \alpha_2\varepsilon^2 + O(\varepsilon^4).$$

Here we chose $\alpha_0 = 1$, which is the frequency of the solution of the unperturbed problem (3.22). Write the solution as a perturbation series in powers of ε^2,

$$\theta(\tau) = \Theta(t) = \Theta_0(t) + \varepsilon\Theta_1(t) + \varepsilon^2\Theta_2(t) + O(\varepsilon^4).$$

From Example 3.3.2, the pendulum equation is

$$\frac{d^2\theta}{d\tau^2} + \frac{\sin(\varepsilon\theta)}{\varepsilon} = 0 \tag{3.27}$$

After expanding

$$\sin(\varepsilon\Theta)/\varepsilon = \Theta - \frac{\varepsilon^2\Theta^3}{6} + O(\varepsilon^4),$$

the differential equation becomes

$$\Omega^2\Theta''(t)+\Theta-\frac{\varepsilon^2}{6}\Theta^3+O(\varepsilon^4)=0.$$

Substituting into the above differential equation leads us to

$$\begin{aligned}\left(1\right. &+ \left.\alpha_1\varepsilon+\alpha_2\varepsilon^2\right)^2\left(\Theta_0(t)+\varepsilon\Theta_1(t)+\varepsilon^2\Theta_2(t)\right)''\\ &+ \left(\Theta_0(t)+\varepsilon\Theta_1(t)+\varepsilon^2\Theta_2(t)\right)-\frac{\varepsilon^2}{6}\left(\Theta_0(t)+\varepsilon\Theta_1(t)+\varepsilon^2\Theta_2(t)\right)^3+O(\varepsilon^4)=0.\end{aligned}$$

By collecting orders, $O(1)$ gives the differential equation

$$\Theta_0''(t)+\Theta_0(t)=0, \qquad \Theta_0(0)=1,\ \Theta_0'(0)=0,$$

that has the solution

$$\Theta_0(t)=\cos(t).$$

As for $O(\varepsilon^1)$ we have

$$\Theta_1''(t)+\Theta_1(t)+2\alpha_1\Theta_0''(t)=0, \qquad \Theta_1(0)=0,\ \Theta_1'(0)=0. \tag{3.28}$$

The above equation has a homogeneous solution

$$\Theta_{1h}=A\cos(t)+B\sin(t),$$

for constants A and B to be determined. Due to the existence of $\cos(t)$, $\sin(t)$ in the homogeneous solution, we must assume a particular solution $\Theta_{1p}(t)$ of the form

$$\Theta_{1p}(t)=c_1t\cos(t)+c_2t\sin(t).$$

Substituting the particular solution into (3.28) leads, after some calculations and then simplification to

$$-2c_1\sin(t)+2c_2\cos(t)=2\alpha_1\cos(t).$$

Equating both sides leads to

$$c_1=0, \quad c_2=\alpha_1.$$

Consequently, the particular solution is

$$\Theta_{1p}(t)=\alpha_1t\sin(t).$$

Hence, the solution of

$$\Theta_1(t)=\Theta_{1h}(t)+\Theta_{1p}(t)=A\cos(t)+B\sin(t)+\alpha_1t\sin(t).$$

Applying the initial conditions $\Theta_1(0)=0$, $\Theta_1'(0)=0$ gives $A=B=0$. Finally, the solution of (3.28) is

$$\Theta_1(t)=\alpha_1t\sin(t).$$

The Poincaré–Lindstedt idea is to choose α_1 so that the existence of the secular term (is a term that grows in time and destroys periodicity) $t\sin(t)$, is gone. Thus, we are forced to set

$$\alpha_1 = 0$$

and with that we have

$$\Theta_1(t) = 0.$$

For $O(\varepsilon^2)$ we have

$$\Theta_2''(t) + \Theta_2(t) + 2\alpha_1\Theta_1''(t) + \alpha_1^2\Theta_0''(t) + 2\alpha_2\Theta_0''(t) - \frac{1}{6}\Theta_0^3(t) = 0.$$

Since $\alpha_1 = 0$, the above equation reduces to

$$\Theta_2''(t) + \Theta_2(t) + 2\alpha_2\Theta_0''(t) - \frac{1}{6}\Theta_0^3(t) = 0. \tag{3.29}$$

Using $\Theta_0(t) = \cos(t)$, and the identity $\cos^3(t) = \frac{1}{4}\Big(3\cos(t) + \cos(3t)\Big)$ in (3.29) we obtain

$$\Theta_2''(t) + \Theta_2(t) = \left(2\alpha_2 + \frac{1}{8}\right)\cos(t) + \frac{1}{24}\cos(3t).$$

Due to the presence of the term $\cos(t)$ in the homogeneous solution of the corresponding homogeneous equation, the particular solution will contain a secular term of the form $t\cos(t)$. This term is eliminated by requiring that

$$2\alpha_2 + \frac{1}{8} = 0,$$

from which we obtain

$$\alpha_2 = -\frac{1}{16}.$$

Thus, we are left with the equation

$$\Theta_2''(t) + \Theta_2(t) = \frac{1}{24}\cos(3t), \qquad \Theta_2(0) = 0,\ \Theta_2'(0) = 0.$$

We obtain the solution by a similar reasoning as we did before. The homogeneous solution is given by

$$\Theta_{2h} = A\cos(t) + B\sin(t),$$

for constants A and B to be determined. On the other hand, the particular solution $\Theta_{2p}(t)$ takes the form

$$\Theta_{2p}(t) = c_1\cos(3t) + c_2\sin(3t).$$

Substituting the particular solution into the differential equation and performing some algebraic manipulations, we simplify the result to

$$-8c_1\cos(3t) + 8c_2\sin(3t) = \frac{1}{24}\cos(3t).$$

Equating both sides yields

$$c_2 = 0, \quad c_1 = -\frac{1}{192}.$$

Consequently, the particular solution is

$$\Theta_{2p}(t) = -\frac{1}{192}\cos(3t).$$

Hence, the solution of

$$\Theta_2(t) = A\cos(t) + B\sin(t) - \frac{1}{192}\cos(3t).$$

Applying the initial condition $\Theta_1(0) = 0$, $\Theta_1'(0) = 0$ gives

$$A = \frac{1}{192}, \quad B = 0.$$

Finally, the solution is

$$\Theta_2(t) = \frac{1}{192}\cos(t) - \frac{1}{192}\cos(3t).$$

Thus, the corrected frequency is

$$\Omega(\varepsilon) = 1 - \frac{\varepsilon^2}{16} + O(\varepsilon^4).$$

In summary, we obtained a second-order, uniformly l valid, perturbation solution of (3.27) that is given by

$$\Theta(t) = \cos(t) + \frac{1}{192}\varepsilon^2\Big(\cos(t) - \cos(3t)\Big) + O(\varepsilon^4),$$

where

$$t = \Omega(\varepsilon)\tau = \tau - \frac{1}{16}\varepsilon^2\tau + O(\varepsilon^4).$$

Finally, the solution to our original problem is

$$\theta(\tau) = \cos[(\tau - \frac{1}{16}\varepsilon^2\tau)] + \frac{1}{192}\varepsilon^2\Big(\cos[(\tau - \frac{1}{16}\varepsilon^2\tau)] - \cos[3(\tau - \frac{1}{16}\varepsilon^2\tau)]\Big) + O(\varepsilon^4).$$

The physical period (in the original τ-time) is $P = 2\pi/\Omega$. Expanding for small ε gives

$$P = \frac{2\pi}{\Omega} = 2\pi\Big(1 + \frac{\varepsilon^2}{16} + O(\varepsilon^4)\Big).$$

Hence, the Poincaré–Lindstedt method both removes the secular term from the perturbation expansion and yields the leading correction to the period. This corrected-period expansion explains why the naive regular expansion developed a secular τ-term: the true oscillation frequency is slightly different from unity by an amount

$O(\varepsilon^2)$. When the regular expansion uses the unperturbed time scale, it cannot absorb that frequency shift, so resonant forcing appears at $O(\varepsilon^2)$ and produces secular growth. Poincaré–Lindstedt (or multiple scales) corrects the time scale and therefore produces a uniformly valid approximation for $\tau = O(\varepsilon^{-2})$ and beyond (to the order of the expansion).

Remark 9 *When dealing with a general second–order nonlinear differential equations of the form*

$$y''(\tau) + \beta^2 y(\tau) = F(\varepsilon, \tau, y, y'),$$

in which Poincaré–Lindstedt method has to be used, assume $t = \Omega(\varepsilon)\,\tau$ *with the frequency expansion*

$$\Omega(\varepsilon) = \beta + \alpha_1 \varepsilon + \alpha_2 \varepsilon^2 + O(\varepsilon^4).$$

3.4.2 Exercises

Exercise 3.24 (Poincaré–Lindstedt method for a weakly nonlinear oscillator) *Consider the weakly nonlinear oscillator*

$$y'' + y + \varepsilon y^2 = 0, \qquad y(0) = 1, \qquad y'(0) = 0,$$

where $0 < \varepsilon \ll 1$.

(a) Use the regular perturbation expansion

$$y(t) = y_0(t) + \varepsilon y_1(t) + O(\varepsilon^2)$$

to obtain the first two terms of the approximation.

(b) Apply the Poincaré–Lindstedt method to determine ω_1 *and obtain a uniformly valid two–term approximation. Determine the corresponding first-order correction to the period of oscillation.*

Exercise 3.25 *Consider the special case of the second–order initial value problem for a weakly nonlinear oscillator discussed in Section 2.3.5:*

$$y'' + \beta^2 y + \varepsilon y^3 = 0, \qquad y(0) = 1, \qquad y'(0) = 0,$$

where

$$\beta = \sqrt{\frac{k}{m}}.$$

Here we assume that the damping constant is $c = 0$ *and that there is no external forcing,* $F(t) = 0$.

(a) Use the regular perturbation method to obtain a two–term perturbation solution.

(b) Show that the correction term contains a secular term and explain why this causes the regular perturbation expansion to fail for large times.

(c) Apply the Poincaré–Lindstedt method to remove the secular term and obtain the first correction to the frequency.

Answer to part (a). *The two–term regular perturbation solution is*

$$y(t) = \cos(\beta t) + \varepsilon\left(-\frac{1}{32\beta^2}\cos(\beta t) + \frac{3}{8\beta}t\sin(\beta t) + \frac{1}{32\beta^2}\cos(3\beta t)\right).$$

Exercise 3.26 *For non–zero constant β, consider*

$$y''(t) + \frac{\beta^2 y(t)}{1+y^2(t)}y(t) = 0, \quad y(0) = A,\ y'(0) = B.$$

Use the Poincaré–Lindstedt method to obtain a two-term uniform expansion of the solution.

Exercise 3.27 *For a nonzero constant β, consider*

$$y''(t) + \beta^2 y(t) = \varepsilon\left(y'(t)\right)^2 y(t), \qquad y(0) = A, \qquad y'(0) = B.$$

Use the Poincaré–Lindstedt method to obtain a two-term uniformly valid approximation of the solution.

Exercise 3.28 *Consider*

$$y''(t) + 4y(t) + \varepsilon y^2(t)y''(t) = 0, \qquad y(0) = 1, \qquad y'(0) = 0.$$

Use the Poincaré–Lindstedt method to obtain a two-term uniformly valid approximation of the solution.

Exercise 3.29 *Consider the weakly nonlinear oscillator with periodic parametric forcing near resonance,*

$$x''(t) + x(t) = \varepsilon\cos[(1+\varepsilon)t]x(t), \qquad x(0) = A, \qquad x'(0) = 0.$$

Explain why the standard Poincaré–Lindstedt method is not the most natural method for removing secular terms in this problem. What perturbation method would be more appropriate?

Exercise 3.30 *Find a two-term regular perturbation approximation for*

$$y''(t) - \varepsilon y'(t) - y(t) = 0, \qquad y(0) = 1, \qquad y'(0) = 0.$$

Compare the approximation with the exact solution and discuss whether the approximation is uniformly valid for large t.

3.5 Multiple Scaling

The Poincaré–Lindstedt method is highly effective for weakly nonlinear oscillators possessing a single dominant frequency. By rescaling time, the method removes secular terms and produces uniformly valid periodic approximations.

However, there are important situations in which this approach is no longer sufficient. In particular, difficulties arise when

- several time scales coexist,
- resonance or near-resonance occurs,
- amplitudes or phases evolve slowly in time, or
- energy transfer occurs between oscillatory modes.

In such problems, the solution evolves simultaneously on both fast and slow time scales. The method of multiple scales captures this behavior by introducing additional slow time variables.

The *method of multiple scales* is necessary when a system's response evolves not only on a fast oscillatory timescale but also on slower timescales due to weak nonlinearity, resonance, or external forcing. This can lead to phenomena such as beats, amplitude death, or complex phase modulation. Problems involving near-resonance or more than one intrinsic time scale (multiple frequencies), such as forced nonlinear oscillators near resonance, parametric resonance, or coupled oscillator systems, require the method of multiple scales. Attempting to use the Poincaré–Lindstedt method in these situations leads to unremovable secular (growing) terms and invalid solutions over long time intervals. The method of multiple scales requires the introduction of a fast time and one or more slow times. Fast time is denoted by $T_0 = t$, which represents the standard time variable, capturing rapid oscillations or dynamics within the system. On the other hand, the **slow time** is denoted by $T_1 = \varepsilon t$, $T_2 = \varepsilon^2 t$, etc. These are auxiliary time variables, scaled by powers of the small parameter ε, evolving much more slowly than the fast time and capturing slow modulations such as changes in amplitude or phase. Thus, fast scale captures oscillations or quick changes over short periods (e.g., vibrations), while, slow scale captures gradual changes that accumulate over many oscillations, often leading to amplitude modulation or phase drift.

Instead of seeking a solution $x(t;\varepsilon)$, the method seeks a function $x(T_0, T_1, T_2, \ldots; \varepsilon)$, treating the new time variables as mathematically independent:

$$x(t;\varepsilon) = x(T_0, T_1, T_2, \ldots; \varepsilon),$$

where

$$T_n = \varepsilon^n t, \quad n = 0, 1, 2, \cdots$$

As t increases, T_0 increases quickly, while each subsequent T_n increases much more slowly.

Suppose we introduce multiple time scales of the form

$$T_0 = t, \quad T_1 = \varepsilon t, \quad T_2 = \varepsilon^2 t, \quad \dots$$

A function $x(t)$ is rewritten as $x(T_0, T_1, T_2, \dots)$, with the new time variables treated as independent.

The total derivatives are expanded using the chain rule

$$\frac{d}{dt} = \frac{\partial}{\partial T_0} + \varepsilon \frac{\partial}{\partial T_1} + \varepsilon^2 \frac{\partial}{\partial T_2} + \cdots$$

For the second derivative:

$$\begin{aligned}\frac{d^2}{dt^2} &= \frac{d}{dt}\left(\frac{d}{dt}\right) \\ &= \left(\frac{\partial}{\partial T_0} + \varepsilon \frac{\partial}{\partial T_1} + \varepsilon^2 \frac{\partial}{\partial T_2} + \cdots\right)\left(\frac{\partial}{\partial T_0} + \varepsilon \frac{\partial}{\partial T_1} + \varepsilon^2 \frac{\partial}{\partial T_2} + \cdots\right)\end{aligned}$$

Expanding, we get:

$$\frac{d^2}{dt^2} = \frac{\partial^2}{\partial T_0^2} + 2\varepsilon \frac{\partial^2}{\partial T_0 \partial T_1} + \varepsilon^2 \left(2\frac{\partial^2}{\partial T_0 \partial T_2} + \frac{\partial^2}{\partial T_1^2}\right) + \cdots$$

Consider the weakly nonlinear oscillator under periodic external forcing near resonance

$$\frac{d^2x}{dt^2} + x = \varepsilon \cos[(1+\varepsilon)t]\, x.$$

Here, the weak nonlinearity and near-resonant forcing lead to slow amplitude modulation of the oscillations. In such a situation, the Poincaré–Lindstedt method fails, but the method of multiple scales succeeds by separating the fast oscillations from the slow amplitude and phase variations, as we shall explore next. In the following example, we illustrate the application of both the Poincaré–Lindstedt method and the method of multiple scales to a single equation.

Example 3.6 *Consider the weakly nonlinear oscillator or (Duffing equation)*

$$u'' + u + \varepsilon u^3 = 0, \quad u(0) = 1, \quad u'(0) = 0$$

where $0 < \varepsilon \ll 1$. *We begin by using the Poincaré–Lindstedt method. We let*

$$\tau = \omega t$$

and expand the solution and frequency in powers of ε. *That is*

$$u(\tau) = u_0(\tau) + \varepsilon u_1(\tau) + \dots$$

$$\omega = 1 + \varepsilon \omega_1 + \dots$$

Under substitution, the derivatives become

$$\frac{d^2u}{dt^2} = \omega^2 u''(\tau)$$

Therefore, the equation transforms into

$$\omega^2 u'' + u + \varepsilon u^3 = 0$$

Collecting terms by powers of ε *we have the leading equation with initial conditions*

$$u_0'' + u_0 = 0, \quad u_0(0) = 1,\ u_0'(0) = 0.$$

Its solution is found to be

$$u_0(\tau) = \cos(\tau)$$

The other equation of order ε^1 *is*

$$u_1'' + u_1 = -2\omega_1 u_0'' - u_0^3, \quad u_1(0) = 0,\ u_1'(0) = 0.$$

Plug in $u_0 = \cos(\tau)$, $u_0'' = -\cos(\tau)$ *to arrive at*

$$u_1'' + u_1 = 2\omega_1 \cos(\tau) - \cos^3(\tau)$$

Using the identity $\cos^3(\tau) = \frac{3}{4}\cos(\tau) + \frac{1}{4}\cos(3\tau)$, *the equation becomes*

$$u_1'' + u_1 = (2\omega_1 - \frac{3}{4})\cos(\tau) - \frac{1}{4}\cos(3\tau)$$

The term proportional to $\cos(\tau)$ *would create secular growth unless its coefficient is zero. Therefore, setting*

$$2\omega_1 - \frac{3}{4} = 0$$

gives

$$\omega_1 = \frac{3}{8}.$$

Thus, after substitution and simplification, we arrive at the equation

$$u_1'' + u_1 = -\frac{1}{4}\cos(3\tau),$$

that has the general solution

$$u_1(\tau) = A\cos(\tau) + B\sin(\tau) + \frac{1}{32}\cos(3\tau).$$

Applying the initial conditions and with the absence of any additional secular or resonant terms, the solution simplifies to

$$u_1(\tau) = \frac{1}{32}\left[\cos(3\tau) - \cos(\tau)\right].$$

Finally, the solution up to first order is

$$u(\tau) = \cos(\tau) + \varepsilon\frac{1}{32}[\cos(3\tau) - \cos(\tau)]$$

Converting back to the original time t using $\tau = (1 + \frac{3}{8}\varepsilon)t$, yields

$$u(t) = \cos\left[(1 + \frac{3}{8}\varepsilon)t\right] + \varepsilon\frac{1}{32}\left[\cos\left(3(1 + \frac{3}{8}\varepsilon)t\right) - \cos\left((1 + \frac{3}{8}\varepsilon)t\right)\right].$$

Next we apply the method of multiple scalings. We consider the same equation

$$\frac{d^2x}{dt^2} + x = -\varepsilon x^3, \qquad x(0) = d, \quad x'(0) = 0, \quad 0 < \varepsilon \ll 1$$

We introduce multiple scales in the form of fast time $T_0 = t$ and the slow time $T_1 = \varepsilon t$. Assume the solution has the form

$$x(t) = x_0(T_0, T_1) + \varepsilon x_1(T_0, T_1) + O(\varepsilon^2).$$

Applying the chain rule for derivatives gives

$$\frac{d}{dt} = \frac{\partial}{\partial T_0} + \varepsilon\frac{\partial}{\partial T_1}$$

$$\frac{d^2}{dt^2} = \frac{\partial^2}{\partial T_0^2} + 2\varepsilon\frac{\partial^2}{\partial T_0 \partial T_1}$$

Substituting into the original equation gives

$$\begin{aligned}\left[\frac{\partial^2}{\partial T_0^2} \right. & + \left. 2\varepsilon\frac{\partial^2}{\partial T_0 \partial T_1}\right]\left(x_0(T_0, T_1) + \varepsilon x_1(T_0, T_1)\right) + x_0(T_0, T_1) + \varepsilon x_1(T_0, T_1) \\ & = -\varepsilon\left(x_0(T_0, T_1) + \varepsilon x_1(T_0, T_1)\right)^3\end{aligned}$$

Equating coefficients gives the following equations

$$\frac{\partial^2 x_0}{\partial T_0^2} + x_0 = 0,$$

$$\frac{\partial^2 x_1}{\partial T_0^2} + x_1 = -2\frac{\partial^2 x_0}{\partial T_0 \partial T_1} - (x_0)^3. \tag{3.30}$$

The solution of the first equation is given by

$$x_0(T_0, T_1) = C(T_1)\cos T_0 + B(T_1)\sin T_0 \tag{3.31}$$

Noticing that at $t = 0$ ($T_0 = 0, T_1 = 0$) we have

$$x_0(0,0) = d, \quad x_0'(0,0) = 0.$$

Thus,

$$0 = x_0(0,0) = C(0) = d$$

and

$$\frac{\partial x_0}{\partial T_0}(0,0) = -B(0)\cos(0) = -B(0) = 0.$$

Thus,

$$C(0) = d, \quad B(0) = 0.$$

We still need to determine $C(T_1)$ and $B(T_1)$, which appear in the expression for $x_0(T_0,T_1)$ given by (3.31)*. Notice that substituting* (3.31) *into* (3.30) *to solve for $x_1(T_0,T_1)$ leads to a rather cumbersome expression. To simplify the process, we will rewrite* (3.31) *using a phase–amplitude (or phase–angle) representation. Recall that*

$$a(T_1)\cos(T_0) + b(T_1)\sin(T_0) = A(T_1)\cos\big(T_0 + \phi(T_1)\big),$$

where

$$A(T_1) = \sqrt{a^2(T_1) + b^2(T_1)}, \quad \tan\big(\phi(T_1)\big) = -\frac{b(T_1)}{a(T_1)}.$$

With this in mind, we rewrite the solution of x_0 given by (3.31) *in the form*

$$x_0(T_0,T_1) = A(T_1)\cos\big(T_0 + \phi(T_1)\big) \tag{3.32}$$

As we have stated before that the initial conditions at $t = 0$ ($T_0 = 0, T_1 = 0$) imply that

$$x_0(0,0) = A(0)\cos(\phi(0)) = d, \quad \frac{\partial x_0}{\partial T_0}(0,0) = -A(0)\sin(\phi(0)) = 0.$$

Thus,

$$\phi(0) = 0, \quad A(0) = d,$$

and we take

$$x_0(T_0,T_1) = A(T_1)\cos(T_0 + \phi(T_1)).$$

Suppressing T_1 from the arguments, we compute the right-hand side of (3.30) *and obtain*

$$\begin{aligned}
\frac{\partial x_0}{\partial T_0} &= -A\sin(T_0+\phi),\\
\frac{\partial^2 x_0}{\partial T_0 \partial T_1} &= -\frac{dA}{dT_1}\sin(T_0+\phi) - A\frac{d\phi}{dT_1}\cos(T_0+\phi),\\
x_0^3 = A^3\cos^3(T_0+\phi) &= \frac{3}{4}A^3\cos(T_0+\phi) + \frac{1}{4}A^3\cos(3(T_0+\phi)).
\end{aligned}$$

So:

$$\frac{\partial^2 x_1}{\partial T_0^2} + x_1 = 2\frac{dA}{dT_1}\sin(T_0+\phi) + \left(2A\frac{d\phi}{dT_1} - \frac{3}{4}A^3\right)\cos(T_0+\phi) - \frac{1}{4}A^3\cos(3(T_0+\phi)).$$

To eliminate secular terms we set coefficients of $\sin(T_0+\phi)$ *and* $\cos(T_0+\phi)$ *to zero and arrive at*

$$2\frac{dA}{dT_1}=0, \quad 2A\frac{d\phi}{dT_1}-\frac{3}{4}A^3=0.$$

The first equation has the solution $A(T_1)=L$, *for some constant* L. *Applying the corresponding initial condition* $A(0)=d$ *and solving the resulting equation, we obtain the solution*

$$A(T_1)=d.$$

Substituting $A(T_1)=d$ *into the second differential equation and solving and then using* $\phi(0)=0$, *we obtain the solution*

$$\phi(T_1)=\frac{3}{8}d^2T_1.$$

By making use of $T_0=t$ *and* $T_1=\varepsilon t$, *the leading–order solution becomes*

$$x_0(t)=d\cos\left(t+\frac{3}{8}d^2\varepsilon t\right)=d\cos(\theta), \quad \theta=t\left(1+\frac{3}{8}d^2\varepsilon\right).$$

The differential equation for x_1 *reduce to*

$$\frac{\partial^2x_1}{\partial T_0^2}+x_1=-\frac{1}{4}d^3\cos(3(T_0+\phi)),$$

and has the particular solution x_{1p} *of the form*

$$x_{1p}=D\cos(3(T_0+\phi)).$$

Substituting into the equation gives

$$-9D\cos(3(T_0+\phi))+D\cos(3(T_0+\phi))=-\frac{1}{4}d^3\cos(3(T_0+\phi)).$$

Equating coefficients on both sides yields to

$$D=\frac{d^3}{32}.$$

Thus,

$$x_{1p}=\frac{d^3}{32}\cos(3\theta).$$

Consequently, the general solution for x_1 *is*

$$x_1=\frac{d^3}{32}\cos(3\theta)+C\cos(\theta)+E\sin(\theta).$$

Applying the initial conditions gives

$$C=-\frac{d^3}{32}, \quad E=0.$$

Finally, the two–term approximate solution is

$$\begin{aligned} x(t) &= d\cos(\theta)+\varepsilon\left[\frac{d^3}{32}\cos(3\theta)-\frac{d^3}{32}\cos(\theta)\right] \\ &= d\cos(\theta)+\frac{\varepsilon d^3}{32}\left[\cos(3\theta)-\cos(\theta)\right], \end{aligned}$$

where

$$\theta = t\left(1+\frac{3}{8}d^2\varepsilon\right).$$

Notice that both answers do match, for $d = 1$.

Remark 10 *Both the Poincaré–Lindstedt and multiple scales methods remove secular terms by modifying the phase of the oscillation. The former achieves this through a rescaling of time, while the latter introduces a slow time variable that captures the gradual phase modulation. The agreement between the two approaches confirms the consistency of the asymptotic expansion and highlights the amplitude-dependent frequency shift characteristic of nonlinear oscillators.*

□

Example 3.7 *Consider the forced, weakly nonlinear oscillator near resonance*

$$\frac{d^2x}{dt^2}+x=\varepsilon\cos\left[(1+\varepsilon)t\right]x, \qquad x(0)=d, \quad x'(0)=0.$$

This is a classic parametric resonance problem, where the forcing frequency is close to the system's natural frequency.

We introduce a fast and a slow time scale

$$T_0 = t, \qquad T_1 = \varepsilon t,$$

and seek a solution of the form

$$x(t;\varepsilon) = X(T_0,T_1).$$

Using the chain rule,

$$\frac{d}{dt}=\frac{\partial}{\partial T_0}+\varepsilon\frac{\partial}{\partial T_1}, \qquad \frac{d^2}{dt^2}=\frac{\partial^2}{\partial T_0^2}+2\varepsilon\frac{\partial^2}{\partial T_0\,\partial T_1}+O(\varepsilon^2).$$

We assume an expansion

$$X(T_0,T_1)=X_0(T_0,T_1)+\varepsilon X_1(T_0,T_1)+O(\varepsilon^2).$$

Substituting into the differential equation and collecting terms up to $O(\varepsilon)$ *gives*

$$\frac{\partial^2X}{\partial T_0^2}+2\varepsilon\frac{\partial^2X}{\partial T_0\,\partial T_1}+X_0+\varepsilon X_1=\varepsilon\cos\left[(1+\varepsilon)T_0\right]X+O(\varepsilon^2).$$

We expand the right-hand side:

$$\cos\left[(1+\varepsilon)T_0\right] = \cos(T_0+\varepsilon T_0) \approx \cos T_0 - \varepsilon T_0 \sin T_0,$$

so, to $O(\varepsilon)$,

$$\cos\left[(1+\varepsilon)T_0\right] X \approx \cos T_0 X_0 + \varepsilon\big(\cos T_0 X_1 - T_0 \sin T_0 X_0\big).$$

Order $\boldsymbol{\varepsilon^0}$**.**

$$\frac{\partial^2 X_0}{\partial T_0^2} + X_0 = 0.$$

The general solution is

$$X_0(T_0, T_1) = A(T_1)\cos T_0 + B(T_1)\sin T_0.$$

Order $\boldsymbol{\varepsilon^1}$**.** *At* $O(\varepsilon)$ *we obtain*

$$\frac{\partial^2 X_1}{\partial T_0^2} + X_1 = -2\frac{\partial^2 X_0}{\partial T_0 \partial T_1} + \cos T_0 X_0.$$

Compute

$$X_0 = A(T_1)\cos T_0 + B(T_1)\sin T_0,$$
$$\frac{\partial X_0}{\partial T_0} = -A(T_1)\sin T_0 + B(T_1)\cos T_0,$$
$$\frac{\partial^2 X_0}{\partial T_0 \partial T_1} = -A'(T_1)\sin T_0 + B'(T_1)\cos T_0.$$

Next,

$$\cos T_0 X_0 = \cos T_0\left[A(T_1)\cos T_0 + B(T_1)\sin T_0\right] = A(T_1)\cos^2 T_0 + B(T_1)\sin T_0 \cos T_0.$$

Using

$$\cos^2 T_0 = \tfrac{1}{2}(1+\cos 2T_0), \qquad \sin T_0 \cos T_0 = \tfrac{1}{2}\sin 2T_0,$$

we obtain

$$\cos T_0 X_0 = \tfrac{A(T_1)}{2}(1+\cos 2T_0) + \tfrac{B(T_1)}{2}\sin 2T_0.$$

Thus,

$$\frac{\partial^2 X_1}{\partial T_0^2} + X_1 = 2A'(T_1)\sin T_0 - 2B'(T_1)\cos T_0 + \tfrac{A(T_1)}{2}(1+\cos 2T_0) + \tfrac{B(T_1)}{2}\sin 2T_0.$$

To avoid secular terms proportional to $\cos T_0$ *and* $\sin T_0$ *in* X_1, *we impose*

$$2A'(T_1) = 0, \qquad -2B'(T_1) = 0.$$

Hence

$$A(T_1) = A, \qquad B(T_1) = B,$$

where A, B are constants (independent of T_1). With this choice,

$$\frac{\partial^2 X_1}{\partial T_0^2} + X_1 = \frac{A}{2} + \frac{A}{2}\cos 2T_0 + \frac{B}{2}\sin 2T_0.$$

A particular solution has the form

$$X_{1p}(T_0) = C + D\cos 2T_0 + E\sin 2T_0.$$

Substituting into

$$\frac{\partial^2 X_1}{\partial T_0^2} + X_1 = \frac{A}{2} + \frac{A}{2}\cos 2T_0 + \frac{B}{2}\sin 2T_0$$

gives

$$(-4D)\cos 2T_0 + (-4E)\sin 2T_0 + C + D\cos 2T_0 + E\sin 2T_0 = \frac{A}{2} + \frac{A}{2}\cos 2T_0 + \frac{B}{2}\sin 2T_0,$$

so

$$C = \frac{A}{2}, \qquad D = -\frac{A}{6} \quad \text{and} \quad E = -\frac{B}{6}.$$

Thus, we may take

$$X_1(T_0, T_1) = \frac{A}{2} - \frac{A}{6}\cos 2T_0 - \frac{B}{6}\sin 2T_0,$$

where we have set the homogeneous part of X_1 to zero (this choice will be used to satisfy the initial conditions). At $t = 0$, we have $T_0 = 0$ and $T_1 = 0$. The initial conditions become

$$x(0) = X(0,0) = X_0(0,0) + \varepsilon X_1(0,0) + O(\varepsilon^2) = d,$$

$$x'(0) = \left(X_{T_0} + \varepsilon X_{T_1}\right)\big|_{(0,0)} = X_{0T_0}(0,0) + \varepsilon\left(X_{1T_0}(0,0) + X_{0T_1}(0,0)\right) + O(\varepsilon^2) = 0.$$

Using

$$X_0(T_0, T_1) = A\cos T_0 + B\sin T_0,$$

we get

$$X_0(0,0) = A, \qquad X_{0T_0}(0,0) = B.$$

Thus, at leading order,

$$A = d, \qquad B = 0.$$

With this choice,

$$X_1(0,0) = \frac{A}{2} - \frac{A}{6}\cos 0 - \frac{B}{6}\sin 0 = \frac{d}{2} - \frac{d}{6} = \frac{d}{3},$$

$$X_{1T_0}(0,0) = \left(\frac{A}{3}\sin 2T_0 - \frac{B}{3}\cos 2T_0\right)\Big|_{(0,0)} = 0,$$

$$X_{0T_1}(0,0) = 0$$

(since A, B are constants). Thus,

$$x(0) = d + \varepsilon\frac{d}{3} + O(\varepsilon^2), \qquad x'(0) = 0 + O(\varepsilon^2).$$

The derivative condition is satisfied to this order, but the value condition is satisfied only to leading order. If one wishes to match $x(0) = d$ to $O(\varepsilon)$ as well, one must retain a homogeneous part in X_1 and choose its constants to enforce $X_1(0,0) = 0$. Here, we accept the leading-order match and regard the $O(\varepsilon)$ discrepancy in the initial value as within the asymptotic error.

With $A = d,\ B = 0$, we have

$$X_0(T_0,T_1) = d\cos T_0, \qquad X_1(T_0,T_1) = \frac{d}{2} - \frac{d}{6}\cos 2T_0.$$

Hence, the multiple scales approximation is

$$\begin{aligned} x(t) &= X_0(T_0,T_1) + \varepsilon X_1(T_0,T_1) + O(\varepsilon^2) \\ &= d\cos t + \varepsilon\left(\frac{d}{2} - \frac{d}{6}\cos 2t\right) + O(\varepsilon^2), \end{aligned}$$

where we have used $T_0 = t$. This approximation is free of secular terms and remains valid for long times $t = O(\varepsilon^{-1})$.

In summary, the method of multiple scales introduces slow and fast time variables to capture both the rapid oscillations and their slow modulation due to near-parametric resonance, thereby preventing secular growth and yielding a uniformly valid approximation. □

3.5.1 Exercises

Exercise 3.31 *(Advanced) Consider the weakly damped, weakly forced Duffing equation*

$$\ddot{x} + \mu\dot{x} + x + \beta x^3 = f_0\cos(\Omega t), \qquad 0 < \mu \ll 1, \quad 0 < |\beta| \ll 1, \quad 0 < f_0 \ll 1.$$

Assume that the forcing frequency is near the linear resonance:

$$\Omega = 1 + \sigma, \qquad |\sigma| \ll 1.$$

(a) Using the method of multiple scales with two time scales

$$T_0 = t, \qquad T_1 = \varepsilon t,$$

where $\mu, \beta, f_0, \sigma = O(\varepsilon)$, derive the first-order modulation equations for the amplitude $A(T_1)$ and phase $\phi(T_1)$ of the leading-order solution

$$x_0(T_0,T_1) = A(T_1)\cos\big(T_0 + \phi(T_1)\big).$$

(b) *From the modulation equations, derive the steady-state amplitude equation relating the steady amplitude a, detuning σ, damping μ, nonlinearity β, and forcing amplitude f_0. Show that, to leading order, this relation may be written in the form*

$$\left(1+\frac{3}{4}\beta a^2-\Omega^2\right)^2 a^2+(\mu\Omega)^2a^2=f_0^2.$$

(c) *Explain qualitatively how a positive β produces a hardening spring, while a negative β produces a softening spring. Describe how this changes the resonance curve and explain when jump phenomena and hysteresis may occur.*

Exercise 3.32 *(Hard) Energy and long-time behavior for the unforced Duffing oscillator . Consider*

$$\ddot{x}+\mu\dot{x}+x+\beta x^3=0, \qquad \mu>0, \qquad \beta\in\mathbb{R}.$$

(a) *Define the mechanical energy*

$$E(t)=\frac{1}{2}\dot{x}^2+\frac{1}{2}x^2+\frac{\beta}{4}x^4, \quad \beta>0.$$

Show that

$$\frac{dE}{dt}=-\mu\dot{x}^2\leq 0.$$

Conclude that $E(t)$ is nonincreasing. Under what condition on β is $E(t)$ bounded below?

(b) *Use the result in part (a), together with LaSalle's invariance principle, to describe the long-time behavior of solutions. State the relevant invariant set and any hypotheses needed on β.*

(c) *Optional challenge: For $\beta>0$ and small $\mu>0$, estimate the time scale for decay of the energy from an initial amplitude a_0 by averaging $\dot{E}$ over one oscillation. That is, compute*

$$\langle\dot{E}\rangle$$

to leading order.

Exercise 3.33 *(Weakly nonlinear oscillator with quadratic nonlinearity)*

Consider

$$\frac{d^2x}{dt^2}+x=\varepsilon\left(\alpha x^2+\beta\frac{dx}{dt}\right), \qquad x(0)=A, \qquad x'(0)=0,$$

where

$$0<\varepsilon\lll 1,$$

and α,β are constants of order one.

Using the method of multiple scales with

$$T_0 = t, \qquad T_1 = \varepsilon t,$$

seek a uniformly valid approximation of the form

$$x(t) = x_0(T_0, T_1) + \varepsilon x_1(T_0, T_1) + O(\varepsilon^2).$$

(a) *Derive the equations satisfied by* x_0 *and* x_1.

(b) *Determine the modulation equations governing the slowly varying amplitude and phase.*

(c) *Eliminate secular terms and obtain a uniformly valid approximation for* $x(t)$ *up to* $O(\varepsilon)$.

(d) *Describe the effect of the parameter* β *on the amplitude of oscillation.*

Exercise 3.34 *(Parametrically excited oscillator)*

Consider the weakly nonlinear Mathieu-type equation

$$\frac{d^2y}{dt^2} + (1 + \varepsilon\cos(\omega t))y = \varepsilon\gamma y^3, \qquad y(0) = B, \qquad y'(0) = 0,$$

where

$$0 < \varepsilon \ll 1,$$

and γ *is a constant of order one.*

(a) *Use the method of multiple scales with*

$$T_0 = t, \qquad T_1 = \varepsilon t,$$

to derive the equations governing the leading-order modulation of the solution.

(b) *Show that parametric resonance occurs when the forcing frequency satisfies*

$$\omega \approx 2.$$

(c) *For*

$$\omega = 2 + \varepsilon\sigma,$$

derive the amplitude and phase modulation equations.

(d) *Discuss qualitatively the stability of the trivial solution and describe the growth of oscillations near resonance.*

Exercise 3.35 *(Advanced: van der Pol oscillator with slow parametric modulation)*

Consider the modified van der Pol oscillator

$$\frac{d^2z}{dt^2} + z = \varepsilon(1 - z^2)\frac{dz}{dt} + \varepsilon^2\mu\cos(\varepsilon t)\, z,$$

with initial conditions

$$z(0) = C, \qquad z'(0) = 0,$$

where

$$0 < \varepsilon \ll 1,$$

and μ is a constant of order one.

Use the method of multiple scales with

$$T_0 = t, \qquad T_1 = \varepsilon t, \qquad T_2 = \varepsilon^2 t,$$

and seek an expansion of the form

$$z(t) = z_0(T_0,T_1,T_2) + \varepsilon z_1(T_0,T_1,T_2) + \varepsilon^2 z_2(T_0,T_1,T_2) + \cdots.$$

(a) *Derive the equations satisfied by z_0, z_1, and z_2.*

(b) *Eliminate secular terms and obtain the modulation equations governing the slow evolution of the amplitude and phase.*

(c) *Construct a uniformly valid approximation for $z(t)$ up to $O(\varepsilon^2)$.*

(d) *Discuss qualitatively how the slowly varying parametric term*

$$\varepsilon^2 \mu \cos(\varepsilon t)\, z$$

affects the classical limit-cycle behavior of the van der Pol oscillator.

3.6 A Model for Weakly Nonlinear Standing Waves

In many physical systems, small disturbances of an equilibrium or standing wave may oscillate stably, decay, or grow depending on the interaction between dispersion and nonlinear effects. When the nonlinearity is weak, the dominant behavior is often governed by the linearized equation, while the nonlinear terms produce slow corrections that accumulate over long timescales. These corrections may lead to frequency shifts, amplitude modulation, resonance, or instability.

To illustrate these ideas, we consider the nonlinear wave equation

$$u_{tt} - u_{xx} - u = u^3, \tag{3.33}$$

subject to the initial conditions

$$u(x,0) = \varepsilon\cos(kx), \qquad u_t(x,0) = 0, \tag{3.34}$$

where $0 < \varepsilon \ll 1$ is a small parameter measuring the amplitude of the initial disturbance. The smallness of ε suggests seeking an asymptotic expansion of the form

$$u(x,t) = \varepsilon u_1(x,t) + \varepsilon^2 u_2(x,t) + \varepsilon^3 u_3(x,t) + \cdots. \tag{3.35}$$

and employ techniques such as the method of multiple scales or averaging. These methods remove the secular terms and yield an evolution equation describing the slow modulation of the wave amplitude and phase.

This example illustrates a fundamental principle of weakly nonlinear analysis: even small nonlinearities can produce significant long-term effects that are completely absent in the linear theory.

3.6.1 Exercises

Exercise 3.36 *Consider the system that governs a damped nonlinear oscillator*

$$\frac{d^2y}{dt^2}+y=\varepsilon y\frac{dy}{dt},\quad y(0)=1,\quad \frac{dy}{dt}(0)=0,\quad 0<\varepsilon\ll 1.$$

Use the ***method of multiple scales*** *to obtain a uniformly valid first-order approximation for $y(t)$.*

Exercise 3.37 *Study the weakly nonlinear van der Pol oscillator*

$$x''(t)+x(t)=\varepsilon\left(1-x^2(t)\right)x'(t),\quad \varepsilon\ll 1,$$

with $x(0)=2$ and $x'(0)=0$.

(a) Use Poincaré–Lindstedt method.

(b) Use multiple scales method.

Exercise 3.38 *(Forced oscillator near resonance) Consider the forced oscillator*

$$\frac{d^2x}{dt^2}+x=\varepsilon\cos\left((1+\varepsilon)t\right),\qquad x(0)=0,\quad \dot{x}(0)=0,$$

where $0<\varepsilon\ll 1$.

(a) Use the method of multiple scales to construct a uniformly valid approximation for $x(t)$ accurate up to $O(\varepsilon)$.

(b) Explain why the Poincaré–Lindstedt method cannot remove secular terms in this near-resonant forcing case. In your explanation, discuss the role of the external forcing frequency $1+\varepsilon$ relative to the natural frequency and why a frequency renormalization of the homogeneous solution alone is insufficient.

Exercise 3.39 *(Modulation instability in a weakly coupled oscillator) Consider the weakly coupled (parametrically forced) oscillator*

$$\frac{d^2x}{dt^2}+x=\varepsilon x\cos t,\qquad x(0)=0,\quad \dot{x}(0)=1,$$

where $0<\varepsilon\ll 1$.

(a) *Apply the method of multiple scales to obtain a uniformly valid approximation for $x(t)$ up to $O(\varepsilon)$. Comment on the nature of the modulation (growth/decay or oscillatory modulation).*

(b) *Show that a direct application of the Poincaré–Lindstedt method fails to produce a uniform approximate solution. Explain why the multiple scales approach captures the instability/modulation while Poincaré–Lindstedt does not.*

Exercise 3.40 *Consider the weakly nonlinear wave equation*

$$u_{tt} - u_{xx} + u = \varepsilon u^3, \qquad \varepsilon \ll 1,$$

subject to the initial conditions

$$u(x,0) = a\cos(kx), \qquad u_t(x,0) = 0.$$

(a) *Obtain a first-order perturbation expansion in powers of ε for $u(x,t)$.*

(b) *Discuss whether the expansion obtained is* uniformly valid *for all time.*

Exercise 3.41 *Consider the nonlinear wave equation with small nonlinearity*

$$u_{tt} - c^2 u_{xx} = \varepsilon u u_x, \qquad \varepsilon \ll 1,$$

subject to the initial conditions

$$u(x,0) = f(x) + g(x), \qquad u_t(x,0) = c\,[g'(x) - f'(x)],$$

where $f(x)$ and $g(x)$ are bounded functions of x.

(a) *Determine a first-order* straightforward *expansion for small ε.*

(b) *Discuss the* uniform approximation *of the obtained approximation.*

3.7 Averaging Method

In many physical, biological, and engineering systems, the governing differential equations contain small parameters that lead to multiple time scales or slowly varying oscillations. When a system exhibits such features, the direct integration of the equations may be analytically intractable, and standard perturbation methods may fail to capture the essential long-term behavior. The *method of averaging* is a powerful perturbation technique that provides an approximate description of the slow evolution of the system by averaging out the effects of the fast oscillations.

This method, originally developed in celestial mechanics and later formalized by Bogoliubov and Mitropolsky, allows one to approximate periodic or quasiperiodic solutions and to analyze their stability over long time intervals. Averaging methods are particularly effective for systems that can be written in the form of slowly perturbed oscillator s or weakly nonlinear systems. To explain the basic idea of the Averaging Method, we consider a system of ordinary differential equations of the form

$$\frac{dx}{dt} = \varepsilon f(x,t,\varepsilon), \tag{3.45}$$

where $x \in \mathbb{R}^n$, $t \in \mathbb{R}$, and $0 < \varepsilon \ll 1$ is a small parameter. The function $f(x,t,\varepsilon)$ is assumed to be continuous and T-periodic in t. The small parameter ε represents a weak perturbation that slightly modifies the unperturbed dynamics.

The key observation is that when $\varepsilon = 0$, the system reduces to

$$\frac{dx}{dt} = 0,$$

whose solutions are constant in time. When $\varepsilon \neq 0$, the solution $x(t)$ evolves slowly. The idea of the averaging method is to replace the right-hand side of (3.45) by its average over one period of the fast variable:

$$\bar{f}(x) = \frac{1}{T}\int_0^T f(x,t,0)\,dt. \tag{3.46}$$

We then consider the averaged system

$$\frac{dy}{dt} = \varepsilon \bar{f}(y), \tag{3.47}$$

whose solution $y(t)$ approximates $x(t)$ over time intervals of order $1/\varepsilon$.

The rigorous justification of this approximation is given by the classical *Averaging Theorem*, which states that if $f(x,t,\varepsilon)$ is sufficiently smooth and bounded, then the solutions $x(t)$ of (3.45) and $y(t)$ of (3.47) satisfy

$$\|x(t) - y(t)\| = O(\varepsilon), \quad \text{for} \quad 0 \le t \le \frac{c}{\varepsilon},$$

where $c > 0$ is a constant independent of ε.

For example, if we consider the weakly nonlinear oscillator

$$\ddot{x} + x = \varepsilon f(x,\dot{x}), \tag{3.48}$$

where $0 < \varepsilon \ll 1$. Introducing polar coordinates

$$x = r\cos\theta, \qquad \dot{x} = -r\sin\theta,$$

and differentiating with respect to time, we obtain

$$\dot{r} = \varepsilon F_1(r,\theta), \qquad \dot{\theta} = 1 + \varepsilon F_2(r,\theta),$$

where F_1 and F_2 are 2π-periodic in θ. Averaging over one period of θ, we obtain the averaged equations

$$\dot{r} = \varepsilon \bar{F}_1(r),$$
$$\dot{\theta} = 1 + \varepsilon \bar{F}_2(r), \tag{3.49}$$

where

$$\bar{F}_i(r) = \frac{1}{2\pi} \int_0^{2\pi} F_i(r, \theta)\, d\theta, \quad i = 1, 2.$$

The averaged equation for $r(t)$ determines the slow evolution of the amplitude, while the phase $\theta(t)$ evolves primarily with frequency 1 and a slow correction of order ε

The method of averaging differs from other perturbation approaches, such as the *regular perturbation method*, the *method of multiple scales*, and *matched asymptotic expansions*, in both philosophy and application.

- **Regular perturbation :** In a regular perturbation expansion, one assumes that the solution can be expanded as a power series in ε, i.e.,

  $$x(t) = x_0(t) + \varepsilon x_1(t) + \varepsilon^2 x_2(t) + \cdots.$$

  This approach works well when the perturbation does not alter the qualitative behavior of the solution and when no secular (unbounded) terms appear in the expansion. However, for oscillatory systems, secular terms typically appear and invalidate the approximation for large times.

- **Method of Multiple Scales and the Poincaré–Lindstedt Method:** These methods are used to eliminate secular terms and capture slow modulations in oscillatory systems. Although highly effective, they can become algebraically complex. The averaging method often achieves similar results in a more compact and geometrically transparent manner.

- **Matched Asymptotic Expansions:** This method is designed primarily for singular perturbation problems, where solutions exhibit rapid variations in narrow regions (boundary layers). Averaging, by contrast, is suited for weakly nonlinear oscillatory systems with smooth slow dynamics.

In summary, the averaging method provides a conceptually clear and computationally efficient framework for studying systems with slowly varying amplitudes and phases. Its strength lies in reducing a rapidly oscillating system to a lower-dimensional slow system that governs the essential dynamics.

Higher-order averaging techniques can be developed by retaining additional terms in the expansion of $f(x,t,\varepsilon)$ with respect to ε. These refinements improve the accuracy of the approximation and are useful in analyzing bifurcations, resonance phenomena, and the long-term stability of periodic orbits.

Applications of the averaging method are widespread: from analyzing nonlinear electrical circuits and mechanical vibrations to studying planetary motion and biological rhythms. In all such systems, the method provides an elegant bridge between fast oscillations and slow modulations, revealing the essential dynamical behavior without requiring a full numerical solution.

In conclusion, the perturbation method by averaging is distinguished from other asymptotic methods by its focus on the slow dynamics of oscillatory systems. Rather than expanding the solution in powers of a small parameter, averaging removes the dependence on the fast oscillatory variable, yielding a simpler system that approximates the long-term evolution. Its conceptual simplicity, broad applicability, and geometric insight makes it a central tool in nonlinear dynamics and applied mathematics.

Example 3.8 *(Van der Pol equation in the weakly nonlinear regime) We consider the (nondimensional) Van der Pol equation in the weakly nonlinear regime:*

$$\ddot{x} - \varepsilon(1-x^2)\dot{x} + x = 0, \qquad 0 < \varepsilon \ll 1. \tag{3.50}$$

Equivalently, moving the small term to the right-hand side,

$$\ddot{x} + x = \varepsilon(1-x^2)\dot{x}. \tag{3.51}$$

We will apply the method of averaging to obtain an approximate equation governing the slow evolution of the amplitude and phase of oscillation, accurate on time intervals of order $O(1/\varepsilon)$. Introduce the first-order system by letting $y = \dot{x}$:

$$\begin{aligned} \dot{x} &= y, \\ \dot{y} &= -x + \varepsilon(1-x^2)y. \end{aligned}$$

Define the (time-dependent) energy-like amplitude

$$r(t) = \sqrt{x(t)^2 + y(t)^2},$$

and an angle $\theta(t)$ so that we may write

$$x = r\cos\theta, \qquad y = -r\sin\theta.$$

(We choose this sign convention so that, for the unperturbed oscillator $\ddot{x} + x = 0$, one has $\theta \approx t + const.$)

Differentiate $r^2 = x^2 + y^2$ and use the system to obtain an exact expression for r':

$$\begin{aligned} \frac{d}{dt}(r^2) &= 2x\dot{x} + 2y\dot{y} = 2xy + 2y\big(-x + \varepsilon(1-x^2)y\big) \\ &= 2xy - 2xy + 2\varepsilon(1-x^2)y^2 = 2\varepsilon(1-x^2)y^2. \end{aligned}$$

Hence

$$r' = \frac{1}{2r}\frac{d}{dt}(r^2) = \varepsilon\,\frac{(1-x^2)y^2}{r}. \tag{3.52}$$

Substituting $x = r\cos\theta,\ y = -r\sin\theta$ *we obtain*

$$(1-x^2)y^2 = \left(1-r^2\cos^2\theta\right)\cdot r^2\sin^2\theta,$$

implies that

$$r' = \varepsilon r\,\sin^2\theta\left(1-r^2\cos^2\theta\right).$$

This is an exact *expression for* r' *(no approximations yet). It shows that changes in* r *are* $O(\varepsilon)$ *and depend periodically on* θ.

For the phase evolution, one can compute (exactly) the quantity

$$r^2\theta' = xy' - yx',$$

which follows from differentiating $\tan\theta = -y/x$ *or from standard identities. Using the system,*

$$\begin{aligned} xy' - yx' &= x\big(-x+\varepsilon(1-x^2)y\big) - y\cdot y \\ &= -x^2 - y^2 + \varepsilon xy(1-x^2) \\ &= -r^2 + \varepsilon xy(1-x^2). \end{aligned}$$

Therefore

$$\theta' = -1 + \varepsilon\,\frac{xy(1-x^2)}{r^2}. \tag{3.53}$$

(The apparent negative sign is a consequence of our choice of sign in $y = -r\sin\theta$*; if one defines* $\phi = -\theta$ *then* $\phi' \approx 1+\cdots$*. What matters for averaging is that the* ε*-term in* θ' *is* $O(\varepsilon)$ *and periodic in* θ*.)*

Because r' *is* $O(\varepsilon)$ *and depends periodically on the fast angle* θ*, averaging over one period of* θ *gives the slow dynamics of* r*. Replace the right-hand side of the exact* r' *expression by its average with respect to* θ *(holding* r *constant during the averaging). Compute*

$$\overline{\sin^2\theta\left(1-r^2\cos^2\theta\right)} = \overline{\sin^2\theta} \;-\; r^2\overline{\sin^2\theta\cos^2\theta}.$$

Evaluate the averages over $\theta \in [0,2\pi]$*:*

$$\overline{\sin^2\theta} = \frac{1}{2\pi}\int_0^{2\pi}\sin^2\theta\,d\theta = \frac{1}{2},$$

and

$$\sin^2\theta\cos^2\theta = \frac{1}{4}\sin^2(2\theta), \quad \textit{so} \quad \overline{\sin^2\theta\cos^2\theta} = \frac{1}{4}\,\overline{\sin^2(2\theta)} = \frac{1}{4}\cdot\frac{1}{2} = \frac{1}{8}.$$

Thus

$$\overline{\sin^2\theta\left(1-r^2\cos^2\theta\right)} = \frac{1}{2} - r^2\cdot\frac{1}{8} = \frac{4-r^2}{8}.$$

Averaging (3.52) gives the averaged amplitude equation

$$\frac{dr}{dt} = \varepsilon r\left(\frac{1}{2} - \frac{r^2}{8}\right) + O(\varepsilon^2) = \frac{\varepsilon}{2}r\left(1-\frac{r^2}{4}\right) + O(\varepsilon^2). \tag{3.54}$$

This is an autonomous first-order ODE for the slow amplitude $r(t)$ and is valid on time intervals of order $t = O(1/\varepsilon)$.

Equation (3.54) is separable and has fixed points (constant amplitudes) determined by the zeros of the right-hand side

$$r\left(1-\frac{r^2}{4}\right)=0$$

and are given by

$$r=0 \quad or \quad r=2.$$

Linearizing the averaged ODE about these equilibria:

$$\frac{dr}{dt}=\frac{\varepsilon}{2}r\left(1-\frac{r^2}{4}\right)$$

implies that

$$\frac{d}{dr}\left(\frac{dr}{dt}\right)\bigg|_{r=0}=\frac{\varepsilon}{2}>0,$$

so $r=0$ is unstable *(small perturbations grow). At $r=2$,*

$$\frac{d}{dr}\left(\frac{dr}{dt}\right)\bigg|_{r=2}=\frac{\varepsilon}{2}\left(1-\frac{3r^2}{4}\right)_{r=2}=\frac{\varepsilon}{2}\left(1-\frac{12}{4}\right)=\frac{\varepsilon}{2}(-2)=-\varepsilon<0,$$

so $r=2$ is stable. *Therefore the averaged dynamics predicts that solutions (except the decaying measure zero case) approach an oscillation of amplitude $r=2$. This is the celebrated relaxation/limit-cycle behavior of the Van der Pol oscillator ; for small ε, the limit cycle is approximately sinusoidal with amplitude 2. Next, we explicitly approximate solution for the amplitude by solving the averaged ODE exactly and dropping the $O(\varepsilon^2)$ term.*

$$\frac{dr}{dt}=\frac{\varepsilon}{2}r\left(1-\frac{r^2}{4}\right).$$

Separate variables:

$$\int\frac{dr}{r(1-r^2/4)}=\frac{\varepsilon}{2}\int dt.$$

Computing the integral on the left by partial fractions, we arrive at

$$\frac{1}{r(1-r^2/4)}=\frac{1}{r}+\frac{r}{4(1-r^2/4)}.$$

Then, letting $u=r^2$, we obtain

$$\int\frac{dr}{r(1-r^2/4)}=\int\frac{dr}{r}+\frac{1}{4}\int\frac{rdr}{1-r^2/4}=\ln r-\frac{1}{2}\ln(1-r^2/4)+C.$$

Thus

$$\ln\left(\frac{r}{(1-r^2/4)^{1/2}}\right)=\frac{\varepsilon}{2}t+C.$$

Solving for $r(t)$ *explicitly gives the explicit solution*

$$r(t) = \frac{2Ce^{\varepsilon t/2}}{\sqrt{4 + C^2 e^{\varepsilon t}}},$$

we compute the limit as $t \to \infty$ *(assuming* $\varepsilon > 0$*).*

Factor $e^{\varepsilon t}$ *from the denominator:*

$$r(t) = \frac{2Ce^{\varepsilon t/2}}{\sqrt{C^2 e^{\varepsilon t}\left(1 + \frac{4}{C^2 e^{\varepsilon t}}\right)}}. \tag{3.55}$$

Then

$$r(t) = \frac{2Ce^{\varepsilon t/2}}{|C| e^{\varepsilon t/2}\sqrt{1 + \frac{4}{C^2 e^{\varepsilon t}}}} = \frac{2C}{|C|}\frac{1}{\sqrt{1 + \frac{4}{C^2 e^{\varepsilon t}}}}.$$

Since $e^{\varepsilon t} \to \infty$ *as* $t \to \infty$, *we have*

$$\frac{4}{C^2 e^{\varepsilon t}} \to 0,$$

and therefore

$$\lim_{t \to \infty} r(t) = \frac{2C}{|C|}.$$

Since $r(t)$ *represents the amplitude,*

$$r(t) = \sqrt{x(t)^2 + y(t)^2} \geq 0,$$

we assume the initial amplitude satisfies $r(0) > 0$. *Hence, the constant* C *is positive.*

Therefore,

$$\lim_{t \to \infty} r(t) = 2.$$

Thus the solution approaches the stable equilibrium amplitude

$$r = 2,$$

which is the positive, nonzero root of

$$r\left(1 - \frac{r^2}{4}\right) = 0.$$

Finally, we reconstruct the oscillatory solution. A first-order averaged approximation for the solution $x(t)$ *is obtained by combining the slowly varying amplitude* $r(t)$ *with the fast phase* $\theta \approx t + \theta_0$. *Thus, to leading order,*

$$x(t) \approx r(t)\cos\left(t + \theta_0\right) + O(\varepsilon),$$

where $r(t)$ is given by (3.55) and θ_0 is a constant determined by the initial phase.

The averaged equation shows that the average correction to the phase vanishes at $O(\varepsilon)$, since the $O(\varepsilon)$ term in θ' is a periodic function of θ with zero mean. Therefore, the dominant phase evolution is given by the natural frequency 1.

We draw the following conclusions:

- *The method of averaging reduced the originally non-autonomous, oscillatory system to a simple autonomous ODE for the slowly varying amplitude $r(t)$. The averaged equation captures the long-term behavior (existence and stability of a limit cycle) and yields an explicit approximate formula for the amplitude evolution* (3.55). *The concept of limit cycles will be discussed in detail in Chapter 8.*
- *For the Van der Pol oscillator with small $\varepsilon > 0$, the averaging prediction is the existence of a stable limit cycle of amplitude $r = 2$ (that is, $x(t)$ oscillates with amplitude approximately equal to* 2*). This agrees with the classical result.*
- *Accuracy. The averaging theorem guarantees that the solution of the full system remains $O(\varepsilon)$-close to the averaged approximation on time intervals*

$$0 \le t \le O(1/\varepsilon).$$

- *Comparison with other perturbation methods. The method of multiple scales produces equivalent leading-order amplitude-phase modulation equations, but typically requires the introduction of separate slow time variables and matching expansions. Averaging achieves the same result by computing explicit averages over the fast angle, and is often more direct and geometrically transparent for weakly nonlinear oscillators.*

□

Example 3.9 *(Perturbed logistic growth model)*

Consider the differential equation

$$\dot{x} = \varepsilon\big(x(1-x) + \sin t\big), \qquad x(0) = x_0,$$

where $\varepsilon > 0$ is small ($\varepsilon \ll 1$). Here, x is the normalized population size, t is time, $x(1-x)$ represents logistic growth, and $\sin t$ represents a fast periodic environmental fluctuation.

The system is written in the standard averaging form

$$\dot{x} = \varepsilon F(t,x), \qquad F(t,x) = x(1-x) + \sin t.$$

The averaging method replaces the non-autonomous system by an autonomous one

$$\dot{z} = \varepsilon G(z),$$

where G is the time average of $F(t,x)$ over one period of the fast oscillations, with x treated as constant during the averaging.

Since $\sin t$ *has period* $T = 2\pi$*, the averaged function is*

$$G(z) = \frac{1}{T}\int_0^T F(t,z)\,dt = \frac{1}{2\pi}\int_0^{2\pi}\big(z(1-z)+\sin t\big)\,dt.$$

Computing the integral, we obtain

$$G(z) = \frac{1}{2\pi}\left[\int_0^{2\pi} z(1-z)\,dt + \int_0^{2\pi}\sin t\,dt\right]$$

$$= \frac{1}{2\pi}\left[z(1-z)t\Big|_0^{2\pi} + (-\cos t)\Big|_0^{2\pi}\right]$$

$$= \frac{1}{2\pi}\left[z(1-z)(2\pi) + (-\cos(2\pi)+\cos 0)\right].$$

Since $\cos(2\pi) = \cos 0 = 1$*, the last term vanishes, and we obtain*

$$G(z) = z(1-z).$$

Thus, the averaged system is the logistic equation

$$\dot{z} = \varepsilon z(1-z).$$

We may easily solve the averaged equation by separation of variables, using partial fractions, to obtain

$$z(t) = \frac{z_0}{z_0 + (1-z_0)e^{-\varepsilon t}},$$

which is the standard logistic growth solution.

The averaging theorem guarantees that the solution $z(t)$ approximates the true solution $x(t)$ on long time intervals, specifically for

$$0 \le t \le O(1/\varepsilon),$$

with an error of order $O(\varepsilon)$.

Since the fast oscillatory term $\sin t$ *has zero mean, its effect averages out over time, and the long-term population dynamics are governed by the logistic equation*

$$\dot{z} = \varepsilon z(1-z).$$

□

4

Boundary Layers in Singularly Perturbed Ordinary Differential Equations

Boundary layers represent a fundamental concept in fluid dynamics and singular perturbation theory, describing thin regions near surfaces where rapid changes occur due to the dominance of diffusive processes over convective ones. Their importance extends far beyond theoretical interest, as they critically influence drag forces, heat transfer rates, and separation phenomena in aerodynamic and hydrodynamic systems. In mathematical modeling, boundary layers resolve the singular nature of problems where a small parameter multiplies the highest derivative, allowing solutions to satisfy boundary conditions that would otherwise be incompatible with the reduced outer solution . The analysis of boundary layers through matched asymptotic expansions provides crucial insights into physical systems ranging from aircraft wing design to atmospheric flow patterns, making them indispensable for accurate prediction and optimization in engineering and geophysical applications.

4.1 Singular Perturbations and the Appearance of Boundary Layers

This section explains the origin and structure of boundary layers in singularly perturbed ordinary differential equations (ODEs). We show why the naïve limit $\varepsilon \to 0$ may change the order of the differential equation and force rapid variation in thin regions ("boundary layers "), introduce outer and inner (boundary layer) solutions denoted by $y^o(t)$ and $y^i(t)$, describe the matching procedure, and explain why the composite (uniform) solution is formed by *subtracting* the common limit to avoid double-counting. Finally, we work the classic example

$$\varepsilon y''(t) + y'(t) + y(t) = 0, \qquad y(0) = 0,\ y(1) = 1, \qquad 0 < t < 1,$$

construct the exact solution, determine the boundary layer thickness, and then obtain the outer and inner expansions and the uniformly valid composite approximation.

DOI: 10.1201/9781042009220-4

Consider a family of ODE boundary value problems depending on a small parameter $0 < \varepsilon \ll 1$. A perturbation is called *singular* when setting $\varepsilon = 0$ changes the order (or fundamental type) of the differential equation. A simple manifestation: an ε multiplies the highest-derivative term, so the reduced problem (obtained by $\varepsilon \to 0$) has a lower order and in general, cannot satisfy all boundary conditions of the full problem. Consequently, the full solution must vary rapidly in thin regions (boundary layers) adjacent to one or more boundary points so that as $\varepsilon \to 0$, the reduced (outer) solution and the full solution can both be satisfied by different pieces.

Heuristically: where the highest derivative is multiplied by a small factor, large derivatives are permitted without making the multiplied term large; hence y' or y'' may become $O(1/\varepsilon)$ in a thin region of width $O(\varepsilon)$ (or some power of ε), producing a boundary layer.

Outer and inner solutions and the need to match.

We denote the outer (slow) solution by $y^o(t)$, or $y^{(0)}(t)$, and the inner (fast / boundary layer) solution by $y^i(t)$, or by $y^{(i)}(t)$.The outer solution is found by formally setting $\varepsilon = 0$ in the differential equation and solving the resulting reduced (lower-order) problem. The inner solution is found by introducing a rescaled *stretched* variable (for example, $\tau = t/\delta(\varepsilon)$ where $\delta(\varepsilon) \to 0$) that "zooms into" the thin region where the rapid change occurs. In the inner variable, the derivatives are large enough so that the neglected highest-derivative term becomes leading-order and the boundary condition at the layer end can be enforced (see Fig. 4.1).

Because both the outer and inner approximations aim to describe the same exact solution in their respective regions of validity, we must *match* them in an overlapping intermediate limit. The matching principle expresses the idea that the outer and inner solutions are not independent approximations, but two descriptions of the same exact solution viewed at different spatial scales. Far from the boundary layer, the solution varies slowly and is well approximated by the outer solution , while close to the boundary, it varies rapidly and is described by the inner solution. In an intermediate region far enough from the boundary that the rapid variation has subsided, yet close enough that the outer solution has not fully taken over, both approximations must represent the same behavior. Matching ensures that the inner solution smoothly connects to the outer solution , preventing spurious jumps or mismatches and allowing the two asymptotic descriptions to merge into a single, coherent approximation. Typically

$$\lim_{t \searrow 0} y^o(t) \quad \text{must equal} \quad \lim_{\tau \to \infty} y^i(\tau),$$

or analogous limits at the other boundary. This matching determines integration constants left undetermined by the reduced problems.

Why subtract the common limit when forming a composite solution.

If we simply add the outer and inner approximations, we would double-count the common intermediate behavior. The matched limit (common value) appears as the

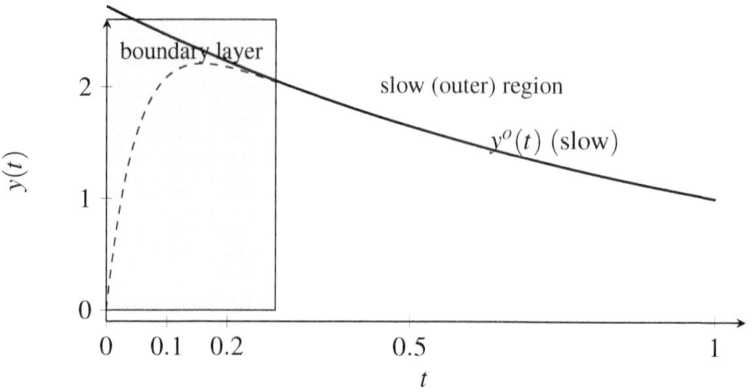

FIGURE 4.1: This is a schematic plot showing a typical situation: a narrow boundary layer of width $O(\varepsilon)$ near $t = 0$, where the solution changes rapidly, and a slow region for $t \gg \varepsilon$. The shaded strip indicates the boundary layer region.

t-independent part of both approximations in the overlap region, so adding them produces that part twice. To produce a uniformly valid approximation, one takes

$$y_{\text{comp}}(t) = y^o(t) + y^i(t) - y^{\text{match}},$$

where

$$y^{\text{match}} = \lim_{t \to \text{layer edge}} y^o(t) = \lim_{\tau \to \infty} y^i(\tau)$$

is the common limit. Subtracting the common limit removes the double-counted portion and yields an approximation that is accurate both inside and outside the layer (to the order of the asymptotics used).

Throughout the remainder of the book, the uniform approximate solution, also called the composite solution, will be denoted by $y^{\text{comp}}(t)$, $y_{\text{comp}}(t)$, or simply $y_u(t)$.

4.1.1 Worked Example

We now present a detailed example illustrating a boundary layer at the left endpoint.

Example 4.1 *Consider the singular perturbed problem*

$$\varepsilon y''(t) + y'(t) + y(t) = 0, \qquad y(0) = 0, \; y(1) = 1, \quad 0 < t < 1, \tag{4.1}$$

with $0 < \varepsilon \ll 1$. *(We use* $'$ *for* d/dt.*)*

We begin by finding the exact solution by considering a solution of the form

$$y(t) = e^{rt}.$$

Substituting into (4.1), we obtain the characteristic equation

$$\varepsilon r^2 + r + 1 = 0,$$

whose two roots are given by

$$r_{1,2} = \frac{-1 \pm \sqrt{1-4\varepsilon}}{2\varepsilon}.$$

Thus the general solution is

$$y(t) = Ae^{r_1 t} + Be^{r_2 t}.$$

Applying $y(0) = 0$ gives $A + B = 0$, so $B = -A$. Applying $y(1) = 1$ gives

$$A\left(e^{r_1} - e^{r_2}\right) = 1, \text{ or } A = \frac{1}{e^{r_1} - e^{r_2}}.$$

Therefore, the exact solution may be written as

$$y(t) = \frac{e^{r_1 t} - e^{r_2 t}}{e^{r_1} - e^{r_2}}. \tag{4.2}$$

Next, we look at the asymptotic form of the roots for small ε.

Asymptotic form of the roots for small ε.

For $0 < \varepsilon \ll 1$, expand

$$\sqrt{1-4\varepsilon} = 1 - 2\varepsilon - 2\varepsilon^2 + O(\varepsilon^3).$$

Then

$$r_1 = \frac{-1 + \sqrt{1-4\varepsilon}}{2\varepsilon} = \frac{-1 + (1 - 2\varepsilon - 2\varepsilon^2 + \cdots)}{2\varepsilon} = -1 + O(\varepsilon),$$

while

$$r_2 = \frac{-1 - \sqrt{1-4\varepsilon}}{2\varepsilon} = \frac{-1 - (1 - 2\varepsilon - 2\varepsilon^2 + \cdots)}{2\varepsilon} = -\frac{1}{\varepsilon} + 1 + O(\varepsilon).$$

Hence, r_1 is the slow root, satisfying

$$r_1 \sim -1,$$

while r_2 is the fast root, satisfying

$$r_2 \sim -\frac{1}{\varepsilon}.$$

Consequently,

$$e^{r_2 t}$$

decays extremely rapidly on the scale $t = O(\varepsilon)$, *whereas*

$$e^{r_1 t}$$

varies on the $O(1)$ *scale. This indicates the presence of a boundary layer near* $t = 0$ *needed to satisfy the condition* $y(0) = 0$, *together with a slowly varying outer solution across the remainder of the interval.*

Outer solution (reduced problem). *Formally set* $\varepsilon = 0$ *in* (4.1) *to obtain the reduced first-order equation*

$$y^{o\prime}(t) + y^o(t) = 0,$$

which has the solution

$$y^o(t) = Ce^{-t}.$$

To determine C *we must use the boundary condition away from the layer. Because the rapid layer occurs near* $t = 0$, *the outer solution should satisfy the boundary condition at* $t = 1$. *Thus*

$$y^o(1) = Ce^{-1} = 1 \quad \text{gives} \quad C = e,$$

and hence the leading-order outer solution is

$$y^o(t) = e^{1-t}. \tag{4.3}$$

Note $y^o(0) = e$, *which does* not *satisfy* $y(0) = 0$; *the discrepancy is resolved by the inner (boundary layer) correction near* $t = 0$.

Inner (boundary layer) solution. *Introduce the stretched inner variable* $\tau = t/\delta(\varepsilon)$ *(so the layer width is* $O(\varepsilon)$*), and define the inner dependent variable*

$$Y(\tau) = y(t) = y(\delta(\varepsilon)\tau).$$

Computing the derivatives, we obtain

$$y' = \frac{1}{\delta(\varepsilon)}Y'(\tau), \qquad y'' = \frac{1}{\delta(\varepsilon)^2}Y''(\tau),$$

where primes now denote $d/d\tau$. *Substituting into* (4.1) *we get*

$$\varepsilon \cdot \frac{1}{\delta(\varepsilon)^2}Y''(\tau) + \frac{1}{\delta(\varepsilon)}Y'(\tau) + Y(\tau) = 0. \tag{4.4}$$

To eliminate ε *from the dominant coefficients multiplying* Y'' *and* Y', *we balance*

$$\varepsilon\delta^{-2}(\varepsilon) \sim \delta^{-1}(\varepsilon).$$

This implies that

$$\delta(\varepsilon) = \varepsilon.$$

Substituting delta with its value into (4.4) and multiply the resulting equation with ε gives

$$Y''(\tau) + Y'(\tau) + \varepsilon Y(\tau) = 0.$$

To leading order as $\varepsilon \to 0$ we obtain

$$Y''(\tau) + Y'(\tau) = 0.$$

Integrate: let $Z(\tau) = Y'(\tau)$. Then $Z' + Z = 0$ so $Z(\tau) = Ke^{-\tau}$ and therefore

$$Y(\tau) = A - Ke^{-\tau},$$

for constants A, K to be determined from matching and the boundary condition at the wall $t = 0$ (i.e., $\tau = 0$):

$$Y(0) = y(0) = 0 \text{ implies that } A - K = 0.$$

So we have

$$K = A.$$

Thus, the leading-order inner solution is

$$Y(\tau) = A\left(1 - e^{-\tau}\right),$$

and in terms of t the inner solution becomes

$$y^i(t) = A\left(1 - e^{-\frac{t}{\varepsilon}}\right). \tag{4.5}$$

To determine the constant A, we must perform matching so that the outer solution and the inner solution agree in an interval where both approximations are simultaneously valid. The inner solution applies when $t = O(\varepsilon)$, placing t within the boundary layer, while the outer solution applies for $t = O(1)$. These two regions do not overlap directly, but we can construct an intermediate region in which both descriptions hold.

A natural choice is to consider values of t satisfying

$$t = O(\sqrt{\varepsilon}),$$

since

$$\varepsilon \ll \sqrt{\varepsilon} \ll 1 \qquad (\varepsilon \to 0).$$

Thus $\sqrt{\varepsilon}$ represents an intermediate scale lying between the inner scale $O(\varepsilon)$ and the outer scale $O(1)$.

This motivates introducing the stretched variable

$$\eta = \frac{t}{\sqrt{\varepsilon}},$$

which is $O(1)$ precisely in the overlap region. It is within this overlap domain, where both the inner and outer expansions can be written in compatible forms, that we *perform the matching and determine the constant A. Substituting* $t = \eta\sqrt{\varepsilon}$ *into the outer and inner expressions* (4.3) *and* (4.5)*, respectively, and then taking the limit as* $\varepsilon \to 0^+$ *allows us to compare their behaviors in the overlap region. This yields*

$$\lim_{\varepsilon \to 0^+} e^{1-\eta\sqrt{\varepsilon}} = \lim_{\varepsilon \to 0^+} A\left(1 - e^{-\eta/\sqrt{\varepsilon}}\right).$$

By evaluating the limits on both sides, we have

$$e = A.$$

Thus,

$$y^i(t) = e\left(1 - e^{-t/\varepsilon}\right).$$

Another way to determine the matching constant is to use the ***matching principle****. The idea is that the inner and outer approximations must represent the same solution in a region where both descriptions are valid.*

The outer solution

$$y^o(t) = e^{1-t}$$

is valid away from the boundary layer, but it should still provide a good approximation as we approach the layer from the outer region. Since the boundary layer occurs near $t = 0$*, we examine the limit*

$$t \to 0^+.$$

On the other hand, the inner solution

$$Y(\tau) = A(1 - e^{-\tau}), \qquad \tau = \frac{t}{\varepsilon},$$

is valid inside the boundary layer. To compare it with the outer solution, we move away from the layer into the overlap region. Because

$$\tau = \frac{t}{\varepsilon},$$

this corresponds to letting

$$\tau \to \infty.$$

Therefore, matching requires

$$\lim_{\tau \to \infty} Y(\tau) = \lim_{t \to 0^+} y^o(t).$$

Since

$$\lim_{\tau \to \infty} A(1 - e^{-\tau}) = A$$

and

$$\lim_{t \to 0^+} e^{1-t} = e,$$

we obtain

$$A = e.$$

Geometrically, the outer solution approaches the boundary layer from the outside, while the inner solution approaches the outer region from within the thin layer. The matching condition ensures that the two asymptotic descriptions connect smoothly in the overlap region.

Thus, if we the common limit is denoted by

$$y^{match} = e,$$

Then, the composite and uniform solution

$$\begin{aligned} y_{comp}(t) &= y^o(t) + y^i(t) - y^{match} \\ &= e^{1-t} + e\left(1 - e^{-t/\varepsilon}\right) - e. \end{aligned}$$

Simplifying,

$$y_{comp}(t) = e^{1-t} - e^{1-t/\varepsilon}. \tag{4.6}$$

Comments about the composite formula.

- *For $t \gg \varepsilon$ the second term is exponentially small and $y_{comp}(t) \sim e^{1-t} = y^o(t)$, the outer solution .*
- *The subtraction of e (the matched value) avoids double-counting the common part e that appears in both y^o and y^i.*

Comparison with the exact solution.

It is instructive to compare the composite approximation (4.6) *with the exact solution* (4.2). *Using the expansions of r_1, r_2 for small ε, one shows (after algebra) that the exact solution admits the decomposition*

$$y(t) = e^{1-t} - e\,e^{-t/\varepsilon} + O(\varepsilon) \tag{4.7}$$

uniformly on $0 \le t \le 1$ (the $O(\varepsilon)$ error may be made precise by a further one-term asymptotic match). Thus, the composite formula gives the leading-order uniform approximation; the exponential layer term $ee^{-t/\varepsilon}$ represents the rapid adjustment from the boundary value at $t = 0$ to the outer profile.

Maximum time reached in the boundary layer interval.

The "maximum time" of the boundary layer refers to the practical time scale over which the rapid exponential transition occurs. For solutions containing a fast-decaying term of the form $e^{-t/\varepsilon}$, it is convenient to introduce the stretched variable

$$\tau = \frac{t}{\varepsilon},$$

so that the exponential becomes $e^{-\tau}$. Since $e^{-\tau}$ decays on the scale $\tau = O(1)$, the boundary layer corresponds to

$$t = O(\varepsilon).$$

To estimate the width more precisely, suppose we want the exponential term to fall below a tolerance γ. We solve

$$e^{-\tau} \leq \gamma.$$

Taking logarithms gives

$$-\tau \leq \ln(\gamma) \text{ implies } \tau \geq -\ln(\gamma),$$

and therefore the corresponding time scale is

$$t \geq \varepsilon(-\ln \gamma).$$

Examples.

- *For 90% of the transition to be completed, the remaining exponential is 10% of its initial value, so $\gamma = 0.1$. Then*

$$\tau = -\ln(0.1) \approx 2.30, \qquad t \approx 2.30\,\varepsilon.$$

- *For 99% completion, take $\gamma = 0.01$. Then*

$$\tau = -\ln(0.01) \approx 4.60, \qquad t \approx 4.60\,\varepsilon.$$

Hence, the boundary layer width is always $O(\varepsilon)$, with the proportionality constant depending only logarithmically on the desired accuracy:

$$0 \leq t \lesssim C\varepsilon, \qquad C = -\ln(\gamma).$$

Values such as $C \approx 2$–5 correspond to 90–99% of the total rapid adjustment, so the boundary layer is typically confined to a small interval $0 \leq t \lesssim 3\varepsilon$.

Summary and final remarks.

1. *A singular perturbation (small factor multiplying the highest derivative) typically produces boundary layers: thin regions where the solution varies rapidly, of width scaling with some power of ε.*
2. *The outer solution solves the reduced problem (set $\varepsilon = 0$). The inner solution uses a stretched variable to resolve the rapid variation in the layer.*
3. *Matching fixes integration constants by equating appropriate limits of the inner and outer solutions in the overlap region.*
4. *The composite solution is formed as $y_{comp} = y^o + y^i - y^{match}$ to avoid double-counting the common part; this yields a uniformly valid approximation.*

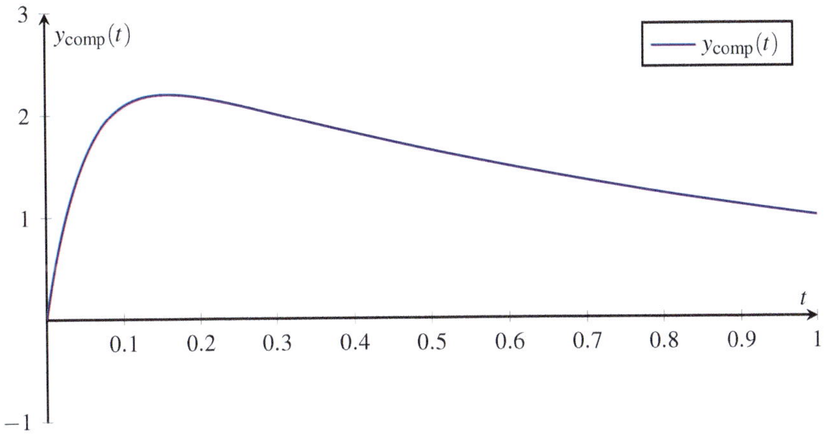

FIGURE 4.2: Graph of $y_{\text{comp}}(t) = e^{1-t} - e^{1-t/\varepsilon}, \varepsilon = 0.05$.

5. *In the worked example, the boundary layer is at* $t = 0$, *of thickness* $O(\varepsilon)$, *the outer solution is* $y^o(t) = e^{1-t}$, *the inner profile is* $y^i(t) = e(1 - e^{-t/\varepsilon})$, *and the composite approximation* $y_{comp}(t) = e^{1-t} - ee^{-t/\varepsilon}$ *matches the exact solution to leading order uniformly on* $[0,1](see\,Fig.4.2)$.

□

4.1.2 Higher–Order Boundary Layer Corrections: O(ε) Terms

One may carry the asymptotic expansion to higher orders by including $O(\varepsilon)$ corrections in the outer expansion (solve y_1^o) and in the inner expansion (solve the inhomogeneous second-order problem obtained after including εY and higher terms). That yields more accurate, uniform approximations and precise error bounds.

In the previous section, we constructed the leading-order outer and inner solutions for the singularly perturbed boundary value problem

$$\varepsilon y''(t) + y'(t) + y(t) = 0, \qquad 0 < t < 1, \qquad y(0) = 0,\ y(1) = 1,$$

and obtained the leading uniform approximation

$$y_{\text{comp}}(t) = e^{1-t} - e\,e^{-t/\varepsilon}.$$

Below, we carry the asymptotic expansion one order further, i.e., we compute the $O(\varepsilon)$ corrections for both outer and inner expansions, perform matching, and produce a uniformly valid approximation

$$y(t) = y^{(o)}(t) + \varepsilon y^{(i)}(t) + O(\varepsilon^2).$$

Order ε^1*:*

Collect $O(\varepsilon)$ terms in the inner equation:

$$Y_1'' + Y_1' + Y_0 = 0.$$

We now substitute $Y_0(\tau) = e(1 - e^{-\tau})$ to obtain

$$Y_1'' + Y_1' = -e\left(1 - e^{-\tau}\right).$$

Set $W(\tau) = Y_1'(\tau)$. Then $W' + W = -e + ee^{-\tau}$.
Multiply by the integrating factor e^{τ} to get

$$\left(e^{\tau}W\right)' = -ee^{\tau} + e.$$

An integration yields

$$e^{\tau}W = -ee^{\tau} + e\tau + B,$$

for some constant B. Consequently,

$$W(\tau) = -e + e\tau e^{-\tau} + Be^{-\tau}.$$

We integrate to get Y_1. Thus,

$$Y_1(\tau) = \int W(\tau)\,d\tau = -e\tau + e\int \tau e^{-\tau}\,d\tau + B\int e^{-\tau}\,d\tau + C.$$

Since $\displaystyle\int \tau e^{-\tau}d\tau = -(\tau+1)e^{-\tau}$, we obtain after some simplifications that

$$Y_1(\tau) = -e\tau - e(\tau+1)e^{-\tau} - Be^{-\tau} + C.$$

Next we apply the inner boundary condition $Y(0) = 0$ at each order. Since $Y_0(0) = 0$, we must have $Y_1(0) = 0$. Hence,

$$Y_1(0) = -e - B + C = 0,$$

which implies that $C = e + B$. Consequently, we arrive at

$$Y_1(\tau) = -e\tau - e(\tau+1)e^{-\tau} - Be^{-\tau} + (e+B). \tag{4.12}$$

Therefore,

$$\begin{aligned} y^{(i)}(t) &= Y_0(\tau) + \varepsilon Y_1(\tau) \\ &= e\left(1 - e^{-\tau}\right) + \varepsilon\left[-e\tau - e(\tau+1)e^{-\tau} - Be^{-\tau} + (e+B).\right] \end{aligned} \tag{4.13}$$

Next we make use of asymptotic **matching principle** or **Van Dyke's principle.** Formally, matching is expressed by Van Dyke's matching principle: the asymptotic expansion of the outer solution, re-expanded in the inner limit, must agree term-by-term with the asymptotic expansion of the inner solution re-expanded in the outer limit. That is, if the boundary layer is located at $t = 0$ and $\tau = t/\delta(\varepsilon)$, then matching requires

$$\lim_{\tau\to\infty} y^i(\tau;\varepsilon) \sim \lim_{t\searrow 0} y^o(t;\varepsilon),$$

with both limits taken as $\varepsilon \to 0^+$. In practice, this condition is enforced by equating the leading-order (and, if desired, higher-order) terms in these limits. The resulting relations determine otherwise undetermined constants in the inner and outer expansions and ensure the existence of an overlap region in which both approximations are simultaneously valid. Now, we use this idea of matching to find the constant B.

We consider an asymptotic solution composed of an outer expansion, valid for $t = O(1)$,

$$y^o(t) \sim y_0(t) + \varepsilon y_1(t) + \cdots,$$

and an inner expansion, valid near a boundary layer at $t = 0$, written in terms of the stretched variable

$$\tau = \frac{t}{\varepsilon},$$

$$y^i(t) \sim Y_0(\tau) + \varepsilon Y_1(\tau) + \cdots.$$

Matching requires that the two expansions agree in their common overlap region, namely, as

$$\varepsilon \to 0, \qquad \tau \to \infty, \quad \text{with } t = \varepsilon\tau \to 0^+.$$

Outer expansion in inner variables

The outer expansion is given by

$$y_0(t) + \varepsilon y_1(t) = e^{1-t} + \varepsilon e^{1-t}(1-t) + \cdots.$$

To match with the inner expansion, we rewrite the outer solution in terms of the inner variable $t - \varepsilon\tau$:

$$y_0(\varepsilon\tau) + \varepsilon y_1(\varepsilon\tau) = e^{1-\varepsilon\tau} + \varepsilon e^{1-\varepsilon\tau}(1-\varepsilon\tau) + \cdots.$$

We now expand for small ε in the overlap region, where $t = \varepsilon\tau \to 0^+$.

$$e^{1-\varepsilon\tau} = e\left(1 - \varepsilon\tau + O(\varepsilon^2)\right).$$

Therefore,

$$\begin{aligned} y^o(\tau) &= y_0 + \varepsilon y_1 = e(1-\varepsilon\tau) + \varepsilon e(1-\varepsilon\tau) + O(\varepsilon^2) \\ &= e + \varepsilon(e - e\tau) + O(\varepsilon^2). \end{aligned} \tag{4.14}$$

Next, we match the $O(\varepsilon)$ terms. Matching is performed in the overlap region where

$$\varepsilon \to 0^+, \qquad \tau \to \infty, \qquad t = \varepsilon\tau \to 0^+.$$

Thus, in the matching process, τ is *not* kept fixed. Instead, the inner expansion is examined in the limit $\tau \to \infty$, while the outer expansion is examined in the limit $t \to 0^+$.

Using (4.13), the inner expansion satisfies

$$y^i(t) = e + \varepsilon\Big[-e\tau + (e+B)\Big] + O\big(\varepsilon e^{-\tau}\big), \qquad \tau \to \infty.$$

On the other hand, from (4.14), the outer expansion written in inner variables is

$$y^o(t) = e + \varepsilon(e - e\tau) + O(\varepsilon^2).$$

Equating the $O(\varepsilon)$ terms gives

$$-e\tau + (e+B) = e - e\tau,$$

and therefore

$$B = 0.$$

Thus, (4.12) becomes

$$Y_1(\tau) = e\big(1 - \tau - (\tau+1)e^{-\tau}\big). \tag{4.15}$$

To form a uniformly valid approximation through $O(\varepsilon)$ we combine the outer and inner expansions and subtract their common (overlap) expansion to avoid double-counting. Using the outer expansion (4.10) and the inner expansions (4.11), (4.15) (with $\tau = t/\varepsilon$), the composite approximation up to $O(\varepsilon^2)$ is

$$\begin{aligned} y_{\text{comp}}(t) &= \big[y_0(t) + \varepsilon y_1(t)\big] + \big[Y_0(\tau) + \varepsilon Y_1(\tau)\big] - \Big[\underbrace{e}_{\text{common } O(1)} + \varepsilon\underbrace{(e - e\tau)}_{\text{common } O(\varepsilon)}\Big] \\ &= \big(e^{1-t} + \varepsilon e^{1-t}(1-t)\big) + \Big(e(1-e^{-\tau}) + \varepsilon e(1 - \tau - (\tau+1)e^{-\tau})\Big) \\ &\quad - \big(e + \varepsilon(e - e\tau)\big) + O(\varepsilon^2), \qquad \tau = \frac{t}{\varepsilon}. \end{aligned}$$

This expression can be rearranged to isolate the leading layer and outer pieces. After algebraic simplification (recall $\tau = t/\varepsilon$) one convenient form is

$$\begin{aligned} y(t) &= e^{1-t} - e\,e^{-t/\varepsilon} \\ &\quad + \varepsilon\Big\{e^{1-t}(1-t) - e\,e^{-t/\varepsilon} - e\frac{t}{\varepsilon}e^{-t/\varepsilon}\Big\} + O(\varepsilon^2), \end{aligned}$$

or, more explicitly,

$$\begin{aligned} y(t) &= e^{1-t} - e\,e^{-t/\varepsilon} \\ &\quad + \varepsilon\Big[ee^{-t}(1-t) - ee^{-t/\varepsilon} - e\frac{t}{\varepsilon}e^{-t/\varepsilon}\Big] + O(\varepsilon^2). \end{aligned} \tag{4.16}$$

We refer to Fig. 4.3.

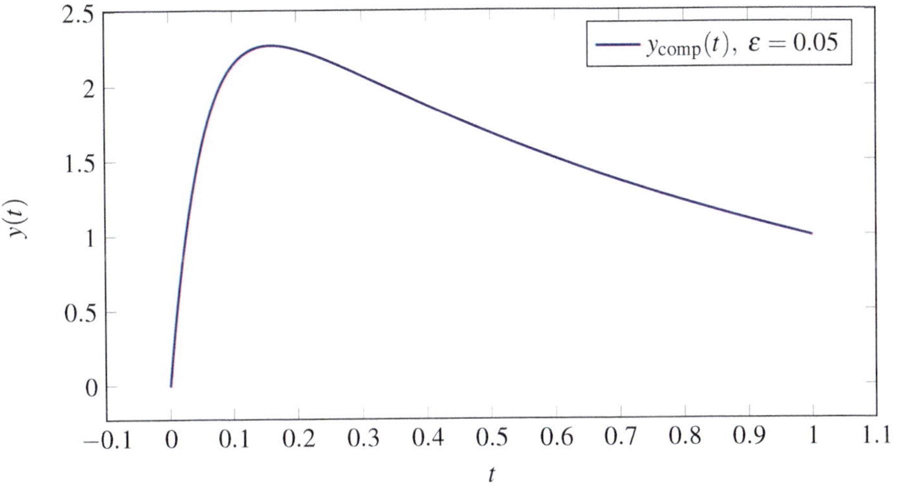

FIGURE 4.3: Graph of $y_{\text{comp}}(t)$ with $\varepsilon = 0.05$.

Remark 11 *A few comments on* (4.16)*:*

- *The leading* $O(1)$ *piece is the same composite found earlier,* $e^{1-t} - ee^{-t/\varepsilon}$.
- *The* $O(\varepsilon)$ *correction contains two kinds of terms:*
  1. *smooth corrections on the outer scale (the term* $ee^{-t}(1-t)$*);*
  2. *correction terms concentrated in the boundary layer (the terms with factors* $e^{-t/\varepsilon}$ *and* $(t/\varepsilon)e^{-t/\varepsilon}$*).*

  These layer-dependent terms are significant only for $t = O(\varepsilon)$ *and are exponentially small for* $t \gg \varepsilon$.
- *In the inner variable* $\tau = t/\varepsilon$ *the inner expansion up to* $O(\varepsilon)$ *reads*

$$Y(\tau) \sim e\left(1 - e^{-\tau}\right) + \varepsilon e\left(1 - \tau - (\tau+1)e^{-\tau}\right),$$

  which for $\tau \to \infty$ *matches the outer expansion* $e + \varepsilon(e - e\tau)$.

In the next example, we pretend the boundary layer occurs at the end point of the interval, and conclude this is not the case.

Example 4.2 *Consider*

$$\varepsilon y'' + (1+\varepsilon^2)y' + (1-\varepsilon^2)y = 0, \quad 0 < t < 1, \quad 0 < \varepsilon \ll 1. \tag{4.17}$$

with boundary conditions

$$y(0) = \alpha, \qquad y(1) = \beta,$$

for constants α, β. *Let us assume that the boundary layer is at the right end, and hence* $y(1) = \beta$ *must be dropped when considering the outer solution . Then we seek an outer expansion in the form*

$$y^o(t) = y_0(t) + \varepsilon y_1(t) + \cdots,$$

which, when substituted into (4.17), *yields*

$$y_0'(t) + y_0(t) = 0, \qquad y_0(0) = \alpha.$$

Thus, the solution is

$$y_0(t) = C_0 e^{-t}.$$

Applying the initial condition $y_0(0) = \alpha$ *we arrive at* $C_0 = \alpha$, *and hence*

$$y^o(t) = \alpha e^{-t}. \tag{4.18}$$

To analyze the behavior of the solution in the assumed boundary layer, we need to magnify the neighborhood of $t = 1$. *As* $\varepsilon \to 0$, *the width of the boundary layer shrinks. If we continue to work in the original variable t, the boundary layer collapses to a point, and the detailed structure of the solution is lost.*

To "zoom in" on this region, we introduce the stretched variable

$$\tau = \frac{1-t}{\varepsilon^\nu}, \qquad \text{where } t < 1 \text{ and } \nu > 0.$$

This magnification allows us to describe the rapid dynamics inside the boundary layer. The numerator $1-t$ *measures the distance from the boundary point* $t = 1$. *The denominator* ε^ν *reflects the unknown thickness of the boundary layer. The exponent* $\nu > 0$ *is left unspecified at this stage because it must be determined by dominant-balance arguments after rewriting the governing equation in terms of* τ.

In other words, ν *is chosen so that the leading terms in the transformed equation remain in balance and yield a nontrivial inner problem.*

Using the chain rule we have

$$\frac{dy}{dt} = \frac{dy}{d\tau}\frac{d\tau}{dt} = -\frac{1}{\varepsilon^\nu}\frac{dy}{d\tau}$$

Differentiating again with respect to t, we get

$$\begin{aligned}\frac{d^2y}{dt^2} &= -\frac{1}{\varepsilon^\nu}\frac{d}{d\tau}\left(\frac{dy}{dt}\right)\\ &= -\frac{1}{\varepsilon^\nu}\frac{d}{d\tau}\left(-\frac{1}{\varepsilon^\nu}\frac{d}{d\tau}\right)\\ &= \frac{1}{\varepsilon^{2\nu}}\frac{d^2y}{d\tau^2}.\end{aligned}$$

Consequently, in terms of τ*, equation* (4.17) *becomes*

$$\varepsilon^{1-2\nu}\frac{d^2y}{d\tau^2} - \varepsilon^{-\nu}(1+\varepsilon^2)\frac{dy}{d\tau} + (1-\varepsilon^2)y = 0. \tag{4.19}$$

As $\varepsilon \to 0$ *in* (4.19)*:*

- *the second-derivative term scales like* $\varepsilon^{1-2\nu}$*,*
- *the first-derivative term scales like* $\varepsilon^{-\nu}$*,*
- *the zeroth-order term scales like* $O(1)$*.*

In a boundary layer, we typically require the highest derivative to balance another leading term. Thus, we balance the first two terms:

$$\varepsilon^{1-2\nu} \sim \varepsilon^{-\nu}.$$

By equating exponents, we have

$$1-2\nu = -\nu,$$

from which we have that $\nu = 1$*. With* $\nu = 1$*, the equation becomes*

$$\varepsilon^{-1}y'' - \varepsilon^{-1}(1+\varepsilon^2)y' + (1-\varepsilon^2)y = 0.$$

Multiplying by ε *leads to the new equation*

$$y'' - (1+\varepsilon^2)y' + \varepsilon(1-\varepsilon^2)y = 0.$$

Assuming that the boundary layer is located at $t = 1$*, which corresponds to* $\tau = 0$*, the inner solution* $y^i(\tau)$ *satisfies*

$$\frac{d^2y^i}{d\tau^2} - \frac{dy^i}{d\tau} = 0, \qquad y^i(0) = \beta.$$

Expressed in terms of the original variable t, the inner solution is

$$y^i(t) = \beta - A + Ae^{\frac{1-t}{\varepsilon}},$$

where A is a constant determined by the matching condition. Sine $\sqrt{\varepsilon}$ *goes to zero slower than* ε*, does, or* $\sqrt{\varepsilon} \ll \varepsilon$ *we use the intermediate scale by letting*

$$\eta = \frac{t}{\sqrt{\varepsilon}}.$$

The condition for matching is that the inner approximation, written in terms of the intermediate variable η*, should agree with the outer approximation, written in terms of the intermediate variable* η*, in the limit as* $\varepsilon \to 0^+$*. Thus,*

$$\lim_{\varepsilon \to 0^+} y^0(\eta\sqrt{\varepsilon}) = \lim_{\varepsilon \to 0^+} y^i(\eta\sqrt{\varepsilon}).$$

That is

$$\lim_{\varepsilon \to 0^+} \alpha e^{-\eta\sqrt{\varepsilon}} = \lim_{\varepsilon \to 0^+} \left(\beta - A + Ae^{\left(\frac{1-\eta\sqrt{\varepsilon}}{\varepsilon}\right)}\right).$$

However,

$$\lim_{\varepsilon \to 0^+} \left(\beta - A + Ae^{\left(\frac{1-\eta\sqrt{\varepsilon}}{\varepsilon}\right)}\right)$$

does not exist, since

$$\lim_{\varepsilon \to 0^+} e^{\frac{1}{\varepsilon}} = \infty.$$

Hence, no matching can be done and we may suspect that the boundary layer is at the point $t = 0$.

Since the inner solution associated with a boundary layer at $t = 1$ fails to admit a finite limit in any intermediate scaling, no matching with the outer solution is possible. We therefore conclude that no boundary layer occurs at $t = 1$, and we are led to investigate the possibility of a boundary layer at $t = 0$ instead. □

4.1.3 Boundary Layers in Nonlinear Equations

Boundary layer theory is a fundamental asymptotic technique used to analyze differential equations containing a small parameter $\varepsilon \ll 1$ multiplying the highest-order derivative. Such problems are called *singularly perturbed* because setting $\varepsilon = 0$ reduces the order of the equation and generally prevents all boundary conditions from being satisfied simultaneously.

In nonlinear differential equations, the presence of small parameters often leads to solutions that exhibit rapid variations in narrow regions of the domain, known as *boundary layers* , while remaining smooth elsewhere. Outside these layers, the solution is well approximated by an *outer solution,* obtained by neglecting the small parameter and solving the reduced (lower-order) nonlinear equation. Near the boundary layer, this approximation breaks down, and an *inner solution* must be constructed by introducing a stretched variable that magnifies the region of rapid variation.

The inner and outer solutions are then connected using a *matching principle*, which ensures that both approximations represent the same asymptotic behavior in their common region of validity. For nonlinear problems, this matching process is especially important, as nonlinear terms can significantly influence the structure and width of the boundary layer.

Finally, a *uniform approximation* is formed by combining the inner and outer solutions and subtracting their common limit. This composite solution provides an accurate approximation across the entire domain and captures both the smooth behavior away from the boundary and the sharp transitions within the boundary layer. We illustrate the idea by providing the following example.

Example 4.3 *Consider the following nonlinear boundary value problem:*

$$\varepsilon y'' + y' + e^{-y} = 0, \qquad y(0) = 2, \quad y(1) = 0. \tag{4.20}$$

In the outer region, letting $\varepsilon = 0$*, we obtain the first-order differential equation*

$$y' + e^{-y} = 0. \tag{4.21}$$

The general solution of (4.21) *is*

$$y^o(t) = \ln(C - t), \tag{4.22}$$

where C is an integration constant.

Suppose the outer region is adjacent to $t = 1$*. Applying the boundary condition* $y(1) = 0$*, we have*

$$y(1) = \ln(C - 1) = 0,$$

from which we have that

$$C = 2.$$

Therefore

$$y^o(t) = \ln(2 - t). \tag{4.23}$$

In the inner region, which is adjacent to $t = 0$*, we introduce the inner stretched variable* $\tau = t/\varepsilon$*. Then,*

$$\frac{dy}{dt} = \frac{dy}{d\tau}\frac{d\tau}{dt} = \frac{1}{\varepsilon}\frac{dy}{d\tau}$$

Differentiating again with respect to t, we get

$$\frac{d^2y}{dt^2} = \frac{1}{\varepsilon^2}\frac{d^2y}{d\tau^2}.$$

Substituting into the original differential equation (4.20) *and then multiplying the resulting equation by* ε*, we obtain*

$$y'' + y' + \varepsilon e^{-y} = 0, \tag{4.24}$$

where $(\prime = \frac{d}{d\tau})$*. From the order* ε^0 *(setting* $\varepsilon = 0$*), we have*

$$y'' + y' = 0,$$

which has the general solution

$$y^i(\tau) = C_1 e^{-\tau} + C_2.$$

where C_1 *and* C_2 *are two integration constants. In terms of the variable t, the solution becomes*

$$y^i(t) = C_1 e^{-\frac{t}{\varepsilon}} + C_2.$$

From the boundary condition at $t = 0$ *gives*

$$C_1 + C_2 = 2,$$

from which we obtain

$$C_2 = 2 - C_1.$$

Hence,

$$y^i(t) = C_1 e^{-t/\varepsilon} - C_1 + 2.$$

To find the constant C_1, we resort, as before, to matching by letting

$$\eta = \frac{t}{\sqrt{\varepsilon}}.$$

Then, requiring

$$\lim_{\varepsilon \to 0^+} y^o(\eta\sqrt{\varepsilon}) = \lim_{\varepsilon \to 0^+} y^i(\eta\sqrt{\varepsilon})$$

we have

$$\lim_{\varepsilon \to 0^+} \ln(2 - \eta\sqrt{\varepsilon}) = \lim_{\varepsilon \to 0^+} \left[C_1 e^{-\eta/\sqrt{\varepsilon}} - C_1 + 2\right].$$

This results into

$$\ln(2) = -C_1 + 2,$$

from which we conclude that

$$C_1 = 2 - \ln(2).$$

Hence

$$y^i(t) = (2 - \ln(2))e^{-t/\varepsilon} + \ln(2).$$

Let the common limit be denoted by

$$y^{match} = \ln(2).$$

Then, the composite and uniform solution

$$\begin{aligned} y_{comp}(t) &= y^o(t) + y^i(t) - y^{match} \\ &= \ln(2-t) + (2 - \ln(2))e^{-t/\varepsilon}. \end{aligned}$$

□

4.1.4 Exercises

Exercise 4.1 *Derive expression* (4.7).

Exercise 4.2 *Find a uniform approximate solution for the singularly perturbed boundary value problem*

$$\varepsilon y''(t) + (1+\varepsilon)y'(t) + y(t) = 0, \qquad y(0) = 0, \qquad y(1) = 1, \qquad 0 < t < 1,$$

where $0 < \varepsilon \ll 1$.

Exercise 4.3 *Show the singularly perturbed boundary value problem*

$$\varepsilon y''(t) - (1+\varepsilon)y'(t) + y(t) = 0, \qquad y(0) = 0, \qquad y(1) = 1, \qquad 0 < t < 1,$$

where $0 < \varepsilon \ll 1$*, has a boundary layer at* $t = 1$*, and find a two-term uniform approximate solution.*

Exercise 4.4 *Find a two-term uniform approximate solution of the singularly perturbed boundary value problem*

$$\varepsilon y''(t) + y'(t) + e^{y(t)} = 0, \qquad y(0) = 2, \qquad y(1) = 0, \qquad 0 < t < 1,$$

where $0 < \varepsilon \ll 1$.

Exercise 4.5 *Find a two-term uniform approximate solution of the singularly perturbed boundary value problem*

$$\varepsilon y''(t) - y'(t) + e^{-y(t)} = 0, \qquad y(0) = 2, \qquad y(1) = 0, \qquad 0 < t < 1,$$

where $0 < \varepsilon \ll 1$.

Exercise 4.6 *Find a higher–order boundary layer corrections up to* $O(\varepsilon)$ *terms as we have done in Subsection 4.1.2 for the singularly perturbed problem*

$$\varepsilon y'' + (1+\varepsilon^2)y' + (1-\varepsilon^2)y = 0, \quad 0 < t < 1, \quad 0 < \varepsilon \ll 1.$$

with boundary conditions

$$y(0) = \alpha, \qquad y(1) = \beta,$$

for constants α, β.

Ans:

$$\begin{aligned} y(t) &= y^o(t) + y^i(t) \\ &= \beta e^{1-t} + (\alpha - \beta e)e^{-\frac{t}{\varepsilon}} \\ &+ \varepsilon\Big[\beta(1-t)e^{1-t} - \beta e^{1-\frac{t}{\varepsilon}} + (\alpha - \beta e)\frac{t}{\varepsilon}e^{-\frac{t}{\varepsilon}}\Big]. \end{aligned}$$

Exercise 4.7 *Consider the singularly perturbed boundary value problem*

$$\varepsilon y''(t) + 2(1+\varepsilon)y'(t) + 4y(t) = 0, \qquad y(0) = 0, \qquad y(1) = 1, \qquad 0 < t < 1,$$

where $0 < \varepsilon \ll 1$.

(a) Find the exact solution.

(b) Construct a two-term uniform approximate solution.

4.2 Equations with Variable Coefficients

For real-valued, continuous functions p, q, and f, we consider the following singularly perturbed boundary value problem

$$\varepsilon y'' + p(t)y' + q(t)y = f(t), \qquad 0 < t < 1, \quad 0 < \varepsilon \ll 1, \tag{4.25}$$

subject to the boundary conditions

$$y(0) = \alpha, \qquad y(1) = \beta. \tag{4.26}$$

The small parameter ε multiplies the highest–order derivative, rendering the problem singularly perturbed: in the limit $\varepsilon \to 0$, the differential equation reduces to a first–order equation, and in general, both boundary conditions cannot be satisfied simultaneously. As a consequence, the solution typically exhibits a boundary layer in a narrow region near one endpoint of the interval. The location of this boundary layer is determined by the sign of the convection coefficient $p(t)$: if $p(t) > 0$ on $[0,1]$, the boundary layer forms near $t = 0$, whereas if $p(t) < 0$ on $[0,1]$, the boundary layer occurs near $t = 1$. Away from the layer, the solution is well approximated by the outer (reduced) solution obtained by setting $\varepsilon = 0$, while within the layer, a suitable stretched variable must be introduced to resolve the rapid variation and ensure that the boundary conditions are satisfied.

Singularly perturbed boundary value problems of the form of (4.25) arise naturally in many applied sciences and engineering disciplines where diffusion effects are weak compared to convection or reaction. In *fluid mechanics*, such equations model convection–diffusion processes and viscous boundary layers in high–Reynolds–number flows, where the small parameter ε represents viscosity. In *heat and mass transfer*, they describe heat conduction or species diffusion with dominant advective transport, leading to sharp thermal or concentration gradients near boundaries. In *chemical reaction engineering*, these problems appear in tubular reactor models when reaction and convection dominate diffusion, producing thin reaction layers. Applications also occur in *aerodynamics*, particularly in the analysis of airflow near solid surfaces, and in *electrical engineering*, where they model charge carrier transport in semiconductors under strong electric fields. More generally, such problems serve as canonical models for multiscale phenomena and motivate the development of asymptotic and numerical methods for singular perturbation problems. The next theorem addresses the case of $p(t) > 0$ on $[0,1]$.

Theorem 4.1 *Consider the singularly perturbed nonhomogeneous boundary value problem given by* (4.25) *and* (4.26). *Assume* $p(t) > 0$, *for all* $t \in [0,1]$. *Then there exists a boundary layer at* $t = 0$ *with a uniform approximate solution given by*

$$y_u(t) = y^i(t) + y^o(t) - C, \tag{4.27}$$

where

$$y^i(t) = C + \left(\alpha - C\right) e^{\frac{-p(0)t}{\varepsilon}},$$

$$y^o(t) = \int_0^t \frac{f(u)}{p(u)} e^{-\int_u^t \frac{q(s)}{p(s)}ds} du + \beta e^{\int_t^1 \frac{q(s)}{p(s)}ds} - \int_0^1 \frac{f(u)}{p(u)} e^{-\int_u^t \frac{q(s)}{p(s)}ds} du,$$

and

$$C = \beta e^{\int_0^1 \frac{q(s)}{p(s)}ds} - \int_0^1 \frac{f(u)}{p(u)} e^{\int_0^u \frac{q(s)}{p(s)}ds} du.$$

Proof *We verify that the assumption of a boundary layer at $t = 0$ is consistent and leads to the uniform approximation stated in the theorem.*

outer solution . *We seek a regular expansion of the form*

$$y(t) = y_0(t) + \varepsilon y_1(t) + \cdots. \tag{4.28}$$

Setting $\varepsilon = 0$ in (4.25), the outer solution must satisfy the reduced (outer) problem

$$p(t)\frac{dy_0}{dt} + q(t)y_0 = f(t), \tag{4.29}$$

which is first order and therefore can satisfy only one boundary condition. Since we are assuming a boundary layer at $t = 0$, the outer solution $y^o(t)$ must satisfy the boundary condition at $t = 1$, namely, $y^o(1) = \beta$. Dividing both sides of (4.36) by $p(t)$ and then multiplying (4.36) the resulting equation by the integrating factor

$$\exp\left(\int_0^t \frac{q(s)}{p(s)}ds\right),$$

we obtain

$$\frac{d}{dt}\left(y^o(t)e^{\int_0^t \frac{q(s)}{p(s)}ds}\right) = \frac{f(t)}{p(t)} e^{\int_0^t \frac{q(s)}{p(s)}ds}.$$

Integrating from 0 to t gives

$$y^o(t)e^{\int_0^t \frac{p(s)}{q(s)}ds} = \int_0^t \frac{f(u)}{p(u)} e^{\int_0^u \frac{q(s)}{p(s)}ds} du + C.$$

After simplifying the above equation, we arrive at

$$y^o(t) = \int_0^t \frac{f(u)}{p(u)} e^{-\int_u^t \frac{q(s)}{p(s)}ds} du + Ce^{-\int_0^t \frac{q(s)}{p(s)}ds}, \tag{4.30}$$

where C is a constant. Imposing the boundary condition $y^o(1) = \beta$ yields

$$C = \beta e^{\int_0^1 \frac{q(s)}{p(s)}ds} - \int_0^1 \frac{f(u)}{p(u)} e^{\int_0^u \frac{q(s)}{p(s)}ds} du. \tag{4.31}$$

Substituting this expression for C into (4.30), we obtain the explicit outer solution

$$y^o(t) = \int_0^t \frac{f(u)}{p(u)} e^{-\int_u^t \frac{q(s)}{p(s)}ds} du + \beta e^{\int_t^1 \frac{q(s)}{p(s)}ds} - \int_0^1 \frac{f(u)}{p(u)} e^{-\int_u^t \frac{q(s)}{p(s)}ds} du. \tag{4.32}$$

Inner solution. *To resolve the boundary layer near $t = 0$, we introduce the stretched variable*

$$\tau = \frac{t}{\delta(\varepsilon)},$$

where $\delta(\varepsilon)$ denotes the boundary layer thickness. Writing $y(t) = u(\tau)$, the chain rule gives

$$\frac{dy}{dt} = \frac{1}{\delta(\varepsilon)}\frac{du}{d\tau}, \qquad \frac{d^2y}{dt^2} = \frac{1}{\delta(\varepsilon)^2}\frac{d^2u}{d\tau^2}.$$

Substitution into (4.25) yields

$$\varepsilon\frac{u''}{\delta(\varepsilon)^2} + \frac{p(\tau\delta(\varepsilon))}{\delta(\varepsilon)}u' + q(\tau\delta(\varepsilon))u = f(\tau\delta(\varepsilon)).$$

Balancing the highest derivative with the convection term gives $\delta(\varepsilon) = \varepsilon$, corresponding to a boundary layer of width $O(\varepsilon)$. With this choice, the equation becomes

$$u''(\tau) + p(\varepsilon\tau)u'(\tau) + \varepsilon q(\varepsilon\tau)u(\tau) = \varepsilon f(\varepsilon\tau).$$

Letting $\varepsilon \to 0^+$ yields the leading-order inner problem

$$u''(\tau) + p(0)u'(\tau) = 0, \qquad u(0) = \alpha.$$

Its general solution is

$$u(\tau) = c_1 + c_2 e^{-p(0)\tau}.$$

The condition $u(0) = \alpha$ implies $c_1 = \alpha - c_2$. Hence, the inner approximation in the original variable is

$$y^i(t) = \alpha - c_2\left(1 - e^{-p(0)t/\varepsilon}\right). \tag{4.33}$$

Matching. *To determine c_2, we match the inner and outer solution s. Introducing the intermediate variable $\eta = t/\sqrt{\varepsilon}$ and taking the limit $\varepsilon \to 0^+$, we obtain*

$$\lim_{\varepsilon\to 0^+} y^i(\sqrt{\varepsilon}\eta) = \alpha - c_2,$$

while

$$\begin{aligned}
\lim_{\varepsilon\to 0^+} y^o(\sqrt{\varepsilon}\eta) &= \lim_{\varepsilon\to 0+}\left[\beta e^{\int_{\sqrt{\varepsilon}\,\eta}^{1}\frac{q(s)}{p(s)}ds} + \int_0^{\sqrt{\varepsilon}\,\eta}\frac{f(u)}{p(u)}e^{-\int_u^{\sqrt{\varepsilon}\,\eta}\frac{q(s)}{p(s)}ds}du\right.\\
&\quad - \left.\int_0^1\frac{f(u)}{p(u)}e^{-\int_u^{\sqrt{\varepsilon}\,\eta}\frac{q(s)}{p(s)}ds}du\right]\\
&= \beta e^{\int_0^1\frac{q(s)}{p(s)}ds} - \int_0^1\frac{f(u)}{p(u)}e^{-\int_u^0\frac{q(s)}{p(s)}ds}du = C.
\end{aligned}$$

Equating these limits yields

$$c_2 = \alpha - C.$$

Substitution into (4.33) *gives*

$$\begin{aligned} y^i(t) &= \; = \alpha - (\alpha - C)\left(1 - e^{-p(0)t/\varepsilon}\right) \\ &= \; C + \left(\alpha - C\right) e^{\frac{-p(0)t}{\varepsilon}}, \end{aligned}$$

where C is given by (4.31).

Finally, subtracting the common part of the inner and outer limits, which equals α, *we define the composite (uniform) approximation*

$$y_u(t) = y^i(t) + y^o(t) - C. \tag{4.34}$$

This completes the proof.

The next theorem addresses the situation when $p(t) < 0,\ t \in [0,1]$.

Theorem 4.2 *Consider the singularly perturbed nonhomogeneous boundary value problem given by* (4.25) *and* (4.26). *Assume* $p(t) < 0$, *for all* $t \in [0,1]$. *Then there exists a boundary layer at* $t = 1$ *with a uniform approximate solution given by*

$$y_u(t) = y^i(t) + y^o(t) - L,$$

where

$$y^i(t) = L + \left(\beta - L\right) e^{p(1)\frac{1-t}{\varepsilon}}$$

$$y^o(t) = \int_0^t \frac{f(u)}{p(u)} e^{-\int_u^t \frac{q(s)}{p(s)} ds} du + \alpha e^{-\int_0^t \frac{q(s)}{p(s)} ds}.$$

and

$$L = \alpha e^{\int_0^1 \frac{q(s)}{p(s)} ds} + \int_0^1 \frac{f(u)}{p(u)} e^{-\int_u^1 \frac{q(s)}{p(s)} ds} du. \tag{4.35}$$

Proof *We verify that the assumption of a boundary layer at* $t = 1$ *is consistent and leads to the uniform approximation stated in the theorem.*

outer solution . *We seek a regular expansion of the form*

$$y(t) = y_0(t) + \varepsilon y_1(t) + \cdots.$$

Setting $\varepsilon = 0$ *in* (4.25), *the outer solution must satisfy the reduced (outer) problem*

$$p(t)\frac{dy_0}{dt} + q(t)y_0 = f(t), \tag{4.36}$$

which is first order and therefore can satisfy only one boundary condition. Since we are assuming a boundary layer at $t = 1$, *the outer solution* $y^o(t)$ *must satisfy the boundary condition at* $t = 0$, *namely,* $y^o(0) = \alpha$. *Dividing both sides of* (4.36) *by*

$p(t)$ and then multiplying (4.36) the resulting equation by the integrating factor and then integrating and simplifying as we did before, we obtain

$$y^0(t) = \int_0^t \frac{f(u)}{p(u)} e^{-\int_u^t \frac{q(s)}{p(s)} ds} du + Ce^{-\int_0^t \frac{q(s)}{p(s)} ds}, \tag{4.37}$$

where C is a constant. Imposing the boundary condition $y^o(0) = \alpha$ yields

$$C = \alpha.$$

Substituting this expression for C into (4.37), we obtain the explicit outer solution

$$y^o(t) = \int_0^t \frac{f(u)}{p(u)} e^{-\int_u^t \frac{q(s)}{p(s)} ds} du + \alpha e^{-\int_0^t \frac{q(s)}{p(s)} ds}. \tag{4.38}$$

Inner solution. *To resolve the boundary layer near $t = 1$, we introduce the stretched variable*

$$\tau = \frac{1-t}{\delta(\varepsilon)},$$

where $\delta(\varepsilon)$, a scaling factor that determines the width of the boundary layer is to be determined. Using the chain rules, and letting $u(\tau) = y(t)$

$$\frac{dy}{dt} = \frac{-1}{\delta(\varepsilon)} \frac{du}{d\tau}$$
$$\frac{d^2y}{dt^2} = \frac{1}{\delta(\varepsilon)^2} \frac{d^2u}{d\tau^2}$$

Substituting the above terms in (4.25) we arrive at

$$\frac{\varepsilon}{\delta(\varepsilon)^2} u'' - \frac{p(1-\tau\delta(\varepsilon))}{\delta(\varepsilon)} u' + q(1-\tau\delta(\varepsilon))u = f(1-\tau\delta(\varepsilon))$$

Balancing the highest derivative with the convection term gives $\delta(\varepsilon) = \varepsilon$, corresponding to a boundary layer of width $O(\varepsilon)$. With this choice, the equation becomes

$$u''(\tau) - p(1-\tau\varepsilon)u'(\tau) + \varepsilon q(1-\tau\varepsilon)u(\tau) = \varepsilon f(1-\tau\varepsilon)$$

Letting $\varepsilon \to 0^+$ yields the leading-order inner problem

$$u''(\tau) - p(1)u'(\tau) = 0, \qquad u(0) = \beta.$$

Its general solution is

$$u(\tau) = c_1 + c_2 e^{p(1)\tau}.$$

The condition $u(0) = \beta$ implies $c_1 = \beta - c_2$. Hence, the inner approximation in the original variable is

$$y^i(t) = \beta - c_2\left(1 - e^{p(1)\frac{1-t}{\varepsilon}}\right). \tag{4.39}$$

Matching. *To determine c_2, we match the inner and outer solutions . Introducing the intermediate variable*

$$\eta = \frac{1-t}{\sqrt{\varepsilon}}$$

and taking the limit $\varepsilon \to 0^+$, we obtain

$$\begin{aligned} \lim_{\varepsilon \to 0^+} y^i(1-\sqrt{\varepsilon}\eta) &= \beta - c_2 \lim_{\varepsilon \to 0^+} \left(1 - e^{p(1)\frac{\eta}{\sqrt{\varepsilon}}}\right) \\ &= \beta - c_2. \end{aligned}$$

while

$$\begin{aligned} \lim_{\varepsilon \to 0^+} y^o(1-\sqrt{\varepsilon}\eta) &= \lim_{\varepsilon \to 0+} \left[\int_0^{1-\sqrt{\varepsilon}\,\eta} \frac{f(u)}{p(u)} e^{-\int_u^{1-\sqrt{\varepsilon}\,\eta} \frac{q(s)}{p(s)} ds} du + \alpha e^{-\int_0^{1-\sqrt{\varepsilon}\eta} \frac{q(s)}{p(s)} ds} \right] \\ &= \int_0^1 \frac{f(u)}{p(u)} e^{-\int_u^1 \frac{q(s)}{p(s)} ds} du + \alpha e^{-\int_0^1 \frac{q(s)}{p(s)} ds} = L. \end{aligned}$$

Equating these limits yields

$$c_2 = \beta - L.$$

Substitution into (4.33) *gives*

$$\begin{aligned} y^i(t) &= \beta - (\beta - L)\left(1 - e^{p(1)\frac{1-t}{\varepsilon}}\right) \\ &= L + \Big(\beta - L\Big) e^{p(1)\frac{1-t}{\varepsilon}} \end{aligned}$$

where L is given by (4.35).

Finally, subtracting the common part L of the inner and outer limits, which equals α, we define the composite (uniform) approximation

$$y_u(t) = y^i(t) + y^o(t) - L.$$

This completes the proof.

4.2.1 Exercises

Exercise 4.8 *Consider the singularly perturbed boundary value problem*

$$\varepsilon y'' + (1+t)y' + y = 0, \qquad 0 < t < 1,$$

with boundary conditions $y(0) = 0$ and $y(1) = 1$. Construct a uniformly valid composite approximation.

Exercise 4.9 *For $r > 1$, we consider the singularly perturbed boundary value problem*

$$\varepsilon y'' + p(t)y' + q(t)y^r = 0, \quad 0 < t < 1, \quad 0 < \varepsilon << 1$$

$$y(0) = \alpha, \quad y(1) = \beta.$$

Find a uniform approximate solution for the following cases:

(a) $p(t) > 0$ *for all* $t \in [0,1]$,

(b) $p(t) < 0$ *for all* $t \in [0,1]$.

Exercise 4.10 *Let*

$$\varepsilon y'' - (2-t)y' + y = t, \qquad 0 < t < 1,$$

with boundary conditions $y(0) = 1$ *and* $y(1) = 0$. *Write down the uniform corresponding composite approximation.*

Exercise 4.11 *Determine the value of* α *so that the singular perturbation problem*

$$\varepsilon y'' + \cos(\alpha - t)y = 1, \quad y(0) = 0, \quad y'(\alpha) = 0, \quad 0 < t < 1,$$

has a uniform approximate solution. Find the uniform approximate solution.

Exercise 4.12 *Construct a uniformly valid composite approximation for the singularly perturbed boundary value problem*

$$\varepsilon y'' + y' + y^2 = 0, \qquad 0 < t < 1,$$

with boundary conditions $y(0) = 0$ *and* $y(1) = 1$.

Exercise 4.13 *Construct a uniformly valid composite approximation for the singularly perturbed boundary value problem*

$$\varepsilon y'' + e^t y' = 1, \qquad 0 < t < 1,$$

with boundary conditions $y(0) = 1$ *and* $y(\infty) = 0$.

Exercise 4.14 *Construct a uniformly valid composite approximation for the singularly perturbed boundary value problem*

$$\varepsilon y'' + \varepsilon y' - y = 0, \qquad 0 < t < 1,$$

with boundary conditions $y(0) = 0$ *and* $y(1) = 1$.

Exercise 4.15 *Construct a uniformly valid composite approximation for the singularly perturbed boundary value problem*

$$\varepsilon y'' + \frac{1}{1+t} y' + \varepsilon y = 0, \qquad 0 < t < 1,$$

with boundary conditions $y(0) = 0$ *and* $y(1) = 1$.

Exercise 4.16 *Construct a uniformly valid composite approximation for each of the singularly perturbed boundary value problem.*

(a) $\varepsilon y'' + (1+t)y' = 1, \qquad y(0) = 0. \quad y(1) = 1, \quad 0 < t < 1,$

(b) $\varepsilon y'' + (t^2 - 2)y = -1, \qquad y'(0) = 0, \quad y(1) = 1, \quad 0 < t < 1,$

(c) $\varepsilon y'' + 4(t^2 - \pi)y = \cos(t), \qquad y(0) = 0, \quad y(\pi/2) = 1, \quad 0 < t < \pi/2,$

(d) $\varepsilon y'' + 2y' + e^y = 0, \qquad y(0) = 0, \quad y(1) = 0, \quad 0 < t < 1.$

4.3 Interior Boundary Layers

An *interior boundary layer* in a singularly perturbed boundary value problem arises when the coefficient of the first-derivative term vanishes or changes sign at some interior point $t = t^*$ of the domain. In this case, the reduced (outer) problem becomes singular at t^*, and the outer solution generally fails to remain uniformly valid across the interval. To accommodate the rapid variation that occurs in a thin neighborhood of t^*, an inner scaling centered at this point is introduced, leading to an interior layer that connects two distinct outer solutions defined on either side of t^*. The resulting interior boundary layer ensures a smooth transition and restores a uniformly valid approximation to the full problem. Sometimes I refer to a term as convection, and below is the justification:

The term *convection coefficient* is correct in the theory of singular perturbations, boundary layers, and convection–diffusion equations. Consider the equation

$$\varepsilon y'' + p(t)y' + q(t)y = f(t).$$

This has the same structure as a one-dimensional steady convection–diffusion–reaction equation:

$$\text{diffusion} + \text{convection} + \text{reaction} = \text{forcing}.$$

term	interpretation
$\varepsilon y''$	diffusion term
$p(t)y'$	convection (transport / advection) term
$q(t)y$	reaction term

The first-derivative term represents transport of the quantity y along the interval, analogous to fluid flow carrying heat or mass. Therefore, $p(t)$ plays the role of a velocity field, and we call it the *convection coefficient* (also called the *advection coefficient* in the PDE literature). When $p(t) = 0$, the convection term vanishes and the reduced equation changes type. This loss of the first-derivative term is precisely the mechanism that produces a turning point and leads to the formation of an interior boundary layer.

Theorem 4.3 (Interior boundary layer at a turning point) *Consider the singularly perturbed boundary value problem*

$$\varepsilon y'' + p(t)y' + q(t)y = f(t), \qquad 0 < t < 1, \quad 0 < \varepsilon \ll 1, \tag{4.40}$$

subject to

$$y(0) = \alpha, \qquad y(1) = \beta, \tag{4.41}$$

where p, q, and f are real-valued continuous functions on $[0,1]$. Assume that the convection coefficient $p(t)$ has a simple zero at an interior point $t = t^* \in (0,1)$, satisfying

$$p(t) < 0 \quad \text{for } 0 \le t < t^*, \qquad p(t) > 0 \quad \text{for } t^* < t \le 1, \qquad p'(t^*) \ne 0.$$

Then the solution of (4.40)–(4.41) possesses an interior boundary layer centered at $t = t^*$ of thickness $O(\varepsilon^{1/2})$. Away from t^* the solution is approximated by two outer solutions y_l^0 on $[0,t^*)$ and y_r^0 on $(t^*,1]$, obtained from the reduced equation

$$p(t)y'(t) + q(t)y(t) = f(t), \tag{4.42}$$

while in a neighborhood of t^* the rapid transition is described by an inner solution obtained using the stretched variable

$$\tau = \frac{t - t^*}{\sqrt{\varepsilon}}.$$

Uniformly valid approximations are obtained separately on the left and right subintervals by matching the inner and outer solutions.

Proof *Setting $\varepsilon = 0$ in (4.40) gives the reduced equation (4.42), which is first order and therefore cannot, in general, satisfy both boundary conditions. Since $p(t) \ne 0$ for $t \ne t^*$, two outer solutions exist:*

$$y_l^0(t), \quad 0 \le t < t^*, \qquad y_r^0(t), \quad t^* < t \le 1,$$

determined by

$$y_l^0(0) = \alpha, \qquad y_r^0(1) = \beta.$$

Because $p(t)$ changes sign at t^*, the reduced solutions generally do not agree at this point, indicating the presence of an interior layer.

To resolve the rapid variation near $t = t^*$, introduce the stretched variable

$$\tau = \frac{t - t^*}{\delta(\varepsilon)}, \qquad y(t) = Y(\tau),$$

where $\delta(\varepsilon)$ is to be determined. Expanding $p(t)$ about t^* gives

$$p(t) = p'(t^*)(t - t^*) + O((t - t^*)^2).$$

For the purpose of illustrating how the analysis proceeds, we assume that the function $f(t)$ is identically zero. Substituting into (4.40) yields

$$\varepsilon \frac{1}{\delta^2} Y'' + p'(t^*) \frac{t - t^*}{\delta} Y' + \cdots = 0.$$

Balancing the first two terms requires

$$\frac{\varepsilon}{\delta^2} \sim \frac{\delta}{\delta} \quad \Rightarrow \quad \delta = \sqrt{\varepsilon}.$$

With the scaling

$$\tau = \frac{t - t^*}{\sqrt{\varepsilon}},$$

the leading inner equation becomes

$$Y'' + p'(t^*)\,\tau Y' = 0. \tag{4.43}$$

Let $Z(\tau) = Y'(\tau)$. Then

$$Z' + p'(t^*)\,\tau Z = 0,$$

so that

$$Z(\tau) = C\exp\left(-\tfrac{1}{2}p'(t^*)\,\tau^2\right).$$

Integrating once more,

$$Y(\tau) = C\int_0^{\tau} \exp\left(-\tfrac{1}{2}p'(t^*)s^2\right)\,ds + D. \tag{4.44}$$

The inner solution connects two constant outer states. Matching with the outer solutions uniquely determines C and D, and shows that the inner solution provides a smooth transition between the two outer solutions across a layer of width $O(\varepsilon^{1/2})$ centered at $t = t^*$.

The limits $\tau \to \pm\infty$ correspond to the right and left outer regions, respectively. Matching these limits with y_r^0 and y_l^0 determines the constants and yields composite approximations

$$\begin{aligned} y_{\text{comp},l}(t) &= y_l^0(t) + y^i(t) - y_l^{\text{match}}, && t \in [0, t^*), \\ y_{\text{comp},r}(t) &= y_r^0(t) + y^i(t) - y_r^{\text{match}}, && t \in (t^*, 1]. \end{aligned}$$

Thus, the solution contains an interior boundary layer of thickness $O(\varepsilon^{1/2})$ centered at $t = t^*$.

Remark 12 *Interior boundary layers of this type are often called* turning-point layers, *since they occur at points where the coefficient $p(t)$ of the first-derivative term vanishes, and changes sign. At such points, the direction of the reduced flow reverses, leading to a mismatch between the left and right outer solutions.*

The Gaussian integral appearing in (4.44) *describes the smooth transition between the two outer states across the layer.*

Before we display the next example, we state and prove the following lemma which is important in doing the matching in the next example.

Lemma 1 *For $x \in \mathbb{R}$,*

$$\int_0^{\infty} e^{-x^2}\,dx = \frac{\sqrt{\pi}}{2}.$$

Proof *Let*

$$I = \int_0^\infty e^{-x^2}\,dx.$$

Then, for $y \in \mathbb{R}$,

$$I^2 = I \cdot I = \int_0^\infty e^{-x^2}dx \int_0^\infty e^{-y^2}dy = \int_0^\infty \int_0^\infty e^{-(x^2+y^2)}dxdy.$$

We switch to polar coordinates by letting

$$x = r\cos(\theta), \quad y = r\sin(\theta).$$

Then,

$$\begin{aligned} I^2 &= \int_0^{\pi/2}\int_0^\infty e^{-r^2} rdr\,d\theta = \int_0^{\pi/2}\left(\int_0^\infty e^{-r^2} rdr\right) d\theta \\ &= \int_0^{\pi/2}\left(-\frac{1}{2}e^{-r^2}\Big|_0^\infty\right) d\theta = \int_0^{\pi/2}\frac{1}{2}\,d\theta = \frac{\pi}{4}. \end{aligned}$$

Taking the square root on both sides, we arrive at

$$I = \int_0^\infty e^{-x^2}dx = \frac{\sqrt{\pi}}{2}.$$

This completes the proof.

Example 4.4 *(Interior boundary layer) In this example, we consider a problem in which* $p(t)$ *vanishes at an interior point, leading to an interior boundary layer. Consider*

$$\varepsilon y'' + (t-\tfrac{1}{2})y' - (t-\tfrac{1}{2})y = 0, \qquad 0<t<1, \quad 0<\varepsilon \ll 1, \tag{4.45}$$

with boundary conditions

$$y(0) = \alpha, \qquad y(1) = \beta.$$

Here $p(t) = t - \frac{1}{2}$ *so that* $p(t) > 0$ *for* $t > \frac{1}{2}$ *and* $p(t) < 0$ *for* $t < \frac{1}{2}$. *There are no boundary layers at the endpoints; instead, a boundary layer forms at* $t = \frac{1}{2}$.

Outer solutions. *We look for an outer approximation* $y^0 = y^0(t)$, *obtaining from* (4.45) *the reduced equation*

$$(t-\tfrac{1}{2})y^{0\prime} - (t-\tfrac{1}{2})y^0 = 0,$$

so that

$$y^0(t) = C_0 e^t,$$

valid away from a small neighborhood of $t = \frac{1}{2}$. *In the right interval* $t > \frac{1}{2}$, y^0 *must satisfy* $y^0(1) = \beta$, *hence*

$$C_0 e = \beta \text{ implies that } C_0 = \beta e^{-1},$$

and

$$y_r^0(t) = \beta e^{t-1}, \tag{4.46}$$

where the subscript r denotes the right outer solution. In the left interval $t < \frac{1}{2}$, y^0 *must satisfy* $y^0(0) = \alpha$, *hence* $C_0 = \alpha$ *and*

$$y_l^0(t) = \alpha e^t, \tag{4.47}$$

where the subscript l denotes the left outer solution.

Inner scaling and failure of $\delta(\varepsilon) = \varepsilon$. *We now derive an inner approximation near* $t = \frac{1}{2}$. *Introduce the stretched variable*

$$\tau = \frac{t - \frac{1}{2}}{\delta(\varepsilon)}.$$

Then

$$\frac{d}{dt} = \frac{1}{\delta(\varepsilon)} \frac{d}{d\tau}, \qquad \frac{d^2}{dt^2} = \frac{1}{\delta(\varepsilon)^2} \frac{d^2}{d\tau^2},$$

and since $t - \frac{1}{2} = \tau\delta(\varepsilon)$, *equation* (4.45) *becomes*

$$\frac{\varepsilon}{\delta(\varepsilon)^2} y''(\tau) + \frac{\tau\,\delta(\varepsilon)}{\delta(\varepsilon)} y'(\tau) - \tau\,\delta(\varepsilon)\,y(\tau) = 0, \tag{4.48}$$

i.e.,

$$\frac{\varepsilon}{\delta(\varepsilon)^2} y''(\tau) + \tau y'(\tau) - \tau\,\delta(\varepsilon)\,y(\tau) = 0.$$

If we choose $\delta(\varepsilon) = \varepsilon$, *then*

$$\frac{\varepsilon}{\delta(\varepsilon)^2} = \frac{1}{\varepsilon} \to \infty, \qquad \tau\delta(\varepsilon) = \tau\varepsilon \to 0,$$

and the dominant balance as $\varepsilon \to 0^+$ *leads formally to* $y''(\tau) = 0$, *whose general solution is*

$$y^i(\tau) = c_1\tau + c_2.$$

Rewriting in terms of t gives

$$y^i(t) = c_1 \frac{t - \frac{1}{2}}{\varepsilon} + c_2.$$

However, when we attempt to match this inner approximation with the outer solutions, no consistent matching is possible as $\varepsilon \to 0^+$. *Thus,* $\delta(\varepsilon) = \varepsilon$ *is not an appropriate boundary layer thickness.*

Correct inner scaling. *We now take* $\delta(\varepsilon) = \sqrt{\varepsilon}$, *so that*

$$\frac{\varepsilon}{\delta(\varepsilon)^2} = 1, \qquad \tau\delta(\varepsilon) = \tau\sqrt{\varepsilon} \to 0.$$

To leading order, (4.48) *becomes*

$$y''(\tau) + \tau y'(\tau) = 0. \tag{4.49}$$

Let $u(\tau) = y'(\tau)$. *Then*

$$u'(\tau) + \tau u(\tau) = 0,$$

so that

$$\frac{u'(\tau)}{u(\tau)} = -\tau, \qquad \ln u(\tau) = -\frac{\tau^2}{2} + C_1,$$

and hence

$$y'(\tau) = C_1 e^{-\tau^2/2}.$$

Integrating from $\tau = 0$ *(which corresponds to* $t = \frac{1}{2}$*) gives*

$$y(\tau) = C_2 + C_1 \int_0^{\tau} e^{-s^2/2}\, ds.$$

Thus, the inner approximation can be written as

$$y^i(t) = c_2 + c_1 \int_0^{\tau} e^{-s^2/2}\, ds = c_2 + c_1 \int_0^{\frac{t-\frac{1}{2}}{\sqrt{\varepsilon}}} e^{-s^2/2}\, ds, \tag{4.50}$$

where c_1, c_2 *are constants to be determined by matching.*

Note that this inner solution is obtained solely from the scaled ODE (4.45)*, which is valid on both sides of* $t = \frac{1}{2}$*; the sign of* τ *changes between left and right, but the functional form of* y^i *is the same in both regions. The difference between the left and right will enter through the matching limits* $\tau \to \pm\infty$.

Matching on the right. *To match* y^i *with* y_r^0*, we let*

$$\eta = \frac{t-\frac{1}{2}}{\varepsilon^{1/4}}, \qquad t > \tfrac{1}{2}.$$

Then $t = \frac{1}{2} + \eta\varepsilon^{1/4}$*, and from* (4.46)

$$y_r^0\left(\tfrac{1}{2} + \eta\varepsilon^{1/4}\right) = \beta e^{\eta\varepsilon^{1/4} - 1/2} \longrightarrow \beta e^{-1/2}, \qquad \varepsilon \to 0^+.$$

On the other hand, from (4.50)*,*

$$\tau = \frac{t-\frac{1}{2}}{\sqrt{\varepsilon}} = \frac{\eta}{\varepsilon^{1/4}} \to +\infty,$$

and therefore

$$y_r^i(t) = c_2 + c_1 \int_0^{\frac{\eta}{\varepsilon^{1/4}}} e^{-s^2/2}\, ds \longrightarrow c_2 + c_1 \int_0^{\infty} e^{-s^2/2}\, ds$$

$$= c_2 + c_1 \frac{\sqrt{\pi}}{\sqrt{2}}, \qquad \varepsilon \to 0^+,$$

using Lemma 1. Matching the two limits gives

$$\beta e^{-1/2} = c_2 + c_1 \frac{\sqrt{\pi}}{\sqrt{2}}. \tag{4.51}$$

Matching on the left. *We next match y^i with y_l^0. Set*

$$\eta = \frac{\frac{1}{2} - t}{\varepsilon^{1/4}}, \qquad t < \tfrac{1}{2},$$

so that $t = \frac{1}{2} - \eta\varepsilon^{1/4}$. From (4.47),

$$y_l^0\left(\tfrac{1}{2} - \eta\varepsilon^{1/4}\right) = \alpha e^{1/2 - \eta\varepsilon^{1/4}} \longrightarrow \alpha e^{1/2}, \qquad \varepsilon \to 0^+.$$

From (4.50), *we now have*

$$\tau = \frac{t - \frac{1}{2}}{\sqrt{\varepsilon}} = -\frac{\eta}{\varepsilon^{1/4}} \to -\infty,$$

and hence

$$y_l^i(t) = c_2 + c_1 \int_0^{-\frac{\eta}{\varepsilon^{1/4}}} e^{-s^2/2}\, ds = c_2 - c_1 \int_0^{\infty} e^{-s^2/2}\, ds$$
$$= c_2 - c_1 \frac{\sqrt{\pi}}{\sqrt{2}}, \qquad \varepsilon \to 0^+,$$

where we use the symmetry of the Gaussian and Lemma 1 again. Matching the limits yields

$$\alpha e^{1/2} = c_2 - c_1 \frac{\sqrt{\pi}}{\sqrt{2}}. \tag{4.52}$$

Solving (4.51) *and* (4.52) *for c_1 and c_2 gives*

$$c_1 = \frac{1}{\sqrt{2\pi}}\left(\beta e^{-1/2} - \alpha e^{1/2}\right), \qquad c_2 = \frac{1}{2}\left(\beta e^{-1/2} + \alpha e^{1/2}\right).$$

Piecewise uniform approximations. *In this problem, we cannot obtain a single uniform approximation valid on the whole interval $(0,1)$. Instead, we construct two uniform approximations, $y_{u_l}(t)$ on $[0, \frac{1}{2})$ and $y_{u_r}(t)$ on $(\frac{1}{2}, 1]$, by the usual "outer + inner − common limit" procedure.*

For $t \in [0, \frac{1}{2})$, we set

$$y_{u_l}(t) = y_l^0(t) + y^i(t) - \alpha e^{1/2}$$

$$= \alpha e^t + \frac{1}{2}(\beta e^{-1/2} + \alpha e^{1/2}) + \frac{1}{\sqrt{2\pi}}(\beta e^{-1/2} - \alpha e^{1/2}) \int_0^{\frac{t-\frac{1}{2}}{\sqrt{\varepsilon}}} e^{-s^2/2} ds - \alpha e^{1/2},$$

and for $t \in (\frac{1}{2}, 1]$,

$$y_{u_r}(t) = y_r^0(t) + y^i(t) - \beta e^{-1/2}$$

$$= \beta e^{t-1} + \frac{1}{2}(\beta e^{-1/2} + \alpha e^{1/2}) + \frac{1}{\sqrt{2\pi}}(\beta e^{-1/2} - \alpha e^{1/2}) \int_0^{\frac{t-\frac{1}{2}}{\sqrt{\varepsilon}}} e^{-s^2/2} ds - \beta e^{-1/2}.$$

These approximations are uniformly valid in their respective subintervals and match smoothly through the interior boundary layer at $t = \frac{1}{2}$.

Remark 13 *The location of boundary layers in problems of the form*

$$\varepsilon y'' + p(t)y' + q(t)y = f(t)$$

is governed by the sign and zeros of the convection coefficient $p(t)$. *In the present example,* $p(t) = t - \frac{1}{2}$ *is monotonic and has a single simple zero at* $t = \frac{1}{2}$. *As a result, the direction of information propagation changes at this point, preventing the outer solutions on either side from being matched directly and leading to an interior boundary layer centered at* $t = \frac{1}{2}$. *This behavior is generic: if* $p(t)$ *is monotonic and has exactly one simple zero in* $(0,1)$, *then exactly one interior boundary layer forms at that zero, regardless of whether* $p(t)$ *is increasing or decreasing. Multiple interior layers arise only when* $p(t)$ *possesses multiple simple zeros, while in the absence of zeros, any boundary layers occur only at the endpoints of the interval.*

We may state without proof the following theorem.

Theorem 4.4 (Location of boundary and interior layers) *Consider the singularly perturbed boundary value problem*

$$\varepsilon y'' + p(t)y' + q(t)y = f(t), \qquad 0 < t < 1,$$

subject to the boundary conditions

$$y(0) = \alpha, \qquad y(1) = \beta,$$

where $0 < \varepsilon \ll 1$. *Assume that*

1. $p, q, f \in C^1[0,1]$,
2. $p(t)$ *is monotonic on* $[0,1]$,
3. *all zeros of* $p(t)$ *in* $(0,1)$, *if any, are simple.*

Then the following conclusions hold:

(i) *If $p(t) \neq 0$ for all $t \in [0,1]$, the reduced problem obtained by setting $\varepsilon = 0$ is first order and any boundary layer that forms is located at one endpoint only: at $t = 0$ if $p(t) > 0$ on $[0,1]$, and at $t = 1$ if $p(t) < 0$ on $[0,1]$.*

(ii) *If $p(t)$ has exactly one simple zero $t^* \in (0,1)$, then no boundary layers occur at the endpoints and exactly one interior boundary layer forms in a neighborhood of $t = t^*$.*

(iii) *More generally, if $p(t)$ has k simple zeros in $(0,1)$, then exactly k interior boundary layers occur, each centered at one of the zeros of $p(t)$.*

Remark 14 *The theorem reflects the change in the direction of information propagation induced by the sign of $p(t)$. Each simple zero of $p(t)$ corresponds to a turning point of the reduced problem, across which outer solutions cannot be smoothly matched, thereby necessitating an interior boundary layer.*

4.3.1 Exercises

Exercise 4.17 *(Identifying an interior layer) Consider*

$$\varepsilon y'' + \left(t - \tfrac{1}{2}\right)y' + y = 0, \quad 0 < t < 1, \quad y(0) = \alpha, \ y(1) = \beta.$$

(a) *Show that the reduced problem has an interior singular point at $t^* = \frac{1}{2}$.*

(b) *Derive the reduced outer problems on $[0, \frac{1}{2})$ and on $(\frac{1}{2}, 1]$ and find the corresponding outer solutions $y_l^0(t)$ and $y_r^0(t)$ satisfying $y_l^0(0) = \alpha$ and $y_r^0(1) = \beta$, respectively.*

(c) *Explain why y_l^0 and y_r^0 cannot, in general, satisfy both boundary conditions simultaneously.*

Exercise 4.18 *(Linear turning point example) Study*

$$\varepsilon y'' + \left(t - \tfrac{1}{2}\right)y' - y = 0, \quad 0 < t < 1, \quad y(0) = 0, \ y(1) = 1,$$

with $0 < \varepsilon \ll 1$.

(a) *Derive the reduced outer equations and solve for y_l^0 and y_r^0.*

(b) *Using the scaling from earlier exercises, determine the correct inner variable near $t^* = \frac{1}{2}$.*

(c) *Solve the leading inner problem and match it with the outer solutions.*

(d) *Construct a uniformly valid composite approximation.*

Exercise 4.19 *(Inhomogeneous interior layer) Consider*

$$\varepsilon y'' + (t-1)y' + y = t, \quad 0 < t < 2, \quad y(0) = \alpha, \ y(2) = \beta.$$

(a) *Show that an interior layer forms at $t^* = 1$ and explain the propagation of information from each boundary.*

(b) *Find outer solutions y_l^0, y_r^0 satisfying the boundary conditions at $t = 0$ and $t = 2$, respectively.*

(c) *Obtain the leading inner problem near $t = 1$, solve it, and construct a composite approximation.*

Exercise 4.20 *(Quadratic convection coefficient) Let*

$$\varepsilon y'' + (t^2 - 1)y' + y = 0, \quad -1 < t < 2, \quad y(-1) = a, \; y(2) = b.$$

(a) *Identify the zero(s) of $p(t) = t^2 - 1$ and determine which ones lie in the domain.*

(b) *Decide which interior point generates a boundary layer and justify your answer using the sign of $p(t)$ on each subinterval.*

(c) *Perform a leading-order matched asymptotic analysis near the relevant interior point, including outer, inner, and composite solutions.*

Exercise 4.21 *Find uniform asymptotic approximations to the solutions of the following problems.*

(a) $\varepsilon y'' + \left(t - \frac{1}{3}\right)y' = 1, \quad 0 < t < 1, \quad y(0) = \alpha, \quad y(1) = \beta,$

(b) $\varepsilon y'' + \left(t - \frac{1}{2}\right)y = 0, \quad 0 < t < 1, \quad y(0) = 1, \quad y(1) = 2,$

(c) $\varepsilon y'' - ty' + 2y = 0, \quad -1 < t < 1, \quad y(-1) = \alpha, \quad y(1) = \beta.$

Exercise 4.22 *Consider the initial value problem (IVP)*

$$x' = ky - x + \varepsilon xy, \quad x(0) = 1,$$
$$\varepsilon y' = x - \varepsilon xy - y, \quad y(0) = 0,$$

where $0 < \varepsilon \ll 1$.

(a) *Use singular perturbation methods to approximate the solution. State clearly both the inner and outer solutions that you derive for $x(t)$ and $y(t)$.*

(b) *Discuss why the convergence to the actual solution differs between the cases $k < 1$ and $k > 1$. Does either case provide a uniformly valid approximation for $t > 0$? Explain.*

Exercise 4.23 *(Piecewise forcing) Consider*

$$\varepsilon y'' + \left(t - \frac{1}{2}\right)y' = \begin{cases} 0, & 0 < t < \frac{1}{2}, \\ 1, & \frac{1}{2} < t < 1, \end{cases} \quad y(0) = 0, \; y(1) = 0.$$

(a) *Find outer solutions on each subinterval, enforcing the boundary conditions.*

(b) *Show that the derivative of the outer solution is discontinuous at $t^* = \frac{1}{2}$.*

(c) *Construct an interior layer near t^* that restores smoothness, and derive the associated composite approximation.*

Exercise 4.24 *(Nonlinear interior layer) Let*

$$\varepsilon y'' + (t - t^*)y' + y^2 = 0, \quad 0 < t < 1, \quad y(0) = 1, \; y(1) = 0,$$

with a fixed $t^* \in (0,1)$.

(a) Derive the reduced outer equations and obtain y_l^0 *and* y_r^0.

(b) Determine the inner scaling near t^* *and write the leading inner equation for* $y^i(\tau)$.

(c) Discuss qualitatively the behavior of $y^i(\tau)$ *needed to connect the two outer solutions.*

Exercise 4.25 *(Multiple interior layers) Assume that* $p(t)$ *has two simple zeros* $t_1, t_2 \in (0,1)$ *with* $t_1 < t_2$.

(a) Describe how the interval $[0,1]$ *splits into three regions and where interior layers may occur.*

(b) Outline the construction of outer and inner solution s and the corresponding composite approximations in this case.

Exercise 4.26 *(Interior layer generated by a zero of the convection coefficient) Consider*

$$\varepsilon y''(t) + (t - \tfrac{1}{2})^{1/3} y'(t) = 0, \quad 0 < t < 1,$$

with boundary conditions

$$y(0) = 0, \quad y(1) = 1.$$

(a) Show that the convection coefficient vanishes at $t = \frac{1}{2}$ *and changes sign there.*

(b) Write down the reduced equation and discuss its singularity structure.

(c) Determine whether the reduced problem can satisfy both boundary conditions.

(d) Identify the location of any interior layer.

(e) Introduce a suitable inner variable centered at $t = \frac{1}{2}$ *and discuss the balance of terms in the inner equation.*

4.4 Existence and Location of Boundary and Interior Layers

We consider the singularly perturbed boundary value problem

$$\varepsilon y''(t) + p(t)y'(t) + q(t)y(t) = f(t), \qquad 0 < t < 1, \quad 0 < \varepsilon \ll 1, \tag{4.53}$$

subject to the boundary conditions

$$y(0) = \alpha, \qquad y(1) = \beta, \tag{4.54}$$

where p, q, and f are real-valued continuous functions on $[0,1]$. The small parameter ε multiplies the highest derivative, and hence the problem is *singularly perturbed*. As $\varepsilon \to 0$, the differential equation formally reduces to the first-order problem

$$p(t)y'(t) + q(t)y(t) = f(t), \tag{4.55}$$

which can satisfy at most one boundary condition. The inability of the reduced problem to satisfy both boundary conditions typically leads to the formation of narrow regions of rapid variation known as *boundary layers* or *interior layers*. The structure and location of these layers depend critically on the behavior of the convection coefficient $p(t)$.

Case I: $p(0) = 0$, $p(1) \neq 0$, and $p(t) \neq 0$ for $0 < t < 1$

In this case, the reduced equation (4.55) is singular at $t = 0$. Near this point, the convective term $p(t)y'$ becomes weak, and the second-derivative term $\varepsilon y''$ may dominate, giving rise to a *boundary layer at the left endpoint*. Away from $t = 0$, where $p(t) \neq 0$, the reduced equation is well posed and yields a smooth outer solution that typically satisfies the boundary condition at $t = 1$. Hence, a single boundary layer of width $O(\varepsilon)$ (or a related scaling depending on the order of vanishing of p) is expected near $t = 0$.

Case II: $p(0) = 0$, $p(1) \neq 0$, and p has a simple zero at $t = t^* \in (0,1)$

Here, the reduced problem is singular both at the left endpoint and at the interior point $t = t^*$. The vanishing of $p(t)$ at $t = t^*$ causes a change in the qualitative behavior of the reduced equation and typically leads to an *interior layer* centered at $t = t^*$. In addition, since $p(0) = 0$, a boundary layer may also form at $t = 0$. Thus, the solution may exhibit a combination of a left boundary layer and an interior layer, whose interaction must be carefully analyzed using matched asymptotic expansions.

Case III: $p(0) \neq 0$, $p(1) = 0$, and $p(t) \neq 0$ for $0 < t < 1$

In this situation, the reduced equation is regular at $t = 0$ but becomes singular at $t = 1$. Consequently, the reduced problem can typically satisfy the boundary condition at $t = 0$, while the condition at $t = 1$ must be enforced through a *right boundary layer*. The solution therefore, consists of a smooth outer solution on $(0,1)$ together with a thin boundary layer near $t = 1$.

Case IV: $p(0) = p(1) = 0$ and $p(t) \neq 0$ for $0 < t < 1$

When $p(t)$ vanishes at both endpoints but has no zeros in the interior, the reduced problem (4.55) is singular at $t = 0$ and $t = 1$. As a result, neither boundary condition can generally be satisfied by the outer solution alone. This configuration typically leads to the formation of *boundary layers at both endpoints*. The solution exhibits rapid variations near $t = 0$ and $t = 1$, while remaining smooth in the interior of the interval.

In summary, zeros of the convection coefficient $p(t)$ determine the location and nature of layers in (4.53)–(4.54). Zeros at the boundary lead to boundary layers , while simple zeros in the interior give rise to interior layers. The subsequent sections are devoted to the systematic construction and matching of inner and outer solutions for each of these cases.

Example 4.5 *Consider the singularly perturbed boundary value problem*

$$\varepsilon y''(t) + t y'(t) - t y(t) = 0, \qquad 0 < t < 1,$$

subject to the boundary conditions

$$y(0) = \alpha, \qquad y(1) = \beta.$$

Here, the convection coefficient is $p(t) = t$, so that $p(0) = 0$. Consequently, a boundary layer is expected to occur at the left endpoint $t = 0$.

As $\varepsilon \to 0^+$, the differential equation formally reduces to

$$t y'(t) - t y(t) = 0.$$

Since the outer solution is required to satisfy the boundary condition at $t = 1$, where $t \neq 0$, we may divide by t to obtain the reduced (outer) problem

$$y'(t) - y(t) = 0, \qquad y(1) = \beta.$$

The general solution of this first–order equation is

$$y(t) = Ce^t,$$

where C is a constant. Imposing the boundary condition $y(1) = \beta$ yields

$$C = \beta e^{-1}.$$

Hence, the outer solution is given by

$$y^0(t) = \beta e^{t-1}.$$

Let

$$\tau = \frac{t}{\delta(\varepsilon)}.$$

Then the differential equation becomes

$$\frac{\varepsilon}{\delta(\varepsilon)^2} y'' + \tau y' - \tau \delta(\varepsilon) y = 0.$$

First scaling: $\delta(\varepsilon) = \varepsilon$*. Then*

$$\frac{1}{\varepsilon}y'' + \tau y' - \varepsilon^2 \tau y = 0,$$

or equivalently,

$$y'' + \varepsilon \tau y' - \varepsilon^2 \tau y = 0.$$

Letting $\varepsilon \to 0^+$ *yields*

$$y'' = 0,$$

so that

$$y(\tau) = c_1 + c_2 \tau,$$

and hence

$$y_1(t) = c_1 + c_2 \frac{t}{\varepsilon},$$

which is unbounded as $\varepsilon \to 0$*. Therefore, this scaling is not appropriate.*

Second scaling: $\delta(\varepsilon) = \sqrt{\varepsilon}$*. Introduce*

$$\tau = \frac{t}{\sqrt{\varepsilon}}.$$

Then

$$\frac{\varepsilon}{\delta^2}y'' + \tau y' - \tau\sqrt{\varepsilon}y = 0,$$

implies that

$$y'' + \tau y' - \tau\sqrt{\varepsilon}y = 0.$$

Thus, as $\varepsilon \to 0^+$*, we have that*

$$y'' + \tau y' = 0.$$

Let $v = y'$*. Then*

$$\frac{dv}{d\tau} + \tau v = 0,$$

which gives

$$\frac{dv}{v} = -\tau d\tau,$$

implies that

$$\ln v = -\frac{\tau^2}{2} + C_1,$$

and hence

$$v(\tau) = C_1 e^{-\tau^2/2}.$$

Integrating once more,

$$y(\tau) = C_2 + C_1 \int_0^{\tau} e^{-s^2/2}\, ds.$$

Thus, the inner solution is

$$y^i(t) = C_2 + C_1 \int_0^{t/\sqrt{\varepsilon}} e^{-s^2/2}\, ds.$$

Next, we match y^i *with* $y^0(t)$.

Let

$$\eta = \frac{t}{\varepsilon^{1/4}}.$$

Then, the outer solution becomes

$$y^o(\eta) = \beta e^{\eta\varepsilon^{1/4}-1},$$

and

$$y^i(\eta) = C_2 + C_1 \int_0^{\eta/\varepsilon^{1/4}} e^{-s^2/2}\, ds.$$

As $\varepsilon \to 0^+$ *from the matching requirement, we have*

$$C_2 + C_1 \int_0^{\infty} e^{-s^2/2}\, ds = \beta e^{-1}.$$

Since

$$\int_0^{\infty} e^{-s^2/2}\, ds = \sqrt{\frac{\pi}{2}},$$

we obtain

$$C_2 + C_1\sqrt{\frac{\pi}{2}} = \beta e^{-1}.$$

Boundary condition at $t = 0$. *From* $y^i(0) = \alpha$,

$$\alpha = C_2,$$

and therefore

$$C_1 = (\beta e^{-1} - \alpha)\sqrt{\frac{2}{\pi}}.$$

Composite (uniform) approximation. *The uniformly valid approximation is*

$$y(t) = y^o(t) + y^i(t) - \beta e^{-1},$$

that is,

$$y(t) = \beta e^{t-1} + \alpha + (\beta e^{-1} - \alpha)\sqrt{\frac{2}{\pi}} \int_0^{t/\sqrt{\varepsilon}} e^{-s^2/2}\, ds - \beta e^{-1}.$$

□

4.4.1 Exercises

Exercise 4.27 *(Reduced problem and missing boundary conditions) Consider the singularly perturbed problem*

$$\varepsilon y''(t) + p(t)y'(t) + q(t)y(t) = f(t), \qquad 0 < t < 1, \quad y(0) = \alpha,\ y(1) = \beta,$$

with continuous p, q, f and $0 < \varepsilon \ll 1$.

(a) *Derive the reduced equation obtained by setting $\varepsilon = 0$ and explain why it can satisfy at most one boundary condition.*

(b) *Show that if $p(t) \neq 0$ on $[0,1]$, then no interior layers occur; only boundary layers are possible.*

(c) *Explain how the sign of $p(t)$ determines at which endpoint a boundary layer may form.*

Exercise 4.28 *(Classification via zeros of $p(t)$) Consider the singularly perturbed boundary value problem*

$$\varepsilon y'' + p(t)y' + q(t)y = r(t), \quad 0 < t < 1,$$

where $0 < \varepsilon \ll 1$, and p, q, r are continuous on $[0,1]$. Assume that $p(t)$ has only simple zeros on $[0,1]$.

(a) *Show that if $p(t^*) = 0$ for some $t^* \in (0,1)$, then the reduced equation*

$$p(t)y' + q(t)y = r(t)$$

degenerates at t^*, and explain why this suggests the possible presence of an interior (turning-point) layer near t^*.

(b) *Show that if $p(0) = 0$ or $p(1) = 0$, then the reduced equation fails to determine a solution satisfying both boundary conditions, and explain why this indicates a possible boundary layer at the corresponding endpoint.*

(c) *Give explicit examples of functions $p(t)$ realizing each of the following cases:*

- *Case I: $p(0) = 0$, $p(1) \neq 0$, no other zeros;*
- *Case II: $p(0) = 0$, simple zero at $t^* \in (0,1)$;*
- *Case III: $p(1) = 0$, $p(0) \neq 0$, no other zeros;*
- *Case IV: $p(0) = p(1) = 0$, $p(t) \neq 0$ for $0 < t < 1$.*

Exercise 4.29 *(Case I: left endpoint degeneracy) Consider*

$$\varepsilon y'' + ty' + y = 0, \qquad 0 < t < 1, \qquad y(0) = \alpha,\ y(1) = \beta.$$

(a) *Identify $p(t)$ and verify that the problem fits Case I.*

(b) *Derive the reduced equation and its general solution.*

(c) Argue which boundary condition can be imposed on the reduced problem and explain why a boundary layer is expected at $t = 0$.

Exercise 4.30 *(Case II: interior layer and left boundary layer) Consider*

$$\varepsilon y'' + t(t - \tfrac{1}{2})y' + y = 0, \qquad 0 < t < 1, \qquad y(0) = \alpha,\ y(1) = \beta.$$

(a) Show that $p(0) = 0$, *and that* $p(t)$ *has a simple zero at* $t^* = \frac{1}{2}$; *identify the case.*

(b) Determine on which subintervals $(0,t^*)$ *and* $(t^*,1)$ *the information propagates from left or right in the reduced equation.*

(c) Explain why a left boundary layer at $t = 0$ *and an interior layer near* $t = t^*$ *are expected.*

Exercise 4.31 *(Case II: approximate locations of layers) For the problem in the previous exercise:*

(a) Sketch the qualitative behavior of $p(t)$ *on* $[0,1]$ *and indicate the regions where* $p(t) > 0$ *and* $p(t) < 0$.

(b) Using the sign structure of $p(t)$, *argue where the outer solution is primarily influenced by* $y(0)$ *and where it is primarily influenced by* $y(1)$.

(c) Mark on the sketch the approximate positions and thicknesses of the boundary layer at $t = 0$ *and the interior layer at* $t = \frac{1}{2}$.

Exercise 4.32 *(Case III: right boundary layer) Consider*

$$\varepsilon y'' + (1 - t)y' + y = 0, \qquad 0 < t < 1, \qquad y(0) = \alpha,\ y(1) = \beta,$$

for a smooth $f(t)$.

(a) Verify that this problem fits Case III.

(b) Solve the reduced problem and identify which boundary condition it can satisfy.

(c) Explain why a boundary layer must form near $t = 1$ *in order to enforce the remaining boundary condition.*

Exercise 4.33 *(Case III: strength of the right layer) Using the problem in the previous exercise:*

(a) Introduce a scaling $1 - t = \delta(\varepsilon)\sigma$ *and balance the terms* $\varepsilon y''$ *and* $p(t)y'$ *near* $t = 1$ *to find* $\delta(\varepsilon)$.

(b) Explain why no interior layers are expected when $p(t) \neq 0$ *for* $0 < t < 1$.

(c) how the thickness of the layer depends on the order of vanishing of $p(t)$ *at* $t = 1$.

Exercise 4.34 *(Case IV: two endpoint degeneracies) Consider*

$$\varepsilon y'' + t(1 - t)y' + y = 0, \qquad 0 < t < 1,$$

with

$$y(0) = \alpha, \qquad y(1) = \beta.$$

(a) Show that $p(0) = p(1) = 0$ and that $p(t) > 0$ for $0 < t < 1$. Identify the case.

(b) Explain why both endpoints are degenerate points of the reduced equation.

(c) Determine which boundary condition is naturally satisfied by the outer solution.

(d) Predict where boundary layers may occur and describe the expected behavior.

Exercise 4.35 *(Case IV: comparing left and right layers) For the problem in the previous exercise:*

(a) Use Taylor expansions of $p(t)$ near $t = 0$ and $t = 1$ to deduce the leading behaviors of the inner equations at each endpoint.

(b) Show that the layer thickness at $t = 0$ and $t = 1$ is of the same order in ε (assuming simple zeros).

(c) Discuss how asymmetry in $p'(0)$ and $p'(1)$ can lead to different amplitudes in the two boundary layers.

Exercise 4.36 *(Locating layers from graphs of $p(t)$) Given four sketches of functions $p(t)$ on $[0,1]$:*

(a) $p(t) = t$,

(b) $p(t) = 1 - t$,

(c) $p(t) = t(1-t)$,

(d) $p(t) = t(t - 1/2)$,

classify each into Cases I–IV. For each case, specify:

(a) the expected type and location of layers (left boundary, right boundary, interior, or both ends);

(b) whether the reduced problem can satisfy the left or right boundary condition;

(c) a qualitative sketch of the anticipated solution profile for small ε.

Exercise 4.37 *(Multiple zeros and multiple layers) Suppose $p(t)$ has two simple zeros in $(0,1)$ at $t = t_1$ and $t = t_2$ with $0 < t_1 < t_2 < 1$.*

(a) Describe qualitatively how the interval $(0,1)$ splits into regions where information propagates from left or right.

(b) Indicate where boundary and/or interior layers can arise.

(c) Propose a general strategy, in terms of outer and inner approximations, for constructing a uniformly valid solution in this situation.

Exercise 4.38 *(Left boundary layer with weak convection) Consider the singularly perturbed problem*

$$\varepsilon y''(t) + t^{1/3} y'(t) + y(t) = 0, \qquad 0 < t < 1,$$

subject to

$$y(0) = 0, \qquad y(1) = 1,$$

where $0 < \varepsilon \ll 1$.

(a) Identify the reduced problem obtained by setting $\varepsilon = 0$.

(b) Explain why the reduced problem is singular at $t = 0$.

(c) Determine which boundary condition is satisfied by the outer solution.

(d) Predict the location and type of boundary layer.

(e) Determine the appropriate stretched variable near the layer and estimate the layer width.

Exercise 4.39 *(Right boundary layer with fractional vanishing) Consider the boundary value problem*

$$\varepsilon y''(t) + (1-t)^{1/3} y'(t) + y(t) = 0, \qquad 0 < t < 1,$$

with

$$y(0) = 0, \qquad y(1) = 0.$$

(a) Show that the reduced equation is singular at $t = 1$.

(b) Determine which boundary condition is satisfied by the outer solution.

(c) Predict the existence and location of a boundary layer.

(d) Compare the expected layer thickness with the classical $O(\varepsilon)$ *scaling.*

(e) Briefly explain how the fractional power in $p(t)$ *affects the layer structure.*

Exercise 4.40 *(Combined left boundary and interior layers) Let*

$$\varepsilon y''(t) + t^{1/3}(t - \tfrac{2}{3}) y'(t) = 1, \qquad 0 < t < 1,$$

with

$$y(0) = 0, \qquad y(1) = 1.$$

(a) Identify all points in $[0,1]$ *where* $p(t) = t^{1/3}(t - \frac{2}{3})$ *vanishes.*

(b) Classify each zero as a boundary or interior zero.

(c) Explain which types of layers are expected in this problem.

(d) Discuss qualitatively how the presence of multiple layers influences the construction of a uniform approximation.

Exercise 4.41 *(Two boundary layers) Consider*

$$\varepsilon y''(t) + t^{1/3}(1-t)^{1/3} y'(t) = 0, \qquad 0 < t < 1,$$

subject to

$$y(0) = 1, \qquad y(1) = 0.$$

(a) Show that $p(t)$ vanishes at both endpoints but nowhere in the interior.

(b) Write down the reduced problem and explain why it is singular at both $t = 0$ and $t = 1$.

(c) Predict the qualitative structure of the solution.

(d) Indicate where boundary layers form and describe their expected thickness.

4.5 Two Boundary Layers

In the preceding sections, we considered problems having one boundary layer. In this section, we consider a problem with two boundary layers. The appearance of two boundary layers is closely related to the order of the differential equation and the number of boundary conditions imposed. In particular, fourth-order singularly perturbed differential equations often exhibit this behavior. When a small parameter multiplies the highest (fourth) derivative, the reduced problem is of second order and can satisfy only two boundary conditions. The remaining boundary conditions must be enforced through boundary layers , which typically form near both ends of the interval. Such equations arise naturally in applications including beam and plate theory, fluid mechanics, and phase-field models. Thus, singular perturbation problems with two boundary layers are not only common but also structurally linked to fourth-order differential equations. Their analysis requires careful asymptotic techniques to capture both the global behavior of the solution and the localized effects near the boundaries. For the sake of matching, we explore the following idea.

Van Dyke's matching principle is a convenient formulation of the idea that inner and outer asymptotic expansions must represent the *same* solution in an overlap region. Suppose a singularly perturbed problem admits

$$y(t;\varepsilon) \sim y^{\text{out}}(t;\varepsilon) = \sum_{n=0}^{N} \varepsilon^n y_n^{\text{out}}(t), \qquad t = O(1),$$

and an inner expansion, written in the stretched variable $\tau = \dfrac{t}{\varepsilon^\alpha}$,

$$y(t;\varepsilon) \sim y^{\text{in}}(\tau;\varepsilon) = \sum_{n=0}^{N} \varepsilon^n y_n^{\text{in}}(\tau), \qquad \tau = O(1),$$

for some exponent $\alpha > 0$. Assume there is an *overlap region* where both descriptions are valid: this corresponds to letting $t \to 0$ in the outer scale while simultaneously letting $\tau \to \infty$ with $t = \varepsilon^\alpha \tau$.

Van Dyke's principle (matching rule).

Write the outer expansion in inner variables and re-expand in powers of ε:

$$y^{\text{out}}(\varepsilon^{\alpha}\tau;\varepsilon) \sim \sum_{n=0}^{N} \varepsilon^{n}\tilde{y}_{n}^{\text{out}}(\tau), \qquad \tau \to \infty.$$

Similarly, write the inner expansion in outer variables and re-expand:

$$y^{\text{in}}\left(\frac{t}{\varepsilon^{\alpha}};\varepsilon\right) \sim \sum_{n=0}^{N} \varepsilon^{n}\tilde{y}_{n}^{\text{in}}(t), \qquad t \to 0.$$

Van Dyke's matching principle states that, order by order in ε, the two re-expanded series must agree in the overlap region. Equivalently, one requires

$$\lim_{\tau\to\infty} y^{\text{in}}(\tau;\varepsilon) \sim \lim_{t\to 0} y^{\text{out}}(t;\varepsilon) \quad \text{as } \varepsilon \to 0,$$

with both sides expanded consistently in powers of ε. This condition determines the unknown constants (or functions) in the inner and outer solutions and ensures that they can be patched together into a uniformly valid composite approximation. In practice, Van Dyke's rule is often summarized as: "*Express the outer expansion in inner variables and the inner expansion in outer variables, expand each in powers of ε, and equate the two truncated series term by term.*" This approach avoids introducing an explicit intermediate variable and provides a systematic way to perform asymptotic matching in boundary layer and other singular perturbation problems. We furnish the following example.

Example 4.6 *Consider*

$$\varepsilon^{2}y^{(4)} - (1+t)^{2}y'' = 2, \qquad 0 < t < 1, \qquad 0 < \varepsilon \ll 1, \tag{4.56}$$

with

$$y(0) = \alpha, \qquad y'(0) = \beta, \qquad y(1) = \gamma, \qquad y'(1) = \delta. \tag{4.57}$$

In this case, the order of the highest derivative is greater than the order of the second term by two, suggesting that there are two boundary layers, one at each end. This suggestion will be verified a posteriori by checking the mathematical consistency of the results. We seek a two-term outer expansion in the form

$$y^{o}(t) = y_{0}(t) + \varepsilon y_{1}(t) + \cdots. \tag{4.58}$$

Substituting (4.58) *into* (4.56) *and equating coefficients of like powers of ε, we obtain*

$$-(1+t)^{2}y_{0}'' = 2, \tag{4.59}$$

$$-(1+t)^{2}y_{1}'' = 0. \tag{4.60}$$

Equation (4.59) *can be rewritten as*

$$y_{0}'' = -\frac{2}{(1+t)^{2}},$$

which can be integrated to give

$$y_0'(t) = \frac{2}{1+t} + A_0.$$

Another integration gives

$$y_0(t) = 2\ln(1+t) + A_0 t + B_0, \tag{4.61}$$

where A_0 and B_0 are arbitrary constants. On the other hand, the general solution of (4.60) is

$$y_1(t) = A_1 t + B_1, \tag{4.62}$$

where A_1 and B_1 are arbitrary constants. Therefore,

$$y^o(t) = 2\ln(1+t) + A_0 t + B_0 + \varepsilon(A_1 t + B_1) + \cdots. \tag{4.63}$$

Since this outer expansion is not expected to satisfy all of the boundary conditions, it must be matched with two boundary layer expansions, one valid near $t = 0$ and the other valid near $t = 1$.

Inner expansion near $t = 0$

We introduce the stretching transformation

$$\tau = \frac{t}{\varepsilon^\nu} \quad or \quad t = \varepsilon^\nu \tau, \qquad \nu > 0. \tag{4.64}$$

In (4.56), denote the inner expansion near $t = 0$ by the superscript i. Then

$$\varepsilon^{2-4\nu}\frac{d^4y^i}{d\tau^4} - \varepsilon^{-2\nu}(1+\varepsilon^\nu\tau)^2\frac{d^2y^i}{d\tau^2} = 2.$$

We choose ν so that there is no small parameter multiplying the highest derivative. Thus, setting $\nu = 1$, we obtain

$$\frac{d^4y^i}{d\tau^4} - (1+2\varepsilon\tau+\varepsilon^2\tau^2)\frac{d^2y^i}{d\tau^2} = 2\varepsilon^2. \tag{4.65}$$

We seek a two-term inner expansion in the form

$$y^i(\tau) = Y_0(\tau) + \varepsilon Y_1(\tau) + \cdots. \tag{4.66}$$

Substituting (4.66) into (4.65) and equating like powers of ε, we obtain

$$Y_0^{(4)} - Y_0'' = 0, \qquad Y_1^{(4)} - Y_1'' = 2\tau Y_0''. \tag{4.67}$$

This inner expansion must satisfy the boundary conditions at $t = 0$, corresponding to $\tau = 0$. Then $y(0) = \alpha$ transforms into

$$y^i(0) = Y_0(0) + \varepsilon Y_1(0) = \alpha,$$

and, since

$$y' = \frac{dy}{dt} = \frac{dy}{d\tau}\frac{d\tau}{dt} = \frac{1}{\varepsilon}\frac{dy}{d\tau},$$

the boundary condition $y'(0) = \beta$ becomes

$$\frac{dy^i}{d\tau}(0) = \varepsilon\beta.$$

Using (4.66), we see that the boundary conditions for Y_0, Y_1 at $\tau = 0$ take the form

$$Y_0(0) = \alpha, \qquad Y_0'(0) = 0, \tag{4.68}$$

$$Y_1(0) = 0, \qquad Y_1'(0) = \beta.$$

The parameter β first influences the solution at order ε, which necessitates retaining at least two terms in the asymptotic expansion. The general solution of the first equation in (4.67) is

$$Y_0(\tau) = a_0 + b_0\tau + c_0 e^{-\tau} + d_0 e^{\tau}.$$

Since $e^{\tau} \to \infty$ as $\tau \to \infty$, the constant d_0 must be zero; otherwise, Y_0 would grow exponentially and could not match the outer solution. Applying (4.68) we obtain

$$a_0 + c_0 = \alpha, \qquad b_0 - c_0 = 0.$$

Hence $b_0 = c_0$ and $a_0 = \alpha - c_0$, so that

$$Y_0(\tau) = \alpha + c_0\left(e^{-\tau} + \tau - 1\right). \tag{4.69}$$

We now perform matching using Van Dyke's principle, matching a one-term outer expansion with a one-term inner expansion. A one-term outer expansion is

$$y_0^o(t) = 2\ln(1+t) + A_0 t + B_0.$$

Rewritten in the inner variable $t = \varepsilon\tau$,

$$y_0^o = 2\ln(1+\varepsilon\tau) + A_0\varepsilon\tau + B_0.$$

Expanding for small ε gives

$$y_0^o = 2\varepsilon\tau + \varepsilon A_0\tau + B_0 + \cdots.$$

Thus, the one-term inner limit of the outer expansion is

$$y_0^o = B_0.$$

Similarly, from (4.69), the one-term outer limit of the inner solution is obtained by rewriting Y_0 in terms of t:

$$Y_0(\tau) = \alpha + c_0\left(e^{-t/\varepsilon} + t/\varepsilon - 1\right).$$

For fixed $t > 0$ and $\varepsilon \to 0^+$, $e^{-t/\varepsilon} \to 0$, so

$$Y_0 \sim \alpha - c_0 + \frac{c_0 t}{\varepsilon} + \cdots .$$

If $c_0 = 0$, then $Y_0 \sim \alpha$, which remains bounded as $\varepsilon \to 0$. If $c_0 \neq 0$, the leading behavior is

$$Y_0(t,\varepsilon) \sim \frac{c_0 t}{\varepsilon},$$

which blows up as $\varepsilon \to 0$. This is why we must set $c_0 = 0$ to obtain an inner solution that can be matched with the outer expansion. Thus, the one-term outer expansion is therefore

$$Y_0 \sim \begin{cases} \alpha, & \text{if } c_0 = 0, \\ \dfrac{c_0 t}{\varepsilon}, & \text{if } c_0 \neq 0. \end{cases} \tag{4.70}$$

Equating the limits of y_0^o and Y_0 requires

$$c_0 = 0, \qquad B_0 = \alpha.$$

Consequently,

$$Y_0(\tau) = \alpha. \tag{4.71}$$

Substituting (4.71) *into the second equation of* (4.67) *gives*

$$Y_1^{(4)} - Y_1'' = 0,$$

whose general solution is

$$Y_1(\tau) = a_1 + b_1\tau + c_1 e^{-\tau} + d_1 e^{\tau}.$$

We must set $d_1 = 0$; otherwise, Y_1 would become unbounded as $\tau \to \infty$. Applying the remaining boundary conditions at $\tau = 0$,

$$a_1 + c_1 = 0, \qquad b_1 - c_1 = \beta.$$

Hence $a_1 = -c_1$ and $b_1 = \beta + c_1$, so that

$$Y_1(\tau) = \beta\tau + c_1\left(e^{-\tau} + \tau - 1\right).$$

Therefore, by denoting the solution with $y_L^i(\tau)$, we have

$$y_L^i(\tau) = Y_0(\tau) + \varepsilon Y_1(\tau) = \alpha + \varepsilon\left[\beta\tau + c_1\left(e^{-\tau} + \tau - 1\right)\right].$$

In terms of t, we have

$$y_L^i(t) = \alpha + \varepsilon\left[\beta\frac{1}{\varepsilon} + c_1\left(e^{-t/\varepsilon} + \frac{t}{\varepsilon} - 1\right)\right]. \tag{4.72}$$

For fixed $t = \varepsilon\tau$ and $\varepsilon \to 0^+$,

$$y_L^i(t) = \alpha + \beta t + c_1 t e^{-t/\varepsilon} - \varepsilon c_1 + (\textit{higher terms}).$$

The two-term outer expansion (4.63), written in the inner variable $t = \varepsilon\tau$, is

$$y^o = 2\ln(1+\varepsilon\tau) + A_0\varepsilon\tau + B_0 + \varepsilon(\varepsilon A_1\tau + B_1).$$

Using $B_0 = \alpha$ and expanding for small ε gives

$$y^o = 2\varepsilon\tau + \varepsilon A_0\tau + \alpha + \varepsilon B_1 + \varepsilon^2 A_1\tau + \cdots.$$

Thus, the two-term inner limit of the outer expansion is

$$\tilde{y}^o = \alpha + \varepsilon(2\tau + A_0\tau + B_1). \tag{4.73}$$

On the other hand, the two-term outer limit of the inner expansion (4.72) is denoted by

$$\tilde{y}^i_{common} = \alpha + \beta t + c_1 t - \varepsilon c_1. \tag{4.74}$$

Expressing (4.73) in terms of t (using $\varepsilon\tau = t$) and equating it to (4.74), we obtain

$$\alpha + 2t + A_0 t + \varepsilon B_1 = \alpha + \beta t + c_1 t - \varepsilon c_1.$$

Hence

$$2 + A_0 = \beta + c_1, \qquad B_1 = -c_1. \tag{4.75}$$

***Inner expansion near** $t = 1$*

We now turn to the boundary layer at $t = 1$ by introducing the stretched variable

$$\zeta = \frac{1-t}{\varepsilon^\nu} \quad or \quad t = 1 - \varepsilon^\nu\zeta, \qquad \nu > 0. \tag{4.76}$$

We denote the inner expansion near $t = 1$ by the superscript i. From (4.56) we obtain

$$\varepsilon^{2-4\nu}\frac{d^4y^i}{d\zeta^4} - \varepsilon^{-2\nu}(2-\varepsilon^\nu\zeta)^2\frac{d^2y^i}{d\zeta^2} = 2.$$

Choosing $\nu = 1$ and simplifying yields

$$\frac{d^4y^i}{d\zeta^4} - (4 - 4\varepsilon\zeta + \varepsilon^2\zeta^2)\frac{d^2y^i}{d\zeta^2} = 2\varepsilon^2. \tag{4.77}$$

We seek a two-term expansion for y^i in the form

$$y^i(\zeta) = \bar{Y}_0(\zeta) + \varepsilon\bar{Y}_1(\zeta) + \cdots. \tag{4.78}$$

Substituting (4.78) into (4.77) and equating coefficients of like powers of ε, we obtain

$$\bar{Y}_0^{(4)} - 4\bar{Y}_0'' = 0, \tag{4.79}$$

$$\bar{Y}_1^{(4)} - 4\bar{Y}_1'' = -4\zeta\bar{Y}_0''. \tag{4.80}$$

This inner expansion must satisfy the boundary conditions at $t = 1$*, corresponding to* $\zeta = 0$*. Then* $y(1) = \gamma$ *transforms into*

$$y^i(0) = \gamma.$$

Since

$$\frac{dy^i}{dt} = \frac{dy^i}{d\zeta}\frac{d\zeta}{dt} = -\frac{1}{\varepsilon}\frac{dy^i}{d\zeta},$$

the boundary condition $y'(1) = \delta$ *becomes*

$$\frac{dy^i}{d\zeta}(0) = -\varepsilon\,\delta. \tag{4.81}$$

Substituting (4.78) *into these conditions and equating coefficients of like powers of* ε*, we obtain*

$$\bar{Y}_0(0) = \gamma, \qquad \bar{Y}_0'(0) = 0, \tag{4.82}$$

$$\bar{Y}_1(0) = 0, \qquad \bar{Y}_1'(0) = -\delta. \tag{4.83}$$

The general solution of (4.79) *is*

$$\bar{Y}_0(\zeta) = \bar{a}_0 + \bar{b}_0\zeta + \bar{c}_0 e^{-2\zeta} + \bar{d}_0 e^{2\zeta}.$$

Since $\bar{d}_0$ *must be zero; otherwise,* $\bar{Y}_0$ *would grow exponentially with* ζ*, we set* $\bar{d}_0 = 0$*. The boundary conditions* (4.82) *then give*

$$\bar{a}_0 + \bar{c}_0 = \gamma, \qquad \bar{b}_0 - 2\bar{c}_0 = 0,$$

so that $\bar{b}_0 = 2\bar{c}_0$ *and* $\bar{a}_0 = \gamma - \bar{c}_0$*. Thus*

$$\bar{Y}_0(\zeta) = \gamma + \bar{c}_0\left(e^{-2\zeta} + 2\zeta - 1\right). \tag{4.84}$$

We match a one-term outer expansion with a one-term inner expansion. The leading outer term is

$$y_0^o(t) = 2\ln(1+t) + A_0 t + \alpha,$$

(using $B_0 = \alpha$*). In the inner variable* $t = 1 - \varepsilon\zeta$ *this becomes*

$$y_0^o = 2\ln(2 - \varepsilon\zeta) + A_0(1 - \varepsilon\zeta) + \alpha.$$

Expanded for small ε*,*

$$y_0^o = 2\ln 2 - \varepsilon\zeta + A_0 - \varepsilon A_0\zeta + \alpha + \cdots. \tag{4.85}$$

Thus, the one-term inner limit of the outer solution is

$$y \sim 2\ln 2 + A_0 + \alpha.$$

On the other hand, the inner solution (4.84)*, written in outer variables, is*

$$y \sim \gamma + \bar{c}_0\left(e^{-2(1-t)/\varepsilon} + \frac{2(1-t)}{\varepsilon} - 1\right).$$

For fixed $t<1$ and $\varepsilon \to 0^+$,

$$y \sim \gamma + \frac{2\bar{c}_0(1-t)}{\varepsilon} - \bar{c}_0 + (\textit{higher terms}).$$

Thus, the one-term outer limit is

$$y \sim \begin{cases} \gamma, & \textit{if } \bar{c}_0 = 0, \\ \dfrac{2\bar{c}_0(1-t)}{\varepsilon}, & \textit{if } \bar{c}_0 \neq 0. \end{cases} \tag{4.86}$$

Equating (4.85) *and* (4.86) *demands that* $\bar{c}_0 = 0$ *and*

$$2\ln 2 + A_0 + \alpha = \gamma, \qquad \textit{that is} \qquad A_0 = \gamma - \alpha - 2\ln 2. \tag{4.87}$$

Hence, at leading order,

$$\bar{Y}_0(\zeta) = \gamma.$$

Substituting $\bar{Y}_0 = \gamma$ into (4.80), *we obtain*

$$\bar{Y}_1^{(4)} - 4\bar{Y}_1'' = 0, \tag{4.88}$$

whose general solution is

$$\bar{Y}_1(\zeta) = \tilde{a}_1 + \tilde{b}_1\zeta + \tilde{c}_1 e^{-2\zeta} + \tilde{d}_1 e^{2\zeta}.$$

Again, $\tilde{d}_1 = 0$ in order that the inner expansion be matchable. The boundary conditions (4.83) *give*

$$\tilde{a}_1 + \tilde{c}_1 = 0, \qquad \tilde{b}_1 - 2\tilde{c}_1 = -\delta.$$

Hence $\tilde{a}_1 = -\tilde{c}_1$ and $\tilde{b}_1 = 2\tilde{c}_1 - \delta$, so that

$$\bar{Y}_1(\zeta) = -\delta\zeta + \tilde{c}_1\left(e^{-2\zeta} + 2\zeta - 1\right). \tag{4.89}$$

Thus, the two–term inner expansion denoted by y_R^i is

$$y_R^i = \bar{Y}_0(\zeta) + \varepsilon\bar{Y}_0(\zeta) = \gamma + \varepsilon\Big[-\delta\zeta + \tilde{c}_1\left(e^{-2\zeta} + 2\zeta - 1\right)\Big]. \tag{4.90}$$

We still need to match the two–term outer expansion given by (4.63) *with the two-term inner expansion given by* (4.90). *Using $B_0 = \alpha$ in* (4.63) *and writing it in terms of the inner variable $t = 1 - \varepsilon\zeta$, we obtain*

$$y_0^o = 2\ln(2-\varepsilon\zeta) + A_0(1-\varepsilon\zeta) + \alpha + \varepsilon\left(A_1 - \varepsilon A_1\zeta + B_1\right).$$

Expanding it for small ε, we arrive at

$$y_0^o = 2\ln 2 - \varepsilon\zeta + A_0 - \varepsilon A_0\zeta + \alpha + \varepsilon A_1 - \varepsilon^2 A_1\zeta + \varepsilon B_1.$$

Finally, only considering two-terms, we get

$$y_0^o = A_0 + \alpha + 2\ln 2 + \varepsilon\left(A_1 + B_1 - \zeta - A_0\zeta\right) \tag{4.91}$$

Next, we turn our attention to the inner solution given by (4.90). *Rewriting it in the outer variables,* $\zeta = \frac{1-t}{\varepsilon}$ *we get*

$$y_R^i(t) = \gamma + \varepsilon\left\{-\delta\frac{1-t}{\varepsilon} + \tilde{c}_1\left[e^{-2(1-t)/\varepsilon} + 2\frac{1-t}{\varepsilon} - 1\right]\right\}. \tag{4.92}$$

Expanding the exponential term for small ε *we obtain*

$$y_R^i(t) = \gamma - \delta(1-t) + 2\tilde{c}_1(1-t) - \varepsilon\tilde{c}_1 + O(\varepsilon^2).$$

That is, a two-term expansion is denoted by

$$y^i_{common} = \gamma - \delta(1-t) + 2\tilde{c}_1(1-t) - \varepsilon\tilde{c}_1. \tag{4.93}$$

Replacing ζ *in* (4.91) *with* $\zeta = \frac{1-t}{\varepsilon}$, *and then equating the resulting equation with* (4.93) *we arrive at the expression*

$$A_0 + \alpha + 2\ln 2 + (-1 - A_0)(1-t) + \varepsilon(A_1 + B_1) = \gamma + (-\delta + 2\tilde{c}_1)(1-t) - \varepsilon\tilde{c}_1.$$

Equating both sides implies that

$$A_0 + \alpha + 2\ln 2 = \gamma, \quad -1 - A_0 = -\delta + 2\tilde{c}_1, \quad A_1 + B_1 = -\tilde{c}_1.$$

Solving the first equation, we obtain

$$A_0 = \gamma - \alpha - 2\ln 2,$$

which is consistent with our earlier result. The second equation gives

$$\tilde{c}_1 = \frac{1}{2}(-1 - \gamma + \alpha + 2\ln 2 + \delta).$$

Invoking (4.75), *we find*

$$B_1 = -c_1 = 2 + A_0 - \beta = -2 - \gamma + \alpha + 2\ln 2 + \beta.$$

Finally, the third equation yields

$$A_1 = \frac{1}{2}(5 - 6\ln 2 + 3\gamma - 2\alpha - 2\beta).$$

Uniformly valid composite expansion

Using (4.63), (4.72), (4.92), (4.74), *and* (4.93), *a uniformly valid composite expansion is then*

$$y_c(t) = y^o(t) + y_L^i(t) + y_R^i(t) - \tilde{y}^i_{common} - y^i_{common}.$$

That is

$$y_c(t) \quad = \quad 2\ln(1+t) + A_0 t + \alpha + \varepsilon(A_1 t + B_1)$$

$$
\begin{aligned}
&+\alpha+\varepsilon\left[\beta\frac{1}{\varepsilon}+c_1\left(e^{-t/\varepsilon}+\frac{t}{\varepsilon}-1\right)\right]\\
&+\gamma+\varepsilon\left\{-\delta\frac{1-t}{\varepsilon}+\tilde{c}_1\left[e^{-2(1-t)/\varepsilon}+2\frac{1-t}{\varepsilon}-1\right]\right\}.\\
&-\left(\alpha+\beta t+c_1 t-\varepsilon c_1\right)\\
&-\left(\gamma-\delta(1-t)+2\tilde{c}_1(1-t)-\varepsilon\tilde{c}_1\right).
\end{aligned}
$$

which is uniformly valid for $0 \le t \le 1$. □

4.5.1 Exercises

Exercise 4.42 *Consider the singularly perturbed fourth–order boundary–value problem*

$$\varepsilon^2 y^{(4)} - y'' = 0, \qquad 0 < t < 1, \qquad 0 < \varepsilon \ll 1,$$

with boundary conditions

$$y(0) = 0, \quad y'(0) = 1, \quad y(1) = 0, \quad y'(1) = -1.$$

Using only dominant–balance arguments (no full solution required),

(a) *Explain why one expects boundary layers to occur at both* $t = 0$ *and* $t = 1$.

(b) *Show that the natural stretched variables are* $\tau = t/\varepsilon$ *near* $t = 0$ *and* $\zeta = (1-t)/\varepsilon$ *near* $t = 1$.

(c) *Write down the reduced (outer) problem obtained by setting* $\varepsilon = 0$ *and state which boundary conditions can be satisfied by the outer solution alone.*

Exercise 4.43 *Consider the singularly perturbed fourth–order boundary–value problem*

$$\varepsilon^2 y^{(4)} - (1+t)y'' = 0, \qquad 0 < t < 1, \qquad 0 < \varepsilon \ll 1,$$

with boundary conditions

$$y(0) = 0, \quad y'(0) = 1, \quad y(1) = 0, \quad y'(1) = -1.$$

(a) *By setting* $\varepsilon = 0$, *derive the reduced (outer) problem and determine how many boundary conditions it can satisfy.*

(b) *Use a dominant–balance argument to explain why boundary layers must appear at both* $t = 0$ *and* $t = 1$.

(c) *Show that the natural stretched variables are* $\tau = t/\varepsilon$ *near* $t = 0$ *and* $\zeta = (1-t)/\varepsilon$ *near* $t = 1$.

Exercise 4.44 *Consider*

$$\varepsilon^2 y^{(4)} - (1+t^2)y'' = 3t, \qquad 0 < t < 1, \qquad 0 < \varepsilon \ll 1,$$

with boundary conditions

$$y(0) = \alpha, \quad y'(0) = \beta, \quad y(1) = \gamma, \quad y'(1) = \delta.$$

(a) *Seek an outer expansion*

$$y^o(t;\varepsilon) \sim y_0(t) + \varepsilon y_1(t),$$

derive the differential equations satisfied by y_0 *and* y_1, *and write their general solutions (with undetermined constants).*

(b) *Introduce the stretched variable* $\tau = t/\varepsilon$ *near* $t = 0$ *and obtain the leading inner equation for* $Y_0(\tau)$ *in an expansion*

$$y^i(\tau;\varepsilon) \sim Y_0(\tau) + \varepsilon Y_1(\tau).$$

Determine the general bounded solution for Y_0.

(c) *Using Van Dyke's matching principle at leading order, write down the outer solution in inner variables and the inner solution in outer variables and state the algebraic relations among the constants that follow from matching.*

Exercise 4.45 *Consider the fourth-order singularly perturbed boundary value problem*

$$\varepsilon^2 y^{(4)} - (2+t)^2 y'' = 3, \quad 0 < t < 1, \quad \varepsilon \ll 1,$$

with boundary conditions

$$y(0) = 1, \quad y'(0) = 0, \quad y(1) = 2, \quad y'(1) = 0.$$

(a) *Determine the two-term outer expansion* $y^o(t)$.

(b) *Introduce appropriate stretching transformations for the boundary layers near* $t = 0$ *and* $t = 1$, *and find the two-term inner expansions* $y^i(\tau)$ *and* $y^i(\zeta)$ *respectively, where* $\tau = t/\varepsilon$ *and* $\zeta = (1-t)/\varepsilon$.

(c) *Use Van Dyke's matching principle to determine all unknown constants in the outer and inner expansions.*

(d) *Construct a uniformly valid composite approximation up to* $O(\varepsilon)$.

Exercise 4.46 *Consider the singular perturbation problem*

$$\varepsilon^2 y^{(4)} - e^t y'' = 1, \quad 0 < t < 1, \quad \varepsilon \ll 1,$$

subject to the boundary conditions

$$y(0) = 0, \quad y'(0) = 1, \quad y(1) = 1, \quad y'(1) = 0.$$

(a) *Find the two-term outer expansion* $y^o(t)$.

(b) *For the boundary layer near* $t = 0$, *use the stretching* $\tau = t/\varepsilon$ *and derive the two-term inner expansion* $y^i(\tau)$. *Similarly, for the boundary layer near* $t = 1$, *use* $\zeta = (1-t)/\varepsilon$ *and find the two-term inner expansion* $y^i(\zeta)$.

(c) *Match the expansions using Van Dyke's principle to determine all arbitrary constants.*

(d) Provide a composite approximation that is uniformly valid on $[0,1]$ *up to* $O(\varepsilon)$.

Exercise 4.47 *Consider the singular perturbation problem*

$$\varepsilon^2 y''' - y' = 1, \quad 0 < t < 1, \quad \varepsilon \ll 1,$$

subject to the boundary conditions

$$y(0) = \alpha, \quad y'(0) = \beta, \quad y(1) = \gamma.$$

(a) Find the exact solution and use it to argue that there is in general, a boundary layer at each of $t = 0$ *and* $t = 1$.

(b) Provide a two-term approximation that is uniformly valid on $[0,1]$ *and compare your solution with the exact solution.*

Exercise 4.48 *Consider the singular perturbation problem*

$$\varepsilon^2 y''' - (t+1)y' = 1, \quad 0 < t < 1, \quad \varepsilon \ll 1,$$

subject to the boundary conditions

$$y(0) = \alpha, \quad y'(0) = \beta, \quad y(1) = \gamma.$$

(a) Find the exact solution and use it to argue that there is in general, a boundary layer at each of $t = 0$ *and* $t = 1$.

(b) Provide a two-term approximation that is uniformly valid on $[0,1]$, *and compare your solution with the exact solution.*

5

Asymptotic Expansions in Integrals

Integral representations play a central role in applied mathematics, mathematical physics, and asymptotic analysis. Many problems involve integrals depending on a large parameter, where exact evaluation is either impossible or unnecessary, and the essential information is captured by the dominant contributions of the integrand. In Example 2.11, we showed how integral equations may arise in the study of nonlinear differential equations. The purpose of this chapter is to develop systematic techniques for constructing asymptotic expansions of such integrals, emphasizing both their theoretical foundation and practical applicability.

We begin with Laplace-type integrals involving a large parameter λ, where the leading behavior is governed by the global maxima of the phase function. These integrals provide a natural entry point into asymptotic analysis, illustrating how local information near critical points determines the global behavior of the integral as $\lambda \to \infty$. Building on this framework, we then study exponentially weighted integrals, introducing powerful approximation techniques that arise in probability theory, statistical mechanics, and special function theory.

A special emphasis is placed on the Gamma function, which serves as a fundamental example of an exactly computable integral whose asymptotic behavior encapsulates many of the ideas developed in this chapter. This example highlights the deep connection between asymptotic expansions, special functions, and scaling arguments.

We also present asymptotic expansions obtained through repeated integration by parts, a versatile method that applies to a wide class of oscillatory and decaying integrals. Finally, the chapter is extended to Laplace-type integrals with power-law prefactors and interior maxima, where competing algebraic and exponential effects lead to richer asymptotic structures.

DOI: 10.1201/9781042009220-5

5.1 Laplace-Type Integrals for Large λ

Before we embark on the discussion of the Laplace-type integrals, we prove the following lemma that we need.

Lemma 2

$$\int_0^\infty t^n e^{-\lambda t}\,dt = \frac{n!}{\lambda^{n+1}}, \qquad n = 0, 1, 2, \ldots,\ \lambda > 0.$$

Proof *The proof is done by induction.*

Base case $n = 0$.

$$\int_0^\infty e^{-\lambda t}\,dt = \left[-\frac{1}{\lambda}e^{-\lambda t}\right]_0^\infty = 0 - \left(-\frac{1}{\lambda}\right) = \frac{1}{\lambda} = \frac{0!}{\lambda^{0+1}}.$$

Inductive step.

Assume that for some $n \geq 0$,

$$\int_0^\infty t^n e^{-\lambda t}\,dt = \frac{n!}{\lambda^{n+1}}.$$

Consider

$$\int_0^\infty t^{n+1} e^{-\lambda t}\,dt.$$

Integrate by parts with

$$u = t^{n+1}, \qquad dv = e^{-\lambda t}\,dt,$$

so

$$du = (n+1)t^n\,dt, \qquad v = -\frac{1}{\lambda}e^{-\lambda t}.$$

Then

$$\int_0^\infty t^{n+1} e^{-\lambda t}\,dt = -\frac{t^{n+1}}{\lambda}e^{-\lambda t}\Big|_0^\infty + \frac{n+1}{\lambda}\int_0^\infty t^n e^{-\lambda t}\,dt.$$

The boundary term vanishes, since

$$\lim_{t\to\infty} t^{n+1}e^{-\lambda t} = 0, \qquad 0^{n+1}e^0 = 0.$$

Thus

$$\int_0^\infty t^{n+1} e^{-\lambda t}\,dt = \frac{n+1}{\lambda}\int_0^\infty t^n e^{-\lambda t}\,dt.$$

Using the induction hypothesis,

$$\int_0^\infty t^{n+1} e^{-\lambda t}\,dt = \frac{n+1}{\lambda}\cdot\frac{n!}{\lambda^{n+1}} = \frac{(n+1)!}{\lambda^{n+2}}.$$

Therefore, by mathematical induction,

$$\int_0^\infty t^n e^{-\lambda t}\,dt = \frac{n!}{\lambda^{n+1}} \quad \textit{for all } n = 0,1,2,\ldots$$

as required.

A large class of problems in applied mathematics leads to integrals depending on a large parameter, whose dominant contributions arise from restricted regions of the integration domain. Asymptotic methods provide systematic tools for extracting accurate approximations of such integrals in the limit of large parameters.

Consider the Laplace-type integral

$$I(\lambda) = \int_0^\infty f(t)\,e^{-\lambda g(t)}\,dt, \qquad \lambda \gg 1,$$

where f and g are sufficiently smooth real-valued functions, with

$$g(t) \geq 0 \quad \text{for } t \geq 0,$$

and $g(t)$ attains its global minimum at some point in the domain.

As $\lambda \to \infty$, the exponential factor $e^{-\lambda g(t)}$ becomes sharply localized near the points where $g(t)$ is minimal. Away from such points, even moderate values of $g(t)$ cause the exponential term to decay rapidly, making negligible contributions to the integral.

- If $g(t)$ has a unique minimum at an interior point $t = t_0$ with $g'(t_0) = 0$ and $g''(t_0) > 0$, then the dominant contribution to $I(\lambda)$ comes from a small neighborhood of t_0. This leads to the classical Laplace method.
- If the minimum occurs at an endpoint of the interval (for example, $t = 0$), then the dominant contribution comes from a boundary layer near that endpoint.

Thus, for large λ, the behavior of $I(\lambda)$ is governed primarily by the local behavior of f and g near the minimum of g.

Approximation for the Case $g(t) = t$

Let

$$I(\lambda) = \int_0^\infty f(t)\,e^{-\lambda t}\,dt, \qquad \lambda \gg 1.$$

Here, $g(t) = t$ is strictly increasing, and its minimum occurs at the endpoint $t = 0$. The exponential term decays rapidly for $t > 0$, so the main contribution to the integral comes from a neighborhood of $t = 0$.

Assume that f is smooth near $t = 0$. Expanding $f(t)$ in a Taylor series about $t = 0$,

$$f(t) = f(0) + f'(0)\,t + \frac{1}{2}f''(0)\,t^2 + \cdots.$$

Substituting this expansion into the integral and integrating term by term, we obtain

$$I(\lambda)=\int_0^\infty \left(f(0)+f'(0)t+\frac{1}{2}f''(0)t^2+\cdots\right)e^{-\lambda t}\,dt.$$

By Lemma 2, which says

$$\int_0^\infty t^n e^{-\lambda t}\,dt=\frac{n!}{\lambda^{n+1}},\qquad n=0,1,2,\ldots,$$

we find the asymptotic expansion

$$I(\lambda)\sim\frac{f(0)}{\lambda}+\frac{f'(0)}{\lambda^2}+\frac{f''(0)}{\lambda^3}+\cdots,\qquad \lambda\to\infty.$$

Keeping only the leading term yields the simplest approximation:

$$I(\lambda)\approx\frac{f(0)}{\lambda},\qquad \lambda\gg 1.$$

This shows explicitly that, for large λ, the integral is dominated by the value of f at the point where $g(t)$ attains its minimum.

Example 5.1 *Consider*

$$I(\lambda)=\int_0^\infty e^{-\lambda t}\ln(1+t^2)\,dt,\qquad \lambda\to\infty.$$

This is a Laplace-type integral. For large λ, the exponential $e^{-\lambda t}$ strongly damps the integrand away from $t=0$, so the main contribution comes from a small neighborhood of $t=0$. Consequently, we expand the slowly varying part $\ln(1+t^2)$ about $t=0$ and integrate term by term.

Using the Taylor expansion at $t=0$,

$$\ln(1+t^2)=t^2-\frac{t^4}{2}+\frac{t^6}{3}-\frac{t^8}{4}+\cdots=\sum_{k=1}^{\infty}(-1)^{k+1}\frac{t^{2k}}{k},\qquad |t|<1.$$

This expansion is not used for convergence on $(0,\infty)$, but only as an asymptotic description near $t=0$, which is sufficient for large λ.

Substitute the expansion into the integral:

$$I(\lambda)\sim\sum_{k=1}^{\infty}(-1)^{k+1}\frac{1}{k}\int_0^\infty e^{-\lambda t}t^{2k}\,dt,\qquad \lambda\to\infty.$$

The Laplace integral is explicit:

$$\int_0^\infty e^{-\lambda t}t^n\,dt=\frac{n!}{\lambda^{n+1}},\qquad n=0,1,2,\ldots$$

Hence,

$$\int_0^\infty e^{-\lambda t} t^{2k}\,dt = \frac{(2k)!}{\lambda^{2k+1}}.$$

Therefore,

$$I(\lambda) \sim \sum_{k=1}^{\infty} (-1)^{k+1} \frac{(2k)!}{k\lambda^{2k+1}}, \qquad \lambda \to \infty.$$

Keeping the first few terms gives

$$\begin{aligned} I(\lambda) &\sim \frac{2!}{\lambda^3} - \frac{4!}{2\lambda^5} + \frac{6!}{3\lambda^7} - \frac{8!}{4\lambda^9} + \cdots \\ &= \frac{2}{\lambda^3} - \frac{12}{\lambda^5} + \frac{240}{\lambda^7} - \frac{5040}{\lambda^9} + \cdots. \end{aligned}$$

Next, we try to interpret the asymptotic expansions.

For each fixed N,

$$I(\lambda) - \sum_{k=1}^{N} (-1)^{k+1} \frac{(2k)!}{k\lambda^{2k+1}} = o\left(\frac{1}{\lambda^{2N+1}}\right), \qquad \lambda \to \infty.$$

Thus,

$$I(\lambda) \sim \frac{2}{\lambda^3} - \frac{12}{\lambda^5} + \frac{240}{\lambda^7} - \frac{5040}{\lambda^9} + \cdots, \qquad \lambda \to \infty,$$

is an asymptotic expansion *in the precise sense discussed earlier.*

The series above is not expected to converge. Its meaning is that each additional term improves the approximation of $I(\lambda)$ for large λ when the series is truncated, which is the hallmark of an asymptotic expansion for Laplace-type integrals. □

5.2 Asymptotic Approximation of an Exponentially Weighted Integral

Consider the integral

$$I(\lambda) = \int_0^\infty \frac{\sin t}{t} e^{\lambda g(t)}\,dt, \qquad \lambda \gg 1.$$

Since the exponential factor is $e^{\lambda g(t)}$ with $\lambda \to +\infty$, the integral is well defined only if $g(t)$ does not grow positively at infinity. We therefore assume:

- g is smooth on $[0,\infty)$,

- $g(t) \leq g(0)$ for all $t \geq 0$,
- $g(0) = 0$ (without loss of generality),
- $g'(0) < 0$.

Thus, $t = 0$ is a (boundary) point of maximum of $g(t)$, and the exponential factor is sharply peaked near $t = 0$ for large λ. As $\lambda \to \infty$, the factor $e^{\lambda g(t)}$ decays rapidly away from $t = 0$. Consequently, the dominant contribution to $I(\lambda)$ comes from a small neighborhood of $t = 0$, and the behavior of the integrand for moderate or large t becomes negligible.

This is an example of the *Laplace method with a boundary maximum.*

We expand both factors of the integrand near $t = 0$. Using the Taylor expansion,

$$\frac{\sin t}{t} = 1 - \frac{t^2}{6} + \frac{t^4}{120} - \cdots, \qquad t \to 0.$$

Since $g(0) = 0$ and $g'(0) < 0$, we write

$$g(t) = g'(0)t + \frac{1}{2}g''(0)t^2 + \cdots, \qquad t \to 0.$$

Let

$$a := -g'(0) > 0.$$

Then, to leading order,

$$g(t) \approx -at.$$

Keeping only the leading terms in both expansions, we obtain

$$I(\lambda) \approx \int_0^\infty e^{-\lambda at}\, dt.$$

Evaluating this integral gives

$$I(\lambda) \sim \frac{1}{\lambda a}, \qquad \lambda \to \infty.$$

To obtain corrections, substitute the full expansions into the integral:

$$I(\lambda) \sim \int_0^\infty \left(1 - \frac{t^2}{6} + \cdots\right) \exp\left[-\lambda at + \frac{\lambda}{2}g''(0)t^2 + \cdots\right] dt.$$

Introduce the stretched variable

$$t = \frac{s}{\lambda},$$

which reflects the fact that the dominant contribution comes from $t = O(\lambda^{-1})$. Then

$$I(\lambda) \sim \frac{1}{\lambda}\int_0^\infty \left(1 - \frac{s^2}{6\lambda^2} + \cdots\right) \exp\left(-as + \frac{g''(0)}{2\lambda}s^2 + \cdots\right) ds.$$

Expanding the exponential and integrating term by term yields an asymptotic series of the form

$$I(\lambda) \sim \frac{1}{\lambda a}\left[1+\frac{c_1}{\lambda}+\frac{c_2}{\lambda^2}+\cdots\right], \qquad \lambda \to \infty,$$

where the constants $c_1, c_2, \ldots$ depend on derivatives of g at $t = 0$.

In conclusion, under the stated assumptions on $g(t)$, the integral

$$I(\lambda) = \int_0^\infty \frac{\sin t}{t} e^{\lambda g(t)}\, dt$$

is dominated by a boundary layer near $t = 0$. The leading-order behavior for large λ is

$$I(\lambda) \sim \frac{1}{\lambda\left(-g'(0)\right)}, \qquad \lambda \to \infty,$$

with higher-order corrections obtainable systematically via the Laplace method.

Example 5.2 *(f depends on the large parameter λ)*

Consider the integral

$$I(\lambda) = \int_0^\infty \frac{\lambda}{t+\lambda} e^{-t}\, dt, \qquad \lambda \to \infty.$$

First rewrite the integrand as

$$\frac{\lambda}{t+\lambda} = \frac{1}{1+t/\lambda}.$$

Hence

$$I(\lambda) = \int_0^\infty \frac{e^{-t}}{1+t/\lambda}\, dt.$$

For fixed t and large λ, we expand the denominator as a geometric series:

$$\frac{1}{1+t/\lambda} = \sum_{n=0}^{\infty} (-1)^n \left(\frac{t}{\lambda}\right)^n, \qquad \lambda \to \infty.$$

Substituting this into the integral and formally integrating term by term, we obtain

$$I(\lambda) \sim \sum_{n=0}^{\infty} \frac{(-1)^n}{\lambda^n} \int_0^\infty t^n e^{-t}\, dt.$$

Since

$$\int_0^\infty t^n e^{-t}\, dt = n!,$$

the formal asymptotic series becomes

$$I(\lambda) \sim \sum_{n=0}^{\infty} (-1)^n \frac{n!}{\lambda^n}, \qquad \lambda \to \infty.$$

Next, we show that the asymptotic series diverges.

The coefficients of the series grow factorially:

$$a_n = (-1)^n n!.$$

Therefore,

$$\left| \frac{a_{n+1}/\lambda^{n+1}}{a_n/\lambda^n} \right| = \frac{n+1}{\lambda} \to \infty \quad \text{as } n \to \infty \quad \text{for any fixed } \lambda.$$

Hence the series

$$\sum_{n=0}^{\infty} (-1)^n \frac{n!}{\lambda^n}$$

diverges for every finite λ. It is therefore not a convergent power series but an asymptotic expansion.

It is helpful sometimes to look at the remainder of the expansion. Let

$$S_N(\lambda) = \sum_{n=0}^{N-1} (-1)^n \frac{n!}{\lambda^n}$$

denote the truncated series, and define the remainder

$$R_N(\lambda) = I(\lambda) - S_N(\lambda).$$

To estimate $R_N(\lambda)$, we write the exact identity

$$\frac{1}{1+t/\lambda} = \sum_{n=0}^{N-1} (-1)^n \left(\frac{t}{\lambda}\right)^n + (-1)^N \frac{(t/\lambda)^N}{1+t/\lambda}.$$

Substituting into the integral yields

$$R_N(\lambda) = (-1)^N \int_0^\infty \frac{t^N}{\lambda^N} \frac{e^{-t}}{1+t/\lambda} \, dt.$$

Next, we examine the asymptotic behavior of the remainder.

Since $1+t/\lambda \geq 1$ for $t \geq 0$, we obtain the bound

$$|R_N(\lambda)| \leq \frac{1}{\lambda^N} \int_0^\infty t^N e^{-t} \, dt = \frac{N!}{\lambda^N}.$$

Therefore, for each fixed N,

$$\lim_{\lambda \to \infty} R_N(\lambda) = 0.$$

This verifies that

$$I(\lambda) \sim \sum_{n=0}^{\infty} (-1)^n \frac{n!}{\lambda^n}, \qquad \lambda \to \infty,$$

in the precise asymptotic sense: for every fixed N,

$$I(\lambda) - \sum_{n=0}^{N-1} (-1)^n \frac{n!}{\lambda^n} = O\left(\frac{1}{\lambda^N}\right), \qquad \lambda \to \infty.$$

This example illustrates a fundamental phenomenon in asymptotic analysis: even though the asymptotic expansion of $I(\lambda)$ diverges, truncating the series after finitely many terms provides an increasingly accurate approximation as $\lambda \to \infty$, with the remainder tending to zero for each fixed truncation order.

The following theorem, often referred to as *Laplace's method,* provides a precise asymptotic approximation for integrals whose exponential term attains a unique interior maximum. It shows that, for large values of the parameter, the integral is governed primarily by the behavior of the integrand in a small neighborhood of the maximizer, and the contribution from the remainder of the interval is exponentially small. Its proof is left as an exercise.

Theorem 5.1 *Consider the integral*

$$I(\lambda) = \int_a^b f(t)\, e^{\lambda g(t)}\, dt, \qquad \lambda \gg 1,$$

where f is continuous, and g is smooth and has a unique maximum at $t = c$ in (a,b). That is

$$g'(c) = 0 \quad \textit{and} \quad g''(c) < 0.$$

Then

$$I(\lambda) \sim f(c) e^{\lambda g(c)} \sqrt{\frac{-2\pi}{\lambda g''(c)}}, \qquad \lambda \gg 1.$$

5.2.1 Gamma Function

The *gamma function* plays an important role in all aspects of applied mathematics.

Definition 5.1 *We denote the Gamma function by Γ, and it is defined by*

$$\Gamma(z) = \int_0^\infty e^{-t} t^{z-1} dt, \quad \mathfrak{Re}(z) > 0. \tag{5.1}$$

By letting $u = t^{z-1}dt$ and $dv = e^{-t}dt$, then a straightforward integration by parts leads to

$$\begin{aligned} \Gamma(z) &= \int_0^\infty e^{-t} t^{z-1} dt \\ &= -e^{-t} t^{z-1}\Big|_0^\infty + \int_0^\infty e^{-t}(z-1)t^{z-2} dt \end{aligned}$$

$$
\begin{aligned}
&= (z-1)\int_0^\infty e^{-t}t^{z-2}dt \\
&= (z-1)\Gamma(z-1).
\end{aligned}
$$

From the above calculations, we arrive at the expression

$$\Gamma(z+1) = z\Gamma(z),$$

which can be used to show that, for every positive integer n

$$\Gamma(n) = (n-1)!, \quad \text{or} \quad \Gamma(n+1) = n!.$$

We will also need the following formula. For a positive integer n, we have

$$\Gamma\big(\frac{1}{2}n\big) = \frac{(n-2)!!\sqrt{\pi}}{2^{(n-1)/2}}, \tag{5.2}$$

where $n!!$ is a *double factorial*. For example,

$$\Gamma\big(\frac{1}{2}\big) = \sqrt{\pi}, \quad \Gamma\big(\frac{3}{2}\big) = \sqrt{\pi}/2, \quad \Gamma\big(\frac{5}{2}\big) = (3\sqrt{\pi})/4, \text{ etc.}$$

Two other important properties of the Gamma function are

$$\Gamma\big(\frac{1}{2}+n\big) = \frac{(2n-1)!!}{2^n}\sqrt{\pi},$$

and

$$\Gamma\big(\frac{1}{2}-n\big) = \frac{(-1)^n 2^n}{(2n-1)!!}\sqrt{\pi}.$$

The next theorem is related to the Γ function, and its proof is left as an exercise.

Another important class of parameter–dependent integrals arises when the exponential factor decays rapidly, and the main contribution comes from an endpoint of the interval rather than from an interior critical point. Such integrals are typical of *Laplace-type-integrals* and occur frequently in asymptotic expansions, special function theory, and boundary layer analysis.

In this setting, the behavior of the integrand near the lower limit of integration determines the leading asymptotics as the parameter becomes large. By exploiting the Taylor expansion of the slowly varying part of the integrand and the strong decay of the exponential term, one can derive a full asymptotic expansion in inverse powers of the large parameter. The following theorem formalizes this procedure and provides an explicit asymptotic series in terms of the derivatives of the integrand at the endpoint. Its proof is left as an exercise.

Theorem 5.2 *For $\beta > -1$, we consider the integral*

$$I(\lambda) = \int_0^b t^\beta f(t)e^{-\lambda t}\,dt, \qquad \lambda \gg 1,$$

where $f(t)$ admits a Taylor series expansion about $t = 0$. If $f(0) \neq 0$, and there are positive constants k and c such that

$$|f(t)| \leq ke^{ct}, \quad 0 < t < b,$$

then

$$I(\lambda) \sim \sum_{n=0}^{\infty} \frac{f^{(n)}(0)\Gamma(\beta+n+1)}{n!\lambda^{\beta+n+1}}, \quad \lambda \gg 1.$$

We furnish the following example.

Example 5.3 *(Asymptotics of* erfc(λ) *for $\lambda \gg 1$)*

Recall the integral representation of the complementary error function

$$\mathrm{erfc}(\lambda) = \frac{2}{\sqrt{\pi}} \int_{\lambda}^{\infty} e^{-u^2}\, du, \qquad \lambda > 0.$$

Introduce the change of variables

$$u = \lambda + t,$$

which yields

$$\mathrm{erfc}(\lambda) = \frac{2}{\sqrt{\pi}} e^{-\lambda^2} \int_0^{\infty} e^{-2\lambda t} e^{-t^2}\, dt.$$

Define

$$I(\lambda) = \int_0^{\infty} e^{-2\lambda t} f(t)\, dt, \qquad f(t) = e^{-t^2}.$$

Verification of the hypotheses of Theorem 5.2. *Here $\beta = 0 > -1$. The function $f(t)$ is smooth and admits a Taylor series expansion about $t = 0$,*

$$f(t) = 1 - t^2 + \frac{t^4}{2!} - \cdots.$$

Moreover, $f(0) = 1 \neq 0$, and the growth condition

$$f(t) \leq ke^{ct}, \qquad t > 0,$$

is satisfied since $e^{-t^2} \leq 1$.

Thus, by Theorem 5.2, we have that

$$I(\lambda) \sim \sum_{n=0}^{\infty} \frac{f^{(n)}(0)\,\Gamma(n+1)}{n!\,(2\lambda)^{n+1}} = \sum_{n=0}^{\infty} \frac{f^{(n)}(0)}{(2\lambda)^{n+1}}, \qquad \lambda \gg 1.$$

Since only even derivatives of f at 0 are nonzero, we compute

$$f(0)=1, \qquad f''(0)=-2, \qquad f^{(4)}(0)=12, \qquad \ldots$$

Hence,

$$I(\lambda)\sim \frac{1}{2\lambda}-\frac{1}{4\lambda^3}+\frac{3}{8\lambda^5}-\cdots, \qquad \lambda \gg 1.$$

Asymptotic expansion of erfc(λ). *Substituting into the representation of* erfc(λ), *we obtain*

$$\operatorname{erfc}(\lambda)\sim \frac{e^{-\lambda^2}}{\sqrt{\pi}\lambda}\left(1-\frac{1}{2\lambda^2}+\frac{3}{(2\lambda^2)^2}-\frac{15}{(2\lambda^2)^3}+\cdots\right), \qquad \lambda \gg 1. \tag{5.3}$$

□

5.2.2 Exercises

Exercise 5.1 *(Warm-up: explicit Laplace-type integrals) For each integral below, compute $I(\lambda)$ exactly and then find its leading asymptotic behavior as $\lambda \to \infty$.*

(a) $I_1(\lambda)=\int_0^\infty e^{-\lambda t}\,dt.$

(b) $I_2(\lambda)=\int_0^\infty t^2 e^{-\lambda t}\,dt.$

Compare your asymptotic expressions with the exact results and verify that they agree to leading order in λ.

Exercise 5.2 *(Interior minimum: classical Laplace method) Consider*

$$I(\lambda)=\int_0^\infty e^{-\lambda(t-1)^2}\,dt, \qquad \lambda \gg 1.$$

(a) Identify $f(t)$ and $g(t)$ and show that $g(t)=(t-1)^2$ has a unique global minimum at $t_0=1$ with $g''(t_0)>0$.

(b) Expand $g(t)$ in a Taylor series about t_0 up to second order and approximate the integral by extending the limits to $(-\infty,\infty)$.

(c) Show that

$$I(\lambda)\sim\sqrt{\frac{\pi}{\lambda}} \quad \text{as } \lambda \to \infty.$$

Exercise 5.3 *Prove Theorem 5.1.*

Exercise 5.4 *Prove Theorem 5.2.*

Exercise 5.5 *(Interior minimum with nontrivial prefactor) Let*

$$I(\lambda) = \int_0^\infty t^2 e^{-\lambda(t-1)^2}\,dt, \qquad \lambda \gg 1.$$

(a) Identify the unique interior minimum of $g(t) = (t-1)^2$ and set $t = 1 + \dfrac{x}{\sqrt{\lambda}}$.

(b) Using the Laplace method, show that

$$I(\lambda) \sim \sqrt{\frac{\pi}{\lambda}} \qquad \text{as } \lambda \to \infty.$$

(c) Explain why the leading term does not depend on the first derivative of $f(t)$ at $t_0 = 1$.

Exercise 5.6 *(Endpoint minimum: boundary layer type) Consider*

$$I(\lambda) = \int_0^\infty e^{-\lambda t^2}\,dt, \qquad \lambda \gg 1.$$

(a) Show that the minimum of $g(t) = t^2$ occurs at the endpoint $t_0 = 0$.

(b) Introduce the scaling $t = \dfrac{x}{\sqrt{\lambda}}$ and rewrite $I(\lambda)$ in terms of x.

(c) that

$$I(\lambda) \sim \frac{1}{2}\sqrt{\frac{\pi}{\lambda}} \qquad \text{as } \lambda \to \infty.$$

Exercise 5.7 *(Endpoint minimum with power prefactor) Consider the Laplace-type integral*

$$I(\lambda) = \int_0^\infty t^\alpha e^{-\lambda t}\,dt, \qquad \alpha > -1, \quad \lambda \gg 1.$$

(a) Identify $f(t) = t^\alpha$ and $g(t) = t$. Note that the minimum of g occurs at the endpoint $t_0 = 0$.

(b) Use the substitution $u = \lambda t$ to evaluate the integral exactly in terms of the Gamma function.

(c) Deduce that

$$I(\lambda) \sim \frac{\Gamma(\alpha+1)}{\lambda^{\alpha+1}} \qquad \text{as } \lambda \to \infty,$$

and interpret this as a boundary layer asymptotic contribution near $t = 0$.

Exercise 5.8 *(Competing endpoints) Let*

$$I(\lambda) = \int_0^1 e^{-\lambda t}\,dt + \int_1^2 e^{-\lambda(2-t)}\,dt, \qquad \lambda \gg 1.$$

(a) Show that in the first integral, the minimum of $g(t) = t$ occurs at $t = 0$, while in the second, the minimum of $g(t) = 2 - t$ occurs at $t = 2$.

(b) Find the leading asymptotic contribution from each integral separately.

(c) Determine which endpoint dominates $I(\lambda)$ as $\lambda \to \infty$ and write the resulting leading-order approximation.

Exercise 5.9 *(Degenerate interior minimum) Consider*

$$I(\lambda) = \int_{-1}^{1} e^{-\lambda t^4}\, dt, \qquad \lambda \gg 1.$$

(a) Show that $g(t) = t^4$ has a global minimum at $t_0 = 0$ with $g'(0) = g''(0) = 0$ but $g^{(4)}(0) > 0$.

(b) Using the scaling $t = \dfrac{x}{\lambda^{1/4}}$, derive the leading asymptotic behavior of $I(\lambda)$ as $\lambda \to \infty$.

(c) Compare the resulting power of λ with the standard Laplace method for a non-degenerate quadratic minimum.

Exercise 5.10 *(Parameter dependence in $f(t)$) Let*

$$I(\lambda, a) = \int_{0}^{\infty} e^{-at} e^{-\lambda(t-1)^2}\, dt, \qquad a > 0, \quad \lambda \gg 1.$$

(a) Identify $f(t) = e^{-at}$ and $g(t) = (t-1)^2$ and locate the minimum of g.

(b) Laplace's method about $t_0 = 1$ to obtain the leading asymptotic approximation of $I(\lambda, a)$ as $\lambda \to \infty$.

(c) Discuss how the parameter a influences only the prefactor and not the λ scaling.

Exercise 5.11 *Let*

$$I(\lambda) = \int_{0}^{1} e^{\lambda(1-(t-\frac{1}{2})^2)}\, dt, \qquad \lambda \gg 1.$$

(a) Identify the point $c \in (0,1)$ at which $g(t) = 1 - (t - \frac{1}{2})^2$ attains its unique maximum.

(b) Verify that $g'(c) = 0$ and $g''(c) < 0$.

(c) Theorem 5.1 to obtain the leading asymptotic approximation of $I(\lambda)$ as $\lambda \to \infty$.

Exercise 5.12 *Consider the integral*

$$I(\lambda) = \int_{0}^{\pi} (1 + \sin t)\, e^{\lambda \cos t}\, dt, \qquad \lambda \gg 1.$$

(a) Show that $g(t) = \cos t$ has a unique maximum at $t = 0$ in $[0, \pi]$.

(b) Compute $g''(0)$ and evaluate $f(0)$.

(c) Apply Theorem 5.1 to derive an asymptotic formula for $I(\lambda)$ as $\lambda \to \infty$.

Exercise 5.13 *Let*

$$I(\lambda) = \int_{-1}^{2} (1+t^2)\, e^{\lambda(-t^2+2t)}\, dt, \qquad \lambda \gg 1.$$

(a) Find the critical point c of $g(t) = -t^2+2t$ and show that it is a point of maximum.

(b) Compute $g(c)$ and $g''(c)$.

(c) Use Theorem 5.1 to obtain the leading-order asymptotic behavior of $I(\lambda)$ as $\lambda \to \infty$.

Exercise 5.14 *Assume f is smooth, and g satisfies the hypotheses of Theorem 5.1. Show that if $f(c) = 0$ but $f'(c) \neq 0$, then the leading-order term in the asymptotic expansion of*

$$I(\lambda) = \int_{a}^{b} f(t)\, e^{\lambda g(t)}\, dt$$

is of order $\lambda^{-3/2} e^{\lambda g(c)}$. (Hint: expand $f(t)$ and $g(t)$ in Taylor series about $t = c$ and apply Laplace's method.)

Exercise 5.15 *Let*

$$I(\lambda) = \int_{0}^{\infty} t^{1/2} e^{-\lambda t} \cos t\, dt, \qquad \lambda \gg 1.$$

(a) Show that $f(t) = \cos t$ satisfies the hypotheses of Theorem 5.2 with $\beta = \frac{1}{2}$.

(b) Compute the first three nonzero terms in the asymptotic expansion of $I(\lambda)$ as $\lambda \to \infty$.

(c) Verify that the leading term is

$$I(\lambda) \sim \frac{\Gamma(\frac{3}{2})}{\lambda^{3/2}}, \qquad \lambda \gg 1.$$

Exercise 5.16 *Consider the Laplace-type integral*

$$I(\lambda) = \int_{0}^{b} t^{\beta} \ln(1+t)\, e^{-\lambda t}\, dt, \qquad \beta > -1,\ \lambda \gg 1.$$

(a) Show that $f(t) = \ln(1+t)$ admits a Taylor expansion about $t = 0$ and satisfies the growth condition required in Theorem 5.2.

(b) Theorem 5.2, derive an asymptotic expansion of $I(\lambda)$ up to and including terms of order $\lambda^{-(\beta+3)}$.

(c) Discuss how the logarithmic singularity of $f'(t)$ at $t = -1$ does not affect the asymptotic expansion for large λ.

5.3 Asymptotic Expansions via Integration by Parts

Integration by parts is one of the most fundamental and systematic tools for deriving asymptotic expansions of integrals containing a large parameter. It is especially effective for *Laplace-type integrals* and *Fourier-type integrals*, where a rapidly varying exponential or oscillatory factor dominates the behavior of the integral. In general, let's consider an integral of the form

$$I(\lambda) = \int_a^b e^{-\lambda g(t)} f(t)\,dt, \qquad \lambda \to \infty,$$

where:

- $g(t) \geq 0$ on $[a,b]$,
- $g(t)$ is smooth and has a unique minimum at $t = a$,
- $g'(t) > 0$ for $t > a$.

For large λ, the exponential $e^{-\lambda g(t)}$ decays rapidly away from $t = a$. The main contribution to the integral comes from the endpoint $t = a$. Integration by parts exploits the identity

$$\frac{d}{dt}\left(e^{-\lambda g(t)}\right) = -\lambda g'(t) e^{-\lambda g(t)},$$

which allows one to systematically extract inverse powers of λ.

Write

$$I(\lambda) = \int_a^b \frac{f(t)}{g'(t)} \left(-\frac{1}{\lambda}\frac{d}{dt} e^{-\lambda g(t)}\right) dt.$$

Applying integration by parts yields

$$I(\lambda) = -\frac{f(t)}{\lambda g'(t)} e^{-\lambda g(t)}\bigg|_a^b + \frac{1}{\lambda}\int_a^b \frac{d}{dt}\left(\frac{f(t)}{g'(t)}\right) e^{-\lambda g(t)}\,dt.$$

Under mild assumptions, the boundary term at $t = b$ vanishes exponentially. The leading contribution comes from $t = a$, and we have that

$$I(\lambda) \sim \frac{f(a)}{\lambda g'(a)} e^{-\lambda g(a)} + \frac{1}{\lambda}\int_a^b \frac{d}{dt}\left(\frac{f(t)}{g'(t)}\right) e^{-\lambda g(t)}\,dt.$$

Repeating this process N times produces an expansion of the form

$$I(\lambda) \sim e^{-\lambda g(a)} \sum_{k=0}^{N-1} \frac{A_k}{\lambda^{k+1}}, \qquad \lambda \to \infty,$$

where the coefficients A_k depend only on f and g and their derivatives at $t = a$. The remainder after N steps is

$$O\left(\frac{e^{-\lambda g(a)}}{\lambda^{N+1}}\right).$$

We notice that integration by parts does not rely on the convergence of a series, and at the same time, it produces a genuine asymptotic expansion,

Remark 15 *The limits of integration can be varied to include integrals of the form*

$$I(\lambda) = \int_{\lambda}^{\infty} e^{-g(t)} f(t)\,dt, \qquad \lambda \to \infty.$$

We explore the idea of integration by parts combined with Taylor series expansion by furnishing the following example.

Example 5.4 *Consider the integral*

$$I(\lambda) = \int_{0}^{\infty} e^{-\lambda t} \ln(1+t)\,dt, \qquad \lambda \to \infty.$$

We derive an asymptotic expansion for large λ. We begin by performing an integration by parts by letting

$$u = \ln(1+t), \qquad dv = e^{-\lambda t}\,dt.$$

Then

$$du = \frac{1}{1+t}\,dt, \qquad v = -\frac{1}{\lambda} e^{-\lambda t}.$$

Hence,

$$I(\lambda) = \left[-\frac{1}{\lambda} e^{-\lambda t} \ln(1+t)\right]_0^{\infty} + \frac{1}{\lambda}\int_0^{\infty} \frac{e^{-\lambda t}}{1+t}\,dt.$$

The boundary term vanishes, and therefore

$$I(\lambda) = \frac{1}{\lambda}\int_0^{\infty} \frac{e^{-\lambda t}}{1+t}\,dt.$$

For $t \geq 0$, we write

$$\frac{1}{1+t} = \sum_{n=0}^{N} (-1)^n t^n + (-1)^{N+1} \frac{t^{N+1}}{1+t}.$$

Substituting into the integral yields

$$I(\lambda) = \frac{1}{\lambda}\sum_{n=0}^{N} (-1)^n \int_0^{\infty} t^n e^{-\lambda t}\,dt + R_N(\lambda),$$

where

$$R_N(\lambda) = \frac{(-1)^{N+1}}{\lambda} \int_0^\infty \frac{t^{N+1}}{1+t} e^{-\lambda t}\, dt.$$

Next, we do term-by-term integration. Using

$$\int_0^\infty t^n e^{-\lambda t}\, dt = \frac{n!}{\lambda^{n+1}},$$

we obtain

$$I(\lambda) = \sum_{n=0}^{N} (-1)^n \frac{n!}{\lambda^{n+2}} + R_N(\lambda).$$

Thus, as $\lambda \to \infty$,

$$I(\lambda) \sim \sum_{n=0}^{\infty} (-1)^n \frac{n!}{\lambda^{n+2}}.$$

Now we try to estimate the remainder. Since $\frac{1}{1+t} \le 1$ *for* $t \ge 0$, *we have*

$$|R_N(\lambda)| \le \frac{1}{\lambda} \int_0^\infty t^{N+1} e^{-\lambda t}\, dt = \frac{(N+1)!}{\lambda^{N+3}}.$$

Hence,

$$R_N(\lambda) = O\left(\frac{(N+1)!}{\lambda^{N+3}}\right), \qquad \lambda \to \infty.$$

We remark that the series is divergent but asymptotic; optimal truncation yields highly accurate approximations for large λ. □

We conclude this section with a classic example that illustrates how repeated integration by parts can be used to derive an asymptotic expansion of a Laplace-type integral. The complementary error function provides a standard and important application, since its large-argument behavior is governed by the rapid decay of the Gaussian exponential. The following example demonstrates, step by step, how successive integrations by parts lead to the well-known asymptotic expansion of $\operatorname{erfc}(\lambda)$ as $\lambda \to \infty$.

Example 5.5 *We consider*

$$\operatorname{erfc}(\lambda) = \frac{2}{\sqrt{\pi}} \int_\lambda^\infty e^{-t^2}\, dt, \qquad \lambda \gg 1.$$

To perform the first integration by parts, we observe that

$$e^{-t^2} = -\frac{1}{2t}\frac{d}{dt}\left(e^{-t^2}\right).$$

Then

$$\int_\lambda^\infty e^{-t^2}\, dt = \int_\lambda^\infty \left(-\frac{1}{2t}\right) d\left(e^{-t^2}\right).$$

Integrating by parts,

$$\int_{\lambda}^{\infty} e^{-t^2}\,dt = \left[-\frac{e^{-t^2}}{2t}\right]_{\lambda}^{\infty} - \int_{\lambda}^{\infty} e^{-t^2}\,d\left(-\frac{1}{2t}\right).$$

Since the boundary term at infinity vanishes, we obtain

$$\int_{\lambda}^{\infty} e^{-t^2}\,dt = \frac{e^{-\lambda^2}}{2\lambda} - \frac{1}{2}\int_{\lambda}^{\infty} \frac{e^{-t^2}}{t^2}\,dt.$$

For the second integration by parts, we apply the same idea to the remaining integral and obtain

$$\int_{\lambda}^{\infty} \frac{e^{-t^2}}{t^2}\,dt = \int_{\lambda}^{\infty} \left(-\frac{1}{2t^3}\right) d\left(e^{-t^2}\right).$$

Then an integration by parts yields

$$\int_{\lambda}^{\infty} \frac{e^{-t^2}}{t^2}\,dt = \frac{e^{-\lambda^2}}{2\lambda^3} - \frac{3}{2}\int_{\lambda}^{\infty} \frac{e^{-t^2}}{t^4}\,dt.$$

Substituting back,

$$\int_{\lambda}^{\infty} e^{-t^2}\,dt = \frac{e^{-\lambda^2}}{2\lambda}\left(1 - \frac{1}{2\lambda^2}\right) + \frac{3}{4}\int_{\lambda}^{\infty} \frac{e^{-t^2}}{t^4}\,dt.$$

Similarly, for the third integration by parts, we write

$$\int_{\lambda}^{\infty} \frac{e^{-t^2}}{t^4}\,dt = \int_{\lambda}^{\infty} \left(-\frac{1}{2t^5}\right) d\left(e^{-t^2}\right).$$

Now performing in an integration by parts and then applying the limits of integration we arrive at

$$\int_{\lambda}^{\infty} \frac{e^{-t^2}}{t^4}\,dt = \frac{e^{-\lambda^2}}{2\lambda^5} - \frac{5}{2}\int_{\lambda}^{\infty} \frac{e^{-t^2}}{t^6}\,dt.$$

Consequently,

$$\int_{\lambda}^{\infty} e^{-t^2}\,dt = \frac{e^{-\lambda^2}}{2\lambda}\left(1 - \frac{1}{2\lambda^2} + \frac{3}{(2\lambda^2)^2}\right) - \frac{15}{8}\int_{\lambda}^{\infty} \frac{e^{-t^2}}{t^6}\,dt. \tag{5.4}$$

By performing the fourth integration by parts, we arrive at

$$\int_{\lambda}^{\infty} \frac{e^{-t^2}}{t^6}\,dt = \frac{e^{-\lambda^2}}{2\lambda^7} - \frac{7}{2}\int_{\lambda}^{\infty} \frac{e^{-t^2}}{t^8}\,dt.$$

Substitution into (5.4) *shows that the remainder is of smaller order* $O\left(e^{-\lambda^2}\lambda^{-7}\right)$ *as* $\lambda \to \infty$.

By multiplying by $2/\sqrt{\pi}$, *we arrive at the asymptotic expansion*

$$\operatorname{erfc}(\lambda) \sim \frac{e^{-\lambda^2}}{\sqrt{\pi}\,\lambda}\left(1 - \frac{1}{2\lambda^2} + \frac{3}{(2\lambda^2)^2} - \frac{15}{(2\lambda^2)^3} + \cdots\right), \qquad \lambda \gg 1.$$

We ask the reader to compare this answer with (5.3). □

In many Laplace-type integrals the main contribution for large values of the parameter comes from one endpoint of the interval of integration. A natural first attempt to extract the large λ behavior is to use integration by parts, since each integration by parts introduces an extra factor of $1/\lambda$. However, when one of the endpoints is singular, or contributes a term that is larger than the integral itself, naive integration by parts can produce expressions involving differences of divergent quantities and therefore must be handled with care. The following example illustrates this phenomenon and shows how to reformulate the problem so that integration by parts yields a genuine asymptotic expansion.

Example 5.6 *In this example we examine the use of integration by parts to determine the asymptotic behavior of*

$$I(\lambda) = \int_0^{\lambda} t^{-1/2} e^{-t}\, dt, \qquad \lambda \to +\infty.$$

A direct application of integration by parts leads to difficulty. Indeed,

$$\int_0^{\lambda} t^{-1/2} e^{-t} dt = \left[-t^{-1/2} e^{-t}\right]_0^{\lambda} - \frac{1}{2}\int_0^{\lambda} t^{-3/2} e^{-t} dt,$$

but the boundary term at $t = 0$ *diverges, so this expression is not useful.*

To proceed, we rewrite $I(\lambda)$ *as*

$$I(\lambda) = \int_0^{\infty} t^{-1/2} e^{-t} dt - \int_{\lambda}^{\infty} t^{-1/2} e^{-t} dt$$

The first integral is finite and equals

$$\Gamma\left(\tfrac{1}{2}\right) = \sqrt{\pi}.$$

We now apply integration by parts to the tail integral:

$$\int_{\lambda}^{\infty} t^{-1/2} e^{-t} dt = \left[-t^{-1/2} e^{-t}\right]_{\lambda}^{\infty} - \frac{1}{2}\int_{\lambda}^{\infty} t^{-3/2} e^{-t} dt.$$

Since the contribution at infinity vanishes, this gives

$$\int_{\lambda}^{\infty} t^{-1/2} e^{-t} dt = \frac{e^{-\lambda}}{\sqrt{\lambda}} - \frac{1}{2}\int_{\lambda}^{\infty} t^{-3/2} e^{-t} dt.$$

Hence,

$$I(\lambda) = \sqrt{\pi} - \frac{e^{-\lambda}}{\sqrt{\lambda}} + \frac{1}{2}\int_{\lambda}^{\infty} t^{-3/2} e^{-t}\, dt.$$

Repeated integration by parts yields the asymptotic expansion

$$I(\lambda) \sim \sqrt{\pi} - \frac{e^{-\lambda}}{\sqrt{\lambda}}\left[1 + \sum_{n=1}^{\infty} \frac{(1\cdot 3\cdot 5\cdots(2n-1))}{(2\lambda)^n}\right], \qquad \lambda \to +\infty.$$

This example illustrates that integration by parts may fail when endpoint contributions dominate the integral. Here, the singular behavior at $t = 0$ produces a divergent boundary term, even though the integral itself is finite. □

5.3.1 Exercises

Exercise 5.17 *(General endpoint formula) Let*

$$I(\lambda) = \int_a^b e^{-\lambda g(t)} f(t)\, dt, \qquad \lambda \to \infty,$$

with $g(t) \geq 0$ on $[a,b]$, $g(a) = 0$, $g'(t) > 0$ for $t > a$ and smooth f, g.

(a) Show that

$$I(\lambda) = -\frac{f(t)}{\lambda g'(t)} e^{-\lambda g(t)}\Big|_a^b + \frac{1}{\lambda}\int_a^b \frac{d}{dt}\left(\frac{f(t)}{g'(t)}\right) e^{-\lambda g(t)}\, dt.$$

(b) Under the assumption that the boundary term at $t = b$ is exponentially small, obtain the first two terms in the asymptotic expansion

$$I(\lambda) \sim \frac{f(a)}{\lambda g'(a)} + \frac{1}{\lambda^2}\left(\frac{f'(a)g'(a) - f(a)g''(a)}{(g'(a))^3}\right), \qquad \lambda \to \infty.$$

Exercise 5.18 *(Integration by parts for a rational prefactor) Let*

$$J(\lambda) = \int_0^{\infty} \frac{e^{-\lambda t}}{(1+t)^2}\, dt, \qquad \lambda \to \infty.$$

(a) Apply integration by parts once with $u = \frac{1}{(1+t)^2}$, $dv = e^{-\lambda t} dt$, and show that

$$J(\lambda) = \frac{1}{\lambda}\left[1 - 2\int_0^{\infty} \frac{e^{-\lambda t}}{(1+t)^3}\, dt\right].$$

(b) Using the previous exercise as a guide, derive the first two terms of the asymptotic expansion of $J(\lambda)$ as $\lambda \to \infty$.

(c) Compare your result with what you would obtain by differentiating the expansion for $\int_0^\infty e^{-\lambda t}\ln(1+t)\,dt$ with respect to λ.

Exercise 5.19 *(Asymptotics of an incomplete gamma function) For fixed $a>0$, consider*

$$\Gamma(a,\lambda)=\int_\lambda^\infty t^{a-1}e^{-t}\,dt, \qquad \lambda\to\infty.$$

(a) Write the integrand as $t^{a-1}e^{-t}=-t^{a-1}\,d(e^{-t})$ and integrate by parts to show that

$$\Gamma(a,\lambda)=\lambda^{a-1}e^{-\lambda}+(a-1)\int_\lambda^\infty t^{a-2}e^{-t}\,dt.$$

(b) Iterate this procedure N times and prove that

$$\Gamma(a,\lambda)=e^{-\lambda}\lambda^{a-1}\left(1+\frac{a-1}{\lambda}+\frac{(a-1)(a-2)}{\lambda^2}+\cdots+\frac{(a-1)\cdots(a-N)}{\lambda^N}\right)+R_N(\lambda),$$

where $R_N(\lambda)=O\big(e^{-\lambda}\lambda^{a-N-2}\big)$.

(c) Interpret this expansion in the framework of the general principle for Laplace-type integrals with a boundary minimum at $t=\lambda$.

Exercise 5.20 *Consider the Laplace-type integral*

$$I(\lambda)=\int_0^\infty e^{-\lambda t}t^\alpha\,dt, \qquad \alpha>-1,\ \lambda\to\infty.$$

(a) Perform one integration by parts and identify the leading-order term of the asymptotic expansion.

(b) By repeating the procedure, derive the first three terms of the asymptotic expansion of $I(\lambda)$.

(c) Compare your result with the exact evaluation of the integral.

Exercise 5.21 *Let*

$$I(\lambda)=\int_0^\infty e^{-\lambda t}\frac{t}{1+t}\,dt, \qquad \lambda\to\infty.$$

(a) Apply integration by parts twice to obtain an asymptotic expansion up to terms of order $O(\lambda^{-3})$.

(b) Show that the resulting expansion is asymptotic but not convergent.

(c) Estimate the remainder term after N integrations by parts.

Exercise 5.22 *Consider the integral*

$$I(\lambda)=\int_0^1 e^{\lambda t}f(t)\,dt, \qquad \lambda\to\infty,$$

where f is smooth on $[0,1]$.

(a) *Perform two integrations by parts and express the result in terms of* $f(0)$ *and* $f'(0)$.

(b) *Show that the main contribution comes from the endpoint* $t = 0$.

(c) *Write down the general form of the asymptotic expansion in inverse powers of* λ.

Exercise 5.23 *Let*

$$I(\lambda) = \int_0^\infty e^{-\lambda t^2} f(t)\,dt, \qquad \lambda \to \infty,$$

where f *is smooth and* $f(0) \neq 0$.

(a) *Use the identity*

$$\frac{d}{dt}\left(e^{-\lambda t^2}\right) = -2\lambda t e^{-\lambda t^2}$$

to perform an integration by parts.

(b) *Derive the leading-order term of the asymptotic expansion.*

(c) *Discuss how repeated integration by parts generates higher-order terms.*

Exercise 5.24 *Consider the oscillatory integral*

$$I(\lambda) = \int_0^1 e^{i\lambda t} f(t)\,dt, \qquad \lambda \to \infty,$$

with $f \in C^1[0,1]$.

(a) *Perform an integration by parts to show that* $I(\lambda) = O(\lambda^{-1})$.

(b) *Repeat the argument to obtain an asymptotic expansion in inverse powers of* λ.

(c) *Identify the contributions coming from the endpoints* $t = 0$ *and* $t = 1$.

5.4 Laplace-Type Integrals with Power Prefactors and Endpoint or Interior Maxima

In this section, we study a class of Laplace-type integrals whose integrands consist of an exponential factor containing a large parameter together with a power-law prefactor. Specifically, we consider integrals of the form

$$I(\lambda) = \int_a^b a_0 (t-a)^\beta e^{\lambda g(t)}\,dt, \qquad \lambda \to \infty,$$

where a_0 is a constant, $\beta > -1$, and g is a smooth function that attains its absolute maximum either at an endpoint of the interval or at an interior point.

By performing a Taylor expansion of g about its point of maximum and introducing an appropriate rescaling of the integration variable, we show that the leading contribution to $I(\lambda)$ is governed by a Gaussian-type integral. This leads to an asymptotic approximation expressed in terms of the Gamma function. We formally establish this result in the following theorems.

Recall that $C^n(I)$ denotes the space of all functions whose first n derivatives exist and are continuous on the interval I. Thus, the assumption $g \in C^3(I)$ means that g, g', g'', and $g^{(3)}$ are all continuous on I.

Theorem 5.3 *We consider*

$$I(\lambda) = \int_a^b a_0(t-a)^\beta e^{\lambda g(t)}\, dt, \qquad b > a, \tag{5.5}$$

where a_0 is a constant, $\beta > -1$, and

$$g'(a) = 0, \qquad g''(a) < 0,$$

with g having a strict (absolute) maximum at $t = a$. Assume that $g \in C^3$ in a neighborhood of a. Then the asymptotic expansion of (5.5)

is given by

$$I(\lambda) \sim \frac{1}{2}\left[-\frac{2}{\lambda g''(a)}\right]^{(\beta+1)/2} a_0 e^{\lambda g(a)} \Gamma\left(\frac{\beta+1}{2}\right), \qquad \lambda \to \infty.$$

Proof *We begin by introducing the change of variables*

$$x = t - a, \qquad t = a + x, \qquad dt = dx,$$

so that

$$I(\lambda) = a_0 e^{\lambda g(a)} \int_0^{b-a} x^\beta \exp(\lambda(g(a+x) - g(a)))\, dx.$$

Since g has a maximum at $t = a$ with $g'(a) = 0$, its Taylor expansion is

$$g(a+x) = g(a) + \frac{g''(a)}{2}x^2 + \frac{g^{(3)}(a)}{3!}x^3 + \frac{g^{(4)}(a)}{4!}x^4 + \cdots.$$

Hence

$$g(a+x) - g(a) = \frac{g''(a)}{2}x^2 + O(x^3) \qquad \text{as } x \to 0.$$

Because $g''(a) < 0$, there exists $\delta > 0$ and a constant $c > 0$ such that

$$g(a+x) - g(a) \le -cx^2, \qquad 0 \le x < \delta,$$

and $g(a+x) - g(a) < 0$ for all $x \in (\delta, b-a]$.

For large λ, the main contribution to the integral comes from $0 \leq x \leq \delta$. Split

$$I(\lambda) = a_0 e^{\lambda g(a)} \left(\int_0^{\delta} x^{\beta} e^{\lambda(g(a+x)-g(a))}\, dx + \int_{\delta}^{b-a} x^{\beta} e^{\lambda(g(a+x)-g(a))}\, dx \right).$$

For $x \in [\delta, b-a]$ we have $g(a+x) - g(a) \leq -\eta$ for some $\eta > 0$, hence

$$\left| \int_{\delta}^{b-a} x^{\beta} e^{\lambda(g(a+x)-g(a))}\, dx \right| \leq \frac{(b-a)^{\beta+1}}{\beta+1} e^{-\lambda\eta} = O(e^{-\lambda\eta}),$$

which is exponentially small compared with the contribution from $x \in [0, \delta]$. Thus

$$I(\lambda) \sim a_0 e^{\lambda g(a)} \int_0^{\delta} x^{\beta} e^{\lambda(g(a+x)-g(a))}\, dx, \qquad \lambda \to \infty.$$

Next, we retain the quadratic term in the Taylor expansion of g. Thus, by Using the Taylor expansion, we write

$$g(a+x) - g(a) = \frac{g''(a)}{2} x^2 + R(x),$$

where $R(x) = O(x^3)$ as $x \to 0$. Then

$$\lambda(g(a+x) - g(a)) = \lambda \frac{g''(a)}{2} x^2 + \lambda R(x).$$

For x of order $\lambda^{-1/2}$, we have $\lambda R(x) = O(\lambda x^3) = O(\lambda^{-1/2}) \to 0$, so $R(x)$ contributes only higher-order corrections. To obtain the leading term, we therefore approximate

$$e^{\lambda(g(a+x)-g(a))} \sim \exp\left(\lambda \frac{g''(a)}{2} x^2 \right).$$

We may also extend the upper limit of the integral to $+\infty$, since for $x \geq \delta$ the integrand is exponentially small:

$$\int_0^{\delta} \cdots dx \sim \int_0^{\infty} \cdots dx \quad \text{to leading order.}$$

Hence

$$I(\lambda) \sim a_0 e^{\lambda g(a)} \int_0^{\infty} x^{\beta} \exp\left(\lambda \frac{g''(a)}{2} x^2 \right) dx.$$

We now transform the integral into Gamma-function form. Since $g''(a) < 0$, write

$$\frac{g''(a)}{2} = -\frac{|g''(a)|}{2}.$$

Define the scaling

$$y=\sqrt{-\frac{\lambda g''(a)}{2}}\,x, \qquad x=\sqrt{-\frac{2}{\lambda g''(a)}}\,y, \qquad dx=\sqrt{-\frac{2}{\lambda g''(a)}}\,dy.$$

Then

$$\lambda\frac{g''(a)}{2}x^2=\lambda\frac{g''(a)}{2}\left(-\frac{2}{\lambda g''(a)}y^2\right)=-y^2,$$

and

$$x^\beta=\left(-\frac{2}{\lambda g''(a)}\right)^{\beta/2}y^\beta.$$

Therefore

$$\int_0^\infty x^\beta \exp\left(\lambda\frac{g''(a)}{2}x^2\right)dx=\left(-\frac{2}{\lambda g''(a)}\right)^{(\beta+1)/2}\int_0^\infty y^\beta e^{-y^2}\,dy.$$

The remaining integral is a standard Gamma-function integral. Using the substitution

$$u=y^2, \quad du=2y\,dy,$$

we obtain

$$\int_0^\infty y^\beta e^{-y^2}\,dy=\frac{1}{2}\int_0^\infty u^{(\beta-1)/2}e^{-u}\,du=\frac{1}{2}\Gamma\left(\frac{\beta+1}{2}\right).$$

Hence

$$\int_0^\infty x^\beta \exp\left(\lambda\frac{g''(a)}{2}x^2\right)dx=\frac{1}{2}\left(-\frac{2}{\lambda g''(a)}\right)^{(\beta+1)/2}\Gamma\left(\frac{\beta+1}{2}\right).$$

Combining this with the prefactor $a_0e^{\lambda g(a)}$ *gives*

$$I(\lambda)\sim a_0e^{\lambda g(a)}\frac{1}{2}\left(-\frac{2}{\lambda g''(a)}\right)^{(\beta+1)/2}\Gamma\left(\frac{\beta+1}{2}\right), \qquad \lambda\to\infty.$$

Therefore,

$$I(\lambda)\sim\frac{1}{2}\left[-\frac{2}{\lambda g''(a)}\right]^{(\beta+1)/2}a_0e^{\lambda g(a)}\Gamma\left(\frac{\beta+1}{2}\right), \qquad \lambda\to\infty,$$

which completes the proof.

The next theorem considers the case where the quadratic term vanishes, so that the first nonzero term in the Taylor expansion occurs at cubic order. Its proof follows the same ideas as the previous theorem and is left as an exercise.

Theorem 5.4 *Consider*

$$I(\lambda) = \int_a^b a_0(t-a)^\beta e^{\lambda g(t)}\,dt, \qquad b > a, \tag{5.6}$$

where a_0 is a constant, $\beta > -1$, and

$$g'(a) = 0, \qquad g''(a) = 0, \qquad g'''(a) < 0,$$

with g having a strict absolute maximum at the endpoint $t = a$. Assume that $g \in C^4[a,b]$. Then

$$I(\lambda) \sim \frac{1}{3}\left(-\frac{6}{\lambda g'''(a)}\right)^{(\beta+1)/3} a_0 e^{\lambda g(a)} \Gamma\left(\frac{\beta+1}{3}\right), \qquad \lambda \to \infty.$$

Remark 16 *The condition $g'''(a) < 0$ implies that the leading decay away from the maximum is cubic rather than quadratic. Consequently, the peak at $t = a$ is flatter than in the Gaussian case $g''(a) < 0$, and the dominant contribution arises from the scaling*

$$t - a = O(\lambda^{-1/3}),$$

instead of the usual Gaussian scaling

$$t - a = O(\lambda^{-1/2}).$$

Unlike the endpoint case, in the next theorem, the dominant contribution now comes from a symmetric neighborhood around the interior maximum $t = c$, producing a full Gaussian integral over $(-\infty, \infty)$.

Theorem 5.5 *We consider*

$$I(\lambda) = \int_a^b a_0|t-c|^\beta e^{\lambda g(t)}\,dt, \qquad b > a, \tag{5.7}$$

where a_0 is a constant, $\beta > -1$, and

$$g'(c) = 0, \qquad g''(c) < 0, \quad a < c < b,$$

with g having a strict (absolute) maximum at $t = c$. Assume that $g \in C^3[a,b]$. Then the asymptotic expansion of (5.7)

is given by

$$I(\lambda) \sim a_0\left[-\frac{2}{\lambda g''(c)}\right]^{(\beta+1)/2} e^{\lambda g(c)} \Gamma\left(\frac{\beta+1}{2}\right), \quad \text{as} \quad \lambda \to \infty.$$

Proof *We study the Laplace-type integral*

$$I(\lambda) = \int_a^b a_0(t-c)^\beta e^{\lambda g(t)}\,dt, \qquad \lambda \to \infty,$$

where $a < b$, $a < c < b$, $\beta > -1$, a_0 is a constant, and

$$g'(c) = 0, \qquad g''(c) < 0,$$

with g attaining its absolute maximum at $t = c$ on $[a,b]$. We assume that g admits a Taylor expansion about $t = c$. Now we examine the behavior of g in the neighborhood $t = c$, at which the absolute maximum of g occurs. Since g has a strict maximum at c and $g''(c) < 0$, there exists $\delta > 0$ such that $c - \delta > a$ and $c + \delta < b$, and

$$g(t) < g(c) \quad \textit{for} \quad t \in [a,b] \setminus (c-\delta, c+\delta).$$

Hence, for t outside $(c-\delta, c+\delta)$, there is a constant $\eta > 0$ with

$$g(t) \le g(c) - \eta.$$

Therefore

$$\int_{[a,b]\setminus(c-\delta,c+\delta)} |a_0(t-c)^{\beta} e^{\lambda g(t)}|\, dt \le C e^{\lambda(g(c)-\eta)} = e^{\lambda g(c)} \cdot C e^{-\lambda\eta},$$

for some $C > 0$ independent of λ. Thus this contribution is $O\left(e^{\lambda g(c)} e^{-\lambda\eta}\right)$ and is exponentially small relative to the main term. It follows that

$$I(\lambda) \sim \int_{c-\delta}^{c+\delta} a_0(t-c)^{\beta} e^{\lambda g(t)}\, dt, \qquad \lambda \to \infty.$$

Since g is smooth and $g'(c) = 0$, $g''(c) < 0$, its Taylor expansion about c is

$$g(t) = g(c) + \frac{g''(c)}{2}(t-c)^2 + R_3(t),$$

where the remainder satisfies

$$R_3(t) = O\left((t-c)^3\right) \quad \textit{as } t \to c.$$

Hence, for $t \in (c-\delta, c+\delta)$,

$$e^{\lambda g(t)} = e^{\lambda g(c)} \exp\left(\lambda \frac{g''(c)}{2}(t-c)^2 + \lambda R_3(t)\right).$$

Because $g''(c) < 0$, the quadratic term is negative, and the factor

$$\exp\left(\lambda \frac{g''(c)}{2}(t-c)^2\right)$$

is sharply peaked at $t = c$.

Next, we introduce a rescaling that transforms the leading-order behavior into a Gaussian integral:

$$x = \sqrt{-\lambda g''(c)}\,(t-c), \quad \textit{so that} \quad t - c = \frac{x}{\sqrt{-\lambda g''(c)}}, \quad dt = \frac{dx}{\sqrt{-\lambda g''(c)}}.$$

Then, for λ large,

$$\begin{aligned}
I(\lambda) &\sim a_0 e^{\lambda g(c)} \int_{c-\delta}^{c+\delta} |t-c|^\beta \exp\left(\lambda \frac{g''(c)}{2}(t-c)^2 + \lambda R_3(t)\right) dt \\
&= a_0 e^{\lambda g(c)} \int_{-\delta\sqrt{-\lambda g''(c)}}^{\delta\sqrt{-\lambda g''(c)}} \left|\frac{x}{\sqrt{-\lambda g''(c)}}\right|^\beta \exp\left[\lambda \frac{g''(c)}{2} \frac{x^2}{-\lambda g''(c)}\right. \\
&\qquad \left. + \lambda R_3\left(c + \frac{x}{\sqrt{-\lambda g''(c)}}\right)\right] \frac{dx}{\sqrt{-\lambda g''(c)}}.
\end{aligned}$$

Using

$$\lambda \frac{g''(c)}{2} \frac{x^2}{-\lambda g''(c)} = -\frac{x^2}{2},$$

we obtain

$$I(\lambda) \sim a_0 e^{\lambda g(c)} \left[-\lambda g''(c)\right]^{-(\beta+1)/2} \int_{-\delta\sqrt{-\lambda g''(c)}}^{\delta\sqrt{-\lambda g''(c)}} |x|^\beta \exp\left(-\frac{x^2}{2} + \lambda R_3\left(c + \frac{x}{\sqrt{-\lambda g''(c)}}\right)\right) dx.$$

We now estimate the remainder term. Since $R_3(t) = O((t-c)^3)$ as $t \to c$, there exists $K > 0$ such that

$$|R_3(t)| \leq K|t-c|^3, \qquad t \in (c-\delta, c+\delta).$$

In terms of x,

$$\lambda R_3\left(c + \frac{x}{\sqrt{-\lambda g''(c)}}\right) = O\left(\lambda \left|\frac{x}{\sqrt{\lambda}}\right|^3\right) = O\left(\frac{|x|^3}{\sqrt{\lambda}}\right), \qquad \lambda \to \infty.$$

Therefore, for fixed x, $\lambda R_3 \to 0$ as $\lambda \to \infty$, and

$$\exp\left(\lambda R_3\left(c + \frac{x}{\sqrt{-\lambda g''(c)}}\right)\right) = 1 + O\left(\frac{|x|^3}{\sqrt{\lambda}}\right).$$

Moreover, the Gaussian factor $e^{-x^2/2}$ ensures integrable decay, so by dominated convergence we may replace the exponential containing R_3 by 1 in the leading term. Also, as $\lambda \to \infty$ the integration limits $\pm\delta\sqrt{-\lambda g''(c)}$ tend to $\pm\infty$, and again the Gaussian decay justifies extending the limits to the whole real line. Thus

$$I(\lambda) \sim a_0 e^{\lambda g(c)} \left[-\lambda g''(c)\right]^{-(\beta+1)/2} \int_{-\infty}^{\infty} |x|^\beta e^{-x^2/2}\, dx.$$

Then the integrand $x^\beta e^{-x^2/2}$ is even, and

$$\int_{-\infty}^{\infty} |x|^\beta e^{-x^2/2}\, dx = 2\int_0^\infty x^\beta e^{-x^2/2}\, dx.$$

Set $u = x^2/2$*, so that* $x = \sqrt{2u}$ *and* $dx = \dfrac{du}{\sqrt{2u}}$*. Then*

$$\int_0^\infty |x|^\beta e^{-x^2/2}\,dx = \int_0^\infty (2u)^{\beta/2} e^{-u} \frac{du}{\sqrt{2u}} = 2^{\beta/2}\int_0^\infty u^{\beta/2} e^{-u} (2^{-1/2} u^{-1/2})\,du$$

$$= 2^{(\beta-1)/2}\int_0^\infty u^{(\beta-1)/2} e^{-u}\,du = 2^{(\beta-1)/2}\Gamma\left(\frac{\beta+1}{2}\right),$$

where we used the definition of the Gamma function $\Gamma(z) = \int_0^\infty u^{z-1}e^{-u}\,du$*. Therefore,*

$$\int_{-\infty}^\infty |x|^\beta e^{-x^2/2}\,dx = 2^{(\beta+1)/2}\Gamma\left(\frac{\beta+1}{2}\right).$$

Substituting this into our expression for $I(\lambda)$ *gives*

$$I(\lambda) \sim a_0 e^{\lambda g(c)} \left[-\lambda g''(c)\right]^{-(\beta+1)/2} 2^{(\beta+1)/2}\Gamma\left(\frac{\beta+1}{2}\right)$$

$$= a_0\left[-\frac{2}{\lambda g''(c)}\right]^{(\beta+1)/2} e^{\lambda g(c)}\Gamma\left(\frac{\beta+1}{2}\right), \qquad \lambda \to \infty.$$

This is precisely the desired asymptotic behavior.

5.4.1 Exercises

Exercise 5.25 *Prove Theorem 5.5*

Exercise 5.26 *Obtain the leading-order asymptotic behavior of* $I(\lambda)$ *as* $\lambda \to \infty$ *for*

$$I(\lambda) = \int_0^1 t^\beta e^{\lambda(1-t^2)}\,dt, \qquad \beta > -1.$$

Exercise 5.27 *Show the leading-order asymptotic expansion of* $I(\lambda)$ *as* $\lambda \to \infty$ *for*

$$I(\lambda) = \int_a^b e^{-\frac{1}{t-a}} e^{-\lambda(t-a)^2}\,dt, \qquad \beta > -1,$$

is given by

$$I(\lambda) \sim \left(\frac{\pi}{3t}\right)^{1/2} e^{-\frac{3}{2}(2t)^{1/3}}.$$

Exercise 5.28 *Consider*

$$I(\lambda) = \int_{-1}^1 |t|^\beta e^{-\lambda t^2}\,dt, \qquad \beta > -1.$$

(a) Show that the dominant contribution comes from $t = 0$*.*

(b) Rewrite the integral in the form required by the theorem for an interior maximum.

(c) Derive the leading asymptotic term as $\lambda \to \infty$*.*

Exercise 5.29 *Let*

$$I(\lambda) = \int_0^1 t^{\beta} e^{\lambda(1-t^3)}\, dt, \qquad \beta > -1.$$

(a) Verify that $g(t) = 1 - t^3$ *satisfies* $g'(0) = g''(0) = 0$ *and* $g'''(0) < 0$.

(b) Explain why the maximum at $t = 0$ *is degenerate.*

(c) Obtain the leading asymptotic behavior of $I(\lambda)$.

Exercise 5.30 *Consider*

$$I(\lambda) = \int_0^b t^{\beta} e^{\lambda(-t^3+t^4)}\, dt, \qquad 0 < b < 1.$$

(a) Show that $t = 0$ *is a strict maximum of* $g(t) = -t^3 + t^4$.

(b) Determine the first nonvanishing derivative of g *at* $t = 0$.

(c) Identify which theorem applies and compute the leading asymptotic term.

Exercise 5.31 *Let*

$$I(\lambda) = \int_{-1}^2 |t-1|^{\beta} e^{\lambda(1-(t-1)^2)}\, dt, \qquad \lambda \to \infty,$$

with $\beta > -1$.

(a) Show that $g(t) = 1 - (t-1)^2$ *has a strict maximum at* $c = 1$ *and that* $g''(1) = -2$.

(b) Identify a_0, β, c, *and* g *so that* $I(\lambda)$ *is of the form in Theorem 5.5.*

(c) Use Theorem 5.5 to obtain the leading asymptotic approximation of $I(\lambda)$ *as* $\lambda \to \infty$.

Exercise 5.32 *Consider*

$$I(\lambda) = \int_0^{\pi} a_0 \left|t - \frac{\pi}{2}\right|^{\beta} \exp\left(\lambda \cos\left(t - \frac{\pi}{2}\right)\right) dt, \qquad \lambda \to \infty,$$

with constant a_0 *and* $\beta > -1$.

(a) Verify that $g(t) = \cos\left(t - \frac{\pi}{2}\right)$ *has a strict maximum at* $c = \frac{\pi}{2}$, *and determine* $g''(c)$.

(b) Check that the assumptions of Theorem 5.5 are satisfied on $(a,b) = (0,\pi)$.

(c) Apply Theorem 5.5 to obtain an explicit leading-order formula for $I(\lambda)$.

Exercise 5.33 *Let*

$$I(\lambda) = \int_{-2}^3 |t-1|^{\beta} e^{-\lambda(t-1)^2}\, dt, \qquad \lambda \to \infty,$$

where $\beta > -1$.

(a) Rewrite $I(\lambda)$ in the form (5.7) *by specifying a_0, $g(t)$, and c.*

(b) Show that $g'(c) = 0$, $g''(c) < 0$, and that g has a strict maximum at c on $[-2,3]$.

(c) Using Theorem 5.5, derive the leading asymptotic approximation of $I(\lambda)$ and simplify it as much as possible.

Exercise 5.34 *Apply Theorem 5.5 to find the leading asymptotic behavior as $\lambda \to \infty$ of*

$$I_1(\lambda) = \int_0^2 |t-1|^3 e^{\lambda(1-(t-1)^2)}\,dt.$$

Verify that the conditions of the theorem are satisfied and explicitly compute the leading term.

Exercise 5.35 *Consider the integral*

$$I_2(\lambda) = \int_{-1}^1 |t|^\beta e^{\lambda\cos(\pi t)}\,dt, \qquad \beta > -1.$$

(a) Determine the point c where $g(t) = \cos(\pi t)$ attains its maximum on $[-1,1]$.

(b) Check that $g''(c) < 0$ and that the conditions of Theorem 5.5 are met.

(c) Use Theorem 5.5 to find the leading asymptotic behavior of $I_2(\lambda)$ as $\lambda \to \infty$.

Exercise 5.36 *Using the rescaling*

$$t = c + \frac{s}{\sqrt{\lambda}},$$

derive the leading asymptotic term of

$$I(\lambda) = \int_a^b |t-c|^\beta e^{\lambda g(t)}\,dt,$$

under the assumptions of Theorem 5.5. Show that the resulting leading-order approximation agrees with the formula stated in the theorem.

Exercise 5.37 *Find the asymptotic expansion as $\lambda \to \infty$ of the integral*

$$I_3(\lambda) = \int_0^\pi \sin^\beta t\, e^{\lambda\cos t}\,dt, \quad \beta > -1.$$

Hint: Use the endpoint maximum at $t = 0$, where $\sin t \sim t$ and $\cos t = 1 - \frac{t^2}{2} + O(t^4)$.

Exercise 5.38 *(a) Find the leading behavior as $\lambda \to +\infty$ of Laplace-type integrals of the form*

$$I(\lambda) = \int_a^b (t-a)^\beta g(t)\, e^{\lambda\phi(t)}\,dt,$$

where $\phi(t)$ has a strict maximum at $t = a$ and $g(a) = 1$. Assume $\beta > -1$ and $\phi'(a) < 0$. Derive an explicit leading-order asymptotic formula for $I(\lambda)$ as $\lambda \to +\infty$.

(b) Repeat the analysis of part (a) *under the assumptions* $\beta > -1$ *and*

$$\phi'(a) = \phi''(a) = \cdots = \phi^{(p-1)}(a) = 0, \qquad \phi^{(p)}(a) < 0,$$

for some integer $p \geq 2$*. Determine how the order of vanishing of* ϕ *at* $t = a$ *modifies the leading asymptotic behavior of* $I(\lambda)$ *as* $\lambda \to +\infty$*.*

Exercise 5.39 *Use a suitable method to find the leading behavior as* $\lambda \to +\infty$ *of each of the following integrals.*

(a)

$$I_a(\lambda) = \int_0^{\pi/2} \sqrt{\sin t}\, e^{-\lambda \sin^4 t}\, dt.$$

(b) For a fixed constant $a > 0$*,*

$$I_b(\lambda) = \int_0^1 \sqrt{t(1-t)}\, (t+a)^{-\lambda}\, dt.$$

(c)

$$I_c(\lambda) = \int_0^{\pi/4} \sqrt{\tan t}\, e^{-\lambda t^2}\, dt.$$

(d)

$$I_d(\lambda) = \int_0^{\pi} \sqrt{t^2+1}\, e^{-\lambda t \sin t}\, dt.$$

(e)

$$I_e(\lambda) = \int_0^{\pi} \sin^2(t)\, e^{-\lambda (e^t - 2t)}\, dt.$$

(f)

$$I_f(\lambda) = \int_{-1}^{1} (1+t)^3\, e^{-\lambda t (\log|t| - 1)}\, dt.$$

(g)

$$I_g(\lambda) = \int_0^{2\pi} \cos(2t)\, e^{-\lambda (\sin t + \cos t)}\, dt.$$

(h)

$$I_h(\lambda) = \int_{-1}^{1} \cosh(t)\, e^{\lambda \cos(\sqrt{t^2+1})}\, dt.$$

5.5 WKB Approximation for Singularly Perturbed Second-Order Equations

In this section, we discuss the Wentzel–Kramers–Brillouin (WKB) method as a perturbation technique for second-order linear differential equations containing a small parameter. We focus on three prototypical families:

$$\varepsilon^2 y''(t) + q(t)y(t) = 0, \qquad 0 < \varepsilon \ll 1, \tag{5.8}$$

$$y''(t) + \left(\lambda^2 p(t) - q(t)\right)y(t) = 0, \qquad \lambda \gg 1, \tag{5.9}$$

$$y''(t) + q(\varepsilon t)^2 y(t) = 0, \qquad 0 < \varepsilon \ll 1, \tag{5.10}$$

subject to appropriate boundary conditions on a finite interval $a \le t \le b$, or on the interval $t \in [a, \infty)$. In all cases, the coefficient in front of y'' is either explicitly small or effectively small for large λ, so that the solutions typically exhibit rapid oscillations or exponential growth/decay on the t–scale. The three equations (5.8)–(5.10) represent equivalent ways of introducing a *large parameter* (or, equivalently, a *small parameter*) into a second–order linear differential equation so that WKB-type asymptotic behavior arises.

If we start from (5.8) and divide by ε^2, we obtain

$$y''(t) + \frac{q(t)}{\varepsilon^2}\, y(t) = 0.$$

Introducing the large parameter $\lambda = \varepsilon^{-1} \gg 1$, this becomes

$$y''(t) + \lambda^2 q(t)\, y(t) = 0,$$

which is a special case of (5.9) with $p(t) = q(t)$ and $q(t) \equiv 0$. Thus, (5.9) can be viewed as a refinement of (5.8) in which a lower-order term $-q(t)y$ is retained. Conversely, given (5.9)

$$y''(t) + \left(\lambda^2 p(t) - q(t)\right)y(t) = 0,$$

set $\varepsilon = \lambda^{-1}$ and multiply the equation by ε^2:

$$\varepsilon^2 y''(t) + \left(p(t) - \varepsilon^2 q(t)\right)y(t) = 0.$$

For $\varepsilon \ll 1$, the term $\varepsilon^2 q(t)$ is a higher-order correction, so at leading order, we recover an equation of the form (5.8). Hence (5.8) represents the leading-order structure of (5.9). In the third equation

$$y''(t) + q(\varepsilon t)^2 y(t) = 0,$$

the small parameter appears through a *slowly varying coefficient*. Introducing the stretched variable

$$\tau = \varepsilon t,$$

and writing $y(t) = Y(\tau)$, we have

$$\frac{d}{dt} = \varepsilon \frac{d}{d\tau}, \qquad \frac{d^2}{dt^2} = \varepsilon^2 \frac{d^2}{d\tau^2}.$$

Substituting into the equation gives

$$\varepsilon^2 Y''(\tau) + q(\tau)^2 Y(\tau) = 0,$$

which is precisely of the form (5.8) (with $q(\tau)^2$ in place of $q(\tau)$). Thus, (5.10) is equivalent to (5.8) under a change of variables.

Conclusion. All three formulations encode the same asymptotic structure: a second-order equation in which either

- the highest derivative is multiplied by a small parameter (5.8),
- or the potential term is scaled by a large parameter (5.9),
- or the coefficients vary on a slow scale (5.10).

Through simple rescalings of parameters and variables, each equation can be reduced to the canonical form (5.8). Therefore, it suffices to analyze (5.8), since the others follow from it by transformation.

The above equations are sub-equations of the Sturm-Liouville problem

$$\left(r(t)y'\right)' + q(t)y + \lambda p(t)y = 0, \tag{5.11}$$

For constants α_1, α_2, β_1, β_2, we impose the boundary conditions

$$\alpha_1 y(a) + \beta_1 y'(a) = 0, \quad \alpha_2 y(b) + \beta_2 y'(b) = 0, \tag{5.12}$$

with

$$\alpha_1^2 + \beta_1^2 \neq 0, \quad \alpha_2^2 + \beta_2^2 \neq 0.$$

Definition 5.2 *The Sturm-Liouville problem given by* (5.11) *and* (5.12) *is said to be* **regular** *when*
(i) $p(t), q(t), r(t)$, *and* r' *are continuous on the closed interval* $a \leq t \leq b$;
(ii) $p(t) > 0$ *and* $r(t) > 0$ *when* $a \leq t \leq b$.

A value of λ for which problem (5.11) and (5.12) has a **nontrivial solution** is called an **eigenvalue**, and the **nontrivial solution** is called an **eigenfunction**.

Corollary 2 *If* λ *is an eigenvalue of the regular Sturm-Liouville problem* (5.11) *and* (5.12), *then* $\lambda \geq 0$ *provided that*

$$q(t) \leq 0 \quad (a \leq t \leq b) \quad \text{and} \quad \alpha_1\beta_1 \leq 0, \quad \alpha_2\beta_2 \geq 0.$$

Theorem 5.6 (Formal WKB approximation for $\varepsilon^2 y'' + q(t)y = 0$) *Let $q \in C^\infty(a,b)$ with $q(t) \neq 0$ for all $t \in (a,b)$, and consider the singularly perturbed differential equation*

$$\varepsilon^2 y''(t) + q(t)y(t) = 0, \qquad 0 < \varepsilon \ll 1. \tag{5.13}$$

Assume a WKB-type ansatz (educated guess)

$$y(t) = \exp\left(\frac{1}{\varepsilon}S(t)\right), \qquad S(t) = S_0(t) + \varepsilon S_1(t) + \varepsilon^2 S_2(t) + \cdots.$$

Then, away from the zeros of $q(t)$, the leading-order behavior of solutions of (5.13) *is given by*

$$y_\pm(t) \sim \frac{1}{(-q(t))^{1/4}} \exp\left(\pm\frac{1}{\varepsilon}\int^t \sqrt{-q(s)}\,ds\right), \qquad \varepsilon \to 0, \tag{5.14}$$

where the choice of branch of $\sqrt{-q}$ is fixed on (a,b).

Proof *Substitute the WKB*

$$y(t) = \exp\left(\frac{1}{\varepsilon}S(t)\right)$$

into (5.13)*. Differentiating gives*

$$y'(t) = \frac{S'(t)}{\varepsilon}\exp\left(\frac{1}{\varepsilon}S(t)\right), \qquad y''(t) = \left(\frac{S''(t)}{\varepsilon} + \frac{(S'(t))^2}{\varepsilon^2}\right)\exp\left(\frac{1}{\varepsilon}S(t)\right).$$

Hence

$$\varepsilon^2 y''(t) = \left(\varepsilon S''(t) + \left(S'(t)\right)^2\right)\exp\left(\frac{1}{\varepsilon}S(t)\right).$$

Substituting into (5.13) *and dividing by the nonvanishing factor* $\exp(S/\varepsilon)$ *yields*

$$\varepsilon S''(t) + \left(S'(t)\right)^2 + q(t) = 0. \tag{5.15}$$

Next, expand S as a formal series

$$S(t) = S_0(t) + \varepsilon S_1(t) + \varepsilon^2 S_2(t) + \cdots$$

and compute

$$S'(t) = S_0'(t) + \varepsilon S_1'(t) + \varepsilon^2 S_2'(t) + \cdots, \qquad S''(t) = S_0''(t) + \varepsilon S_1''(t) + \varepsilon^2 S_2''(t) + \cdots.$$

Then

$$\left(S'(t)\right)^2 = \left(S_0'(t)\right)^2 + 2\varepsilon S_0'(t)S_1'(t) + \varepsilon^2\left(S_1'(t)^2 + 2S_0'(t)S_2'(t)\right) + \cdots.$$

Substituting these expansions into (5.15) *and collecting powers of ε gives*

$$\begin{aligned}
O(1): &\quad \left(S_0'(t)\right)^2 + q(t) = 0,\\
O(\varepsilon): &\quad S_0''(t) + 2S_0'(t)S_1'(t) = 0,\\
O(\varepsilon^2): &\quad S_1''(t) + 2S_0'(t)S_2'(t) + S_1'(t)^2 = 0, \quad \textit{etc.}
\end{aligned}$$

The leading-order equation

$$\left(S_0'(t)\right)^2 = -q(t)$$

is the eikonal equation. Choosing a branch of the square root, we obtain

$$S_0'(t) = \pm\sqrt{-q(t)}.$$

Integrating from a to t, we have that

$$S_0(t) = \pm\int_a^t \sqrt{-q(s)}\,ds + C_0.$$

where C_0 is a constant that can be absorbed into the overall normalization of y.

The $O(\varepsilon)$ equation

$$S_0''(t) + 2S_0'(t)S_1'(t) = 0$$

is the transport equation. Solving for S_1',

$$S_1'(t) = -\frac{S_0''(t)}{2S_0'(t)}. \tag{5.16}$$

Using $\left(S_0'(t)\right)^2 = -q(t)$, differentiate to obtain

$$2S_0'(t)S_0''(t) = -q'(t),$$

or,

$$S_0''(t) = -\frac{q'(t)}{2S_0'(t)}. \tag{5.17}$$

Therefore, after substituting (5.17) into (5.16) we arrive at

$$S_1'(t) = -\frac{-q'(t)/(2S_0'(t))}{2S_0'(t)} = \frac{q'(t)}{4\left(S_0'(t)\right)^2} = \frac{q'(t)}{4(-q(t))} = -\frac{1}{4}\frac{q'(t)}{q(t)}.$$

Integrating, from a to t, we obtain

$$S_1(t) = -\frac{1}{4}\int_a^t \frac{q'(s)}{q(s)}\,ds = -\frac{1}{4}\ln\left(q(t)\right) + C_1,$$

so that

$$e^{S_1(t)} = C\left(q(t)\right)^{-1/4},$$

for some constant C, again absorbable into the overall normalization.

Keeping only S_0 and S_1 in the exponent, the approximate solution takes the form

$$y(t) \approx \exp\left(\frac{1}{\varepsilon}S_0(t) + S_1(t)\right) = \frac{C}{(-q(t))^{1/4}}\exp\left(\pm\frac{1}{\varepsilon}\int^t \sqrt{-q(s)}\,ds\right),$$

where we have used $q(t) < 0$ in the nonoscillatory case so that $(-q(t))^{1/4}$ is real. Dropping the inessential constant C produces the pair of linearly independent WKB solutions

$$y_{\pm}(t) \sim \frac{1}{(-q(t))^{1/4}} \exp\left(\pm\frac{1}{\varepsilon}\int^{t}\sqrt{-q(s)}\,ds\right), \qquad \varepsilon \to 0,$$

valid away from the zeros of $q(t)$. This is the claimed approximation (5.14). *Thus, the general solution is given by*

$$y(t) = \frac{C_1}{(-q(t))^{1/4}} \exp\left(\frac{1}{\varepsilon}\int^{t}\sqrt{-q(s)}\,ds\right) + \frac{C_2}{(-q(t))^{1/4}} \exp\left(-\frac{1}{\varepsilon}\int^{t}\sqrt{-q(s)}\,ds\right)$$

for some constants C_1 and C_2.

As noted, The character of the solutions depends on the sign of $q(t)$.

Oscillatory Case

If $q(t) > 0$ on (a,b), then $-q(t) < 0$ and $\sqrt{-q(t)} = i\sqrt{q(t)}$ is purely imaginary. The leading-order WKB solutions become

$$y_{\pm}(t) \sim \frac{1}{q(t)^{1/4}} \exp\left(\pm\frac{i}{\varepsilon}\int^{t}\sqrt{q(s)}\,ds\right),$$

which are highly oscillatory with slowly varying amplitude. This case is typical for Sturm–Liouville eigenvalue problems and for the stationary Schrödinger equation in classically allowed regions of the potential. In this case, the general solution is given by

$$y(t) = \frac{C_1}{(-q(t))^{1/4}} \exp\left(\frac{i}{\varepsilon}\int_{a}^{t}\sqrt{q(s)}\,ds\right) + \frac{C_2}{(q(t))^{1/4}} \exp\left(-\frac{i}{\varepsilon}\int_{a}^{t}\sqrt{q(s)}\,ds\right) \tag{5.18}$$

for some constants C_1 and C_2. Next, we try to simplify (5.18) *and write it in terms of sine and cosine given by*

$$y(t) = \frac{C_1}{(q(t))^{1/4}} \sin\left(\frac{1}{\varepsilon}\int_{a}^{t}\sqrt{q(s)}\,ds\right) + \frac{C_2}{(q(t))^{1/4}} \cos\left(\frac{1}{\varepsilon}\int_{a}^{t}\sqrt{q(s)}\,ds\right). \tag{5.19}$$

Starting from (5.18)*, write it in a slightly more compact form. Since $q(t) > 0$ on (a,b), we may choose the real fourth root $(q(t))^{1/4} > 0$ and note that*

$$(-q(t))^{1/4} = e^{i\pi/4} q(t)^{1/4},$$

so that the complex phase $e^{i\pi/4}$ can be absorbed into the constant C_1. Renaming the constants, we rewrite (5.18) as

$$y(t) = \frac{A}{q(t)^{1/4}} \exp(i\theta(t)) + \frac{B}{q(t)^{1/4}} \exp(-i\theta(t)), \tag{5.20}$$

where

$$\theta(t) = \frac{1}{\varepsilon} \int_a^t \sqrt{q(s)}\, ds, \qquad A, B \in \mathbb{C}.$$

Next, use Euler's formulas

$$e^{i\theta} = \cos\theta + i\sin\theta, \qquad e^{-i\theta} = \cos\theta - i\sin\theta,$$

to express the exponentials in (5.20) in terms of sine and cosine:

$$\begin{aligned} y(t) &= \frac{1}{q(t)^{1/4}} \Big[A(\cos\theta(t) + i\sin\theta(t)) + B(\cos\theta(t) - i\sin\theta(t))\Big] \\ &= \frac{1}{q(t)^{1/4}} \Big[(A+B)\cos\theta(t) + i(A-B)\sin\theta(t)\Big]. \end{aligned}$$

Define new (complex) constants

$$C_2 = A + B, \qquad C_1 = i(A - B),$$

so that

$$y(t) = \frac{C_1}{q(t)^{1/4}} \sin\theta(t) + \frac{C_2}{q(t)^{1/4}} \cos\theta(t).$$

Substituting back the definition of $\theta(t)$ yields

$$y(t) = \frac{C_1}{(q(t))^{1/4}} \sin\left(\frac{1}{\varepsilon}\int_a^t \sqrt{q(s)}\, ds\right) + \frac{C_2}{(q(t))^{1/4}} \cos\left(\frac{1}{\varepsilon}\int_a^t \sqrt{q(s)}\, ds\right),$$

which is precisely the representation in (5.19). Thus, (5.19) is simply the real trigonometric form of the complex exponential combination in (5.18), obtained via Euler's formula and a redefinition of the constants.

Example 5.7 *(Oscillatory WKB solution: $q(t) > 0$) Consider*

$$\varepsilon^2 y''(t) + q(t)y(t) = 0, \qquad 0 < \varepsilon \ll 1,$$

on (a, b) with

$$q(t) = 1, \qquad a < t < b.$$

Here, $q(t) > 0$ for all t, so the solution is an oscillator y. We apply Theorem 5.13.

Since $q(t) = 1$, we have

$$-q(t) = -1, \qquad \sqrt{-q(t)} = \sqrt{-1} = i,$$

where we choose the branch $\sqrt{-1} = i$. *Hence*

$$(-q(t))^{1/4} = (-1)^{1/4} = e^{i\pi/4},$$

a constant factor that can be absorbed into the normalization. The integral in (5.14) *is*

$$\int^t \sqrt{-q(s)}\, ds = \int^t i\, ds = it + C,$$

where the constant C again only affects an overall complex factor.

Substituting into (5.14) *we obtain*

$$y_\pm(t) \sim \frac{1}{(-1)^{1/4}} \exp\left(\pm\frac{1}{\varepsilon} it\right) = A_\pm \exp\left(\pm i\frac{t}{\varepsilon}\right),$$

where $A_\pm$ *are complex constants. Taking real and imaginary parts gives the familiar oscillatory WKB approximation*

$$y(t) \approx C_1 \cos\left(\frac{t}{\varepsilon}\right) + C_2 \sin\left(\frac{t}{\varepsilon}\right),$$

which agrees with the exact solution of $y'' + (1/\varepsilon^2)y = 0$. *This illustrates the oscillatory case* $q(t) > 0$ *in Theorem 5.6.* □

Example 5.8 *(Nonoscillatory WKB solution:* $q(t) < 0$*) Consider again*

$$\varepsilon^2 y''(t) + q(t)y(t) = 0, \qquad 0 < \varepsilon \ll 1,$$

but now with

$$q(t) = -1, \qquad a < t < b.$$

Here, $q(t) < 0$ *for all t, so the solutions are exponential (nonoscillator y).*

We compute

$$-q(t) = 1, \qquad \sqrt{-q(t)} = \sqrt{1} = 1, \qquad (-q(t))^{1/4} = 1.$$

Thus, the integral in (5.14) *is*

$$\int^t \sqrt{-q(s)}\, ds = \int^t 1\, ds = t + C,$$

and we may absorb C into the normalization. The WKB formula (5.14) *yields*

$$y_\pm(t) \sim \frac{1}{1} \exp\left(\pm\frac{1}{\varepsilon} t\right) = C_+ e^{t/\varepsilon} + C_- e^{-t/\varepsilon},$$

a linear combination of rapidly growing and rapidly decaying exponentials. This matches the exact general solution of $y'' - (1/\varepsilon^2)y = 0$, *demonstrating the nonoscillatory regime* $q(t) < 0$ *in Theorem 5.6.* □

In the next example, we introduce the notion of eigenvalues and the corresponding eigenfunctions arising from specified boundary conditions.

Example 5.9 *Consider again*

$$\varepsilon^2 y''(t) + q(t)y(t) = 0, \qquad 0 < t < \pi, \quad y(0) = 0,\ y(\pi) = 0. \tag{5.21}$$

We assume $q(t) > 0$ *on* $[0,\pi]$. *Dividing equation* (5.21) *with* ε^2 *and setting*

$$\lambda = \frac{1}{\varepsilon^2}$$

we rewrite the equation in the form of Sturm-Liouville problem

$$y''(t) + \lambda\, q(t)y(t) = 0, \qquad 0 < t < \pi, \quad y(0) = 0,\ y(\pi) = 0. \tag{5.22}$$

In terms of λ, *using* (5.19) *the solution is given by*

$$y(t) = \frac{C_1}{\big(-q(t)\big)^{1/4}} \sin\Big(\sqrt{\lambda}\int_0^t \sqrt{q(s)}\,ds\Big) + \frac{C_2}{\big(-q(t)\big)^{1/4}} \cos\Big(\sqrt{\lambda}\int_0^t \sqrt{q(s)}\,ds\Big). \tag{5.23}$$

Applying $y(0) = 0$, *in* (5.23) *gives* $C_2 = 0$. *On the other hand, an application of* $y(\pi) = 0$, *yields*

$$\frac{C_1}{\big(-q(t)\big)^{1/4}} \sin\Big(\sqrt{\lambda}\int_0^\pi \sqrt{q(s)}\,ds\Big) = 0.$$

Now, if $C_1 = 0$, *then we have the trivial solution. Thus, we force*

$$\sin\Big(\sqrt{\lambda}\int_0^\pi \sqrt{q(s)}\,ds\Big) = 0.$$

This is satisfied when

$$\sqrt{\lambda}\int_0^\pi \sqrt{q(s)}\,ds = n\pi, \quad n = 1, 2, \cdots$$

Since $\lambda \gg 1$, *the integer* n *has to be chosen large. That is the eigenvalues denoted by* λ_n *are*

$$\lambda_n = \left(\frac{n\pi}{\int_0^\pi \sqrt{q(s)}\,ds}\right)^2,$$

for large n. *The corresponding eigenfunctions are*

$$y_n(t) = \frac{C_1}{\big(q(t)\big)^{1/4}} \sin\left(n\pi \frac{\int_0^t \sqrt{q(s)}\,ds}{\int_0^\pi \sqrt{q(s)}\,ds}\right), \tag{5.24}$$

for large n. □

Applications

WKB-type equations and their eigenvalue problems appear in many areas of applied mathematics and physics:

- Quantum mechanics: stationary Schrödinger equation, tunneling probabilities, and semiclassical energy spectra of bound states.
- Wave propagation: acoustic and electromagnetic waveguides, where slowly varying media lead to WKB approximations for mode shapes and phase velocities.
- Vibrations and elasticity: high-frequency modes of beams and membranes, treated via Sturm–Liouville theory and WKB eigenvalue estimates.
- Asymptotic analysis of special functions: large-order behaviors of Bessel, Hermite, and Airy functions can be derived using WKB methods applied to their defining differential equations.

In all of these applications, the WKB approximation provides a powerful tool to capture the leading oscillatory or evanescent structure of solutions and to derive accurate asymptotic formulas for eigenvalues and eigenfunctions in the limit of small ε or large λ.

We provide the following example as an application of the stationary Schrödinger equation

Example 5.10 *(WKB quantization for the one-dimensional harmonic oscillator) We illustrate the WKB method by deriving the semiclassical energy levels of the one-dimensional quantum harmonic oscillator. The stationary Schrödinger equation is*

$$-\frac{\hbar^2}{2m}\psi''(x)+\frac{1}{2}m\omega^2x^2\psi(x)=E\psi(x), \qquad x\in\mathbb{R}. \tag{5.25}$$

Here, m is the mass of the particle, ω is the angular frequency, $\hbar$ is the reduced Planck constant, and E is the energy eigenvalue. The potential is

$$V(x)=\frac{1}{2}m\omega^2x^2.$$

Rewriting (5.25) *in the standard WKB form gives*

$$\psi''(x)+\frac{2m}{\hbar^2}\big(E-V(x)\big)\psi(x)=0.$$

Define the classical momentum

$$p(x)=\sqrt{2m\big(E-V(x)\big)}=\sqrt{2mE-m^2\omega^2x^2}.$$

The turning points are determined by $E=V(x)$. Hence

$$E=\frac{1}{2}m\omega^2x^2,$$

so that

$$x_1 = -\sqrt{\frac{2E}{m\omega^2}}, \qquad x_2 = \sqrt{\frac{2E}{m\omega^2}}.$$

The interval

$$x_1 < x < x_2$$

is the classically allowed region, where $E > V(x)$ *and* $p(x)$ *is real. The regions*

$$x < x_1 \qquad \text{and} \qquad x > x_2$$

are the classically forbidden regions, where $E < V(x)$*, the quantity* $p(x)$ *is no longer real, and the WKB solutions become exponentially (growing or decaying).*

To connect solutions across the turning points, we use the standard WKB connection formulas. Near a simple turning point x_0 *satisfying* $E = V(x_0)$*, the local behavior of the WKB solutions changes from oscillatory to exponential.*

Define

$$\kappa(x) = \sqrt{2m(V(x) - E)}.$$

Then the standard connection formulas are

$$\text{Allowed region } (E > V): \quad \psi(x) \sim \frac{2A}{\sqrt{p(x)}} \cos\left(\frac{1}{\hbar}\int_{x_0}^{x} p(s)\,ds - \frac{\pi}{4}\right),$$

$$\text{Forbidden region } (E < V): \quad \psi(x) \sim \frac{A}{\sqrt{\kappa(x)}} \exp\left(-\frac{1}{\hbar}\int_{x_0}^{x} \kappa(s)\,ds\right).$$

Thus a decaying exponential solution in the forbidden region connects to an oscillatory cosine solution in the allowed region with a phase shift of $\pi/4$*.*

Remark 17 (WKB connection principle) *Each simple turning point contributes a phase shift of*

$$\frac{\pi}{4}.$$

Therefore, if a bound-state problem has two turning points, the total phase correction is

$$\frac{\pi}{4} + \frac{\pi}{4} = \frac{\pi}{2}.$$

This additional phase produces the Bohr–Sommerfeld quantization rule

$$\int_{x_1}^{x_2} p(x)\,dx = \left(n + \frac{1}{2}\right)\pi\hbar.$$

You should remember the following general principle:

"Two turning points implies add a phase shift of $\pi/2$."

Imposing that the solution decays as $x \to \pm\infty$ forces consistency between the phases of the oscillatory solutions between the two turning points. This leads to the Bohr–Sommerfeld quantization condition

$$\int_{x_1}^{x_2} p(x)\,dx = \left(n + \frac{1}{2}\right)\pi\hbar, \qquad n = 0, 1, 2, \ldots.$$

The factor $\frac{1}{2}$ arises because each turning point contributes a phase shift of $\frac{\pi}{4}$; with two turning points, the total phase shift is $\frac{\pi}{2}$.

We now evaluate the integral. Since

$$p(x) = \sqrt{2m\left(E - \frac{1}{2}m\omega^2 x^2\right)} = \sqrt{2mE - m^2\omega^2 x^2},$$

we obtain

$$\int_{x_1}^{x_2} p(x)\,dx = \int_{-\sqrt{2E/(m\omega^2)}}^{\sqrt{2E/(m\omega^2)}} \sqrt{2mE - m^2\omega^2 x^2}\,dx.$$

Set

$$a = \sqrt{\frac{2E}{m\omega^2}}.$$

Then

$$2mE = m^2\omega^2 a^2,$$

and hence

$$\sqrt{2mE - m^2\omega^2 x^2} = m\omega\sqrt{a^2 - x^2}.$$

Thus

$$\int_{x_1}^{x_2} p(x)\,dx = m\omega \int_{-a}^{a} \sqrt{a^2 - x^2}\,dx.$$

Using the standard integral

$$\int_{-a}^{a} \sqrt{a^2 - x^2}\,dx = \frac{\pi a^2}{2},$$

we obtain

$$\int_{x_1}^{x_2} p(x)\,dx = m\omega\frac{\pi a^2}{2}.$$

Since

$$a^2 = \frac{2E}{m\omega^2},$$

it follows that

$$\int_{x_1}^{x_2} p(x)\,dx = m\omega\frac{\pi}{2}\frac{2E}{m\omega^2} = \frac{\pi E}{\omega}.$$

Substituting into the quantization condition gives

$$\frac{\pi E}{\omega} = \left(n + \frac{1}{2}\right)\pi\hbar,$$

and therefore

$$E_n = \left(n + \frac{1}{2}\right)\hbar\omega, \qquad n = 0, 1, 2, \ldots.$$

Thus, for the one-dimensional harmonic oscillator, the WKB approximation recovers the exact energy levels.

Remark 18 *It is remarkable that the WKB approximation yields the exact energy spectrum for the harmonic oscillator. This is a special feature of quadratic potentials; for general potentials, the WKB method provides only an approximation.*

□

The preceding example is exceptional in that the WKB method reproduces the exact energy levels of the harmonic oscillator. This exactness is a special consequence of the quadratic form of the potential, for which the semiclassical approximation happens to capture the full quantum behavior.

In general, however, the WKB method is only an asymptotic approximation. Its accuracy improves in the high-energy regime (large quantum numbers), but it does not typically yield exact eigenvalues. Moreover, the form of the quantization condition can depend sensitively on the geometry of the problem, particularly on the number of turning points and the boundary conditions imposed on the wave function.

To illustrate these features, we now consider a nonquadratic potential for which the WKB method leads to an approximate spectrum. In contrast to the harmonic oscillator, this example involves a single turning point together with a boundary condition, resulting in a modified phase shift in the quantization rule.

Example 5.11 *(WKB quantization for a linear potential (approximate spectrum)) To illustrate the genuinely asymptotic nature of the WKB method, we consider a nonquadratic potential for which the WKB approximation is no longer exact.*

Consider the stationary Schrödinger equation

$$-\frac{\hbar^2}{2m}\psi''(x) + Fx\,\psi(x) = E\psi(x), \qquad x \geq 0, \tag{5.26}$$

where $F > 0$ is a constant force (e.g., a uniform gravitational or electric field). We impose the boundary condition

$$\psi(0) = 0,$$

so that the particle is confined to the half-line $x \geq 0$.

The potential is

$$V(x) = Fx,$$

which increases linearly. Rewriting (5.26) *in WKB form gives*

$$\psi''(x) + \frac{2m}{\hbar^2}(E - V(x))\,\psi(x) = 0.$$

Define the classical momentum

$$p(x) = \sqrt{2m(E - Fx)}.$$

The turning point is determined by $E = V(x)$, *giving*

$$x_1 = \frac{E}{F}.$$

The classically allowed region is

$$0 < x < x_1,$$

where $E > V(x)$ *and* $p(x)$ *is real. The region*

$$x > x_1$$

is forbidden, the quantity $p(x)$ *is no longer real in this region, and the WKB solutions transition from oscillatory behavior to exponentially growing or decaying behavior. Physically, this corresponds to a classically forbidden region where the particle cannot propagate as an oscillatory wave. At the boundary* $x = 0$, *we impose the condition* $\psi(0) = 0$, *which selects a sine-type oscillatory solution in the allowed region. Matching this oscillatory solution to a decaying solution beyond the turning point using the standard WKB connection formula leads to the quantization condition*

$$\int_0^{x_1} p(x)\,dx = \left(n + \frac{3}{4}\right)\pi\hbar, \qquad n = 0, 1, 2, \ldots.$$

The factor $\frac{3}{4}$ *arises from:*

- *a phase shift of* $\frac{\pi}{4}$ *at the turning point, and*
- *an additional* $\frac{\pi}{2}$ *shift imposed by the boundary condition* $\psi(0) = 0$.

We now evaluate the integral:

$$\int_0^{x_1} \sqrt{2m(E - Fx)}\,dx.$$

Let

$$u = E - Fx, \qquad du = -F\,dx.$$

Then

$$\int_0^{x_1} \sqrt{2m(E-Fx)}\,dx = \frac{1}{F}\int_0^E \sqrt{2mu}\,du.$$

Evaluating the integral gives

$$\int_0^E \sqrt{2mu}\,du = \sqrt{2m}\int_0^E u^{1/2}\,du = \sqrt{2m}\,\frac{2}{3}E^{3/2}.$$

Hence

$$\int_0^{x_1} p(x)\,dx = \frac{2\sqrt{2m}}{3F}E^{3/2}.$$

Substituting into the quantization condition yields

$$\frac{2\sqrt{2m}}{3F}E^{3/2} = \left(n+\frac{3}{4}\right)\pi\hbar.$$

Solving for E, we obtain the WKB approximation

$$E_n = \left[\frac{3\pi\hbar F}{2\sqrt{2m}}\left(n+\frac{3}{4}\right)\right]^{2/3}, \qquad n = 0,1,2,\ldots.$$

Remark 19 *The exact solutions of* (5.26) *are expressed in terms of Airy functions, and the exact eigenvalues are determined by the zeros of the Airy function. The WKB formula above provides an accurate asymptotic approximation for large n, but unlike the harmonic oscillator, it is not exact.*

□

5.5.1 Exercises

Exercise 5.40 *For large positive eigenvalues* λ*, use the WKB method to find asymptotic approximations to the eigenvalues and corresponding eigenfunctions of the boundary value problem*

$$y''(t) + \lambda(t+\pi)^4 y(t) = 0, \qquad 0 < t < \pi,$$
$$y(0) = 0, \qquad y(\pi) = 0.$$

Exercise 5.41 *Let* $0 < \varepsilon \ll 1$ *and consider*

$$\varepsilon^2 y''(t) + q(t)y(t) = 0, \qquad 0 < t < 1, \tag{5.27}$$
$$y(0) = 0, \qquad y(1) = 0, \tag{5.28}$$

where $q \in C^2[0,1]$ *satisfies* $q(t) > q_0 > 0$ *for all* $t \in [0,1]$.

(a) *Construct the leading-order WKB approximation to solutions of* (5.27).

(b) *Explain why nontrivial solutions satisfying both boundary conditions exist only for special values of* ε.

(c) *Derive the approximate quantization condition*

$$\frac{1}{\varepsilon}\int_0^1 \sqrt{q(t)}\,dt \sim n\pi, \qquad n = 1,2,3,\ldots.$$

Exercise 5.42 *Consider the Sturm–Liouville eigenvalue problem*

$$-y''(t) + q(t)y(t) = \lambda^2 y(t), \qquad a < t < b, \tag{5.29}$$

$$y(a) = 0, \qquad y(b) = 0, \tag{5.30}$$

where $q \in C^2[a,b]$ *and* $q(t) > 0$.

(a) *Rewrite* (5.29) *in the singularly perturbed form*

$$\varepsilon^2 y''(t) + Q(t)y(t) = 0$$

by setting $\varepsilon = 1/\lambda$.

(b) *Use the WKB method to derive an asymptotic formula for the large eigenvalues* λ_n *as* $n \to \infty$.

(c) *Obtain a corresponding leading-order approximation to the eigenfunctions* $y_n(t)$, *normalized so that*

$$\max_{a\le t\le b} |y_n(t)| = 1.$$

Exercise 5.43 *Let* $q(t) \equiv 1$ *on* $[0,\pi]$ *and consider the eigenvalue problem*

$$\varepsilon^2 y''(t) + (1+\lambda)y(t) = 0, \qquad t \in [0,\pi], \tag{5.31}$$

$$y(0) = 0, \qquad y(\pi) = 0,$$

with $0 < \varepsilon \ll 1$.

(a) *For fixed* ε, *determine the exact eigenvalues* $\lambda_n(\varepsilon)$ *and eigenfunctions* $y_n(t)$ *of* (5.31).

(b) *Regard n as large and express* (5.31) *in a WKB form with a small parameter related to* $1/n$.

(c) *Derive an asymptotic formula for the large eigenvalues* λ_n *and compare your result with the exact formula from part (a).*

Exercise 5.44 *Use the WKB method to find a leading-order approximation to the initial value problem*

$$\varepsilon^2 y''(t) - (1+t^2)^2 y(t) = 0, \qquad 0 < t < 1,$$

$$y(0) = 0, \qquad y'(0) = 1,$$

where $0 < \varepsilon \ll 1$.

Exercise 5.45 *(WKB expansion with a slowly varying coefficient) Consider the singularly perturbed ordinary differential equation*

$$y''(t) + q(\varepsilon t)\, y(t) = 0, \qquad 0 < \varepsilon \ll 1,$$

where $q \in C^\infty(\mathbb{R})$ and $q(\tau) > 0$ for all $\tau \in \mathbb{R}$.

(a) Introduce the slow variable $\tau = \varepsilon t$ and rewrite the differential equation in terms of τ.

(b) Seek a WKB-type solution of the form

$$y(t) \sim \exp\left(\frac{1}{\varepsilon} S_0(\tau) + S_1(\tau) + \varepsilon S_2(\tau) + \cdots\right), \qquad \tau = \varepsilon t,$$

and derive the hierarchy of equations for the functions $S_k(\tau)$ by substituting this ansatz into the transformed equation and collecting powers of ε.

(c) Show that the leading-order function $S_0(\tau)$ satisfies

$$(S_0'(\tau))^2 + q(\tau) = 0.$$

Solve this equation formally for $S_0(\tau)$ and explain the significance of the choice of sign.

(d) Derive the equation satisfied by $S_1(\tau)$ and show that

$$S_1'(\tau) = -\frac{1}{4}\frac{q'(\tau)}{q(\tau)}.$$

Deduce an explicit expression for $S_1(\tau)$ up to an additive constant.

Exercise 5.46 *(Leading-order WKB approximation) Under the assumptions and notation of the previous exercise, assume that $q(\tau) > 0$ for all τ in an interval I.*

(a) Use the expressions for S_0 and S_1 to construct a leading-order WKB approximation of the form

$$y(t) \sim A_\pm(\tau) \exp\left(\pm\frac{i}{\varepsilon}\int_{\tau_0}^{\tau} \sqrt{q(s)}\, ds\right), \qquad \tau = \varepsilon t,$$

and determine the amplitude $A_\pm(\tau)$ up to a multiplicative constant.

(b) Show that the approximation can be written in the standard WKB form

$$y(t) \sim \frac{C_1}{q(\varepsilon t)^{1/4}} \exp\left(\frac{i}{\varepsilon}\int_{\tau_0}^{\varepsilon t} \sqrt{q(s)}\, ds\right) + \frac{C_2}{q(\varepsilon t)^{1/4}} \exp\left(-\frac{i}{\varepsilon}\int_{\tau_0}^{\varepsilon t} \sqrt{q(s)}\, ds\right),$$

where C_1 and C_2 are constants.

(c) Rewrite the result in an equivalent real-valued form involving sine and cosine functions.

(d) *Explain briefly under what qualitative conditions on q this WKB approximation is expected to be valid on the interval I. In particular, discuss the role of smoothness and the absence of turning points (points where $q(\tau) = 0$).*

Exercise 5.47 *Consider*

$$y''(t) + (1 + \varepsilon t)\, y(t) = 0, \qquad 0 < \varepsilon \ll 1.$$

(a) *Identify $q(\varepsilon t)$ in this example and verify that $q(\tau) > 0$ for τ in a suitable interval.*

(b) *Apply the leading-order WKB approximation to obtain an explicit asymptotic formula for $y(t)$ in the form*

$$y(t) \sim \frac{1}{(1+\varepsilon t)^{1/4}} \left[C_1 \exp\left(\frac{1}{\varepsilon} \int_0^{\varepsilon t} \sqrt{1+s}\, \mathrm{d}s \right) + C_2 \exp\left(-\frac{1}{\varepsilon} \int_0^{\varepsilon t} \sqrt{1+s}\, \mathrm{d}s \right) \right].$$

(c) *Evaluate the integral explicitly and simplify the exponent as far as possible.*

(d) *Impose initial conditions $y(0) = y_0$, $y'(0) = y_1$ and obtain asymptotic expressions for C_1 and C_2 in terms of y_0, y_1, and ε.*

Exercise 5.48 *Let $q(\tau)$ be smooth and assume there is a simple turning point τ_0 such that*

$$q(\tau_0) = 0, \qquad q'(\tau_0) \neq 0.$$

Consider again

$$y''(t) + q(\varepsilon t)\, y(t) = 0.$$

(a) *Explain why the leading-order WKB approximation constructed away from τ_0 breaks down in a neighborhood of the turning point.*

(b) *Introduce a stretched variable $\xi = \varepsilon^{-\alpha}(t - t_0)$, with $t_0 = \tau_0/\varepsilon$, and determine the exponent α for which the leading-order balance yields an Airy-type equation for the inner solution .*

(c) *Write the resulting leading-order inner problem in terms of ξ and identify the standard Airy equation.*

(d) *Briefly describe (without full derivation) how one would match the inner Airy-type solution to the outer WKB solutions on either side of the turning point.*

Exercise 5.49 *Assume now that $q(\tau) > 0$ and consider the oscillatory WKB form*

$$y(t) \sim \frac{1}{q(\varepsilon t)^{1/4}} \left[A \cos\left(\frac{1}{\varepsilon} \int_{\tau_0}^{\varepsilon t} \sqrt{q(s)}\, \mathrm{d}s + \phi \right) \right].$$

(a) *Derive this cosine form by taking appropriate linear combinations of the exponential WKB solutions.*

(b) *For the specific case* $q(\tau) = 1 + \tau^2$, *compute the phase function*

$$\theta(t) = \frac{1}{\varepsilon} \int_0^{\varepsilon t} \sqrt{1+s^2}\, \mathrm{d}s$$

in closed form using elementary functions and inverse hyperbolic functions.

(c) *Discuss how the slowly varying prefactor* $q(\varepsilon t)^{-1/4}$ *modifies the amplitude of the oscillations as* t *increases.*

Exercise 5.50 *(WKB tunneling probability through a potential barrier) Consider a one-dimensional particle of mass* m *in the potential*

$$V(x) = \begin{cases} 0, & x < 0, \\ V_0, & 0 \le x \le a, \\ 0, & x > a, \end{cases}$$

with $0 < E < V_0$. *The stationary Schrödinger equation is*

$$-\frac{\hbar^2}{2m} \psi''(x) + V(x)\psi(x) = E\psi(x), \qquad x \in \mathbb{R},$$

and the incident wave comes from $x = -\infty$.

(a) *Identify the classically allowed and forbidden regions and write the local classical momentum*

$$p(x) = \sqrt{2m\big(E - V(x)\big)}.$$

(b) *In the barrier region* $0 < x < a$, $E < V_0$ *so* $p(x)$ *is purely imaginary. Use the WKB approximation in the forbidden region to write the general form of the decaying and growing solutions and explain why only the decaying solution is physically acceptable inside the barrier.*

(c) *Using the standard WKB expression for tunneling through a broad barrier,*

$$T \approx \exp\left(-\frac{2}{\hbar} \int_{x_1}^{x_2} \sqrt{2m\big(V(x) - E\big)}\, dx \right),$$

where x_1, x_2 *are the classical turning points, compute the transmission probability* T *explicitly for the square barrier above.*

(d) *Discuss qualitatively how* T *depends on the particle energy* E, *the barrier height* V_0, *and the width* a, *and relate this to the concept of quantum tunneling.*

Exercise 5.51 *(Semiclassical spectrum of a bound state in a general potential well) Let* $V(x)$ *be a smooth potential with a single nondegenerate minimum at* $x = x_0$ *and* $V(x) \to +\infty$ *as* $|x| \to \infty$. *Consider the bound states of the Schrödinger equation*

$$-\frac{\hbar^2}{2m} \psi''(x) + V(x)\psi(x) = E\psi(x),$$

subject to $\psi(x) \to 0$ *as* $|x| \to \infty$.

(a) *For a fixed energy E with* $V_{\min} < E < V_{\max}$, *where* $V_{\min} = V(x_0)$, *define the turning points* $x_1(E) < x_0 < x_2(E)$ *by* $V(x_1) = V(x_2) = E$. *Show that in the classically allowed region* (x_1, x_2) *the WKB solution has the oscillatory form*

$$\psi(x) \approx \frac{1}{\sqrt{p(x)}} \left[A\cos\left(\frac{1}{\hbar}\int_{x_1}^{x} p(s)\,ds\right) + B\sin\left(\frac{1}{\hbar}\int_{x_1}^{x} p(s)\,ds\right)\right],$$

where $p(x) = \sqrt{2m(E - V(x))}$.

(b) *By matching the oscillatory WKB solution across the two turning points to decaying solutions in the forbidden regions* $(E - V < 0)$ *using the standard connection formulas, derive the Bohr–Sommerfeld quantization rule*

$$\int_{x_1(E_n)}^{x_2(E_n)} \sqrt{2m(E_n - V(x))}\,dx = \left(n + \frac{1}{2}\right)\pi\hbar, \qquad n = 0, 1, 2, \ldots.$$

(c) *For the special case* $V(x) = \frac{1}{2}m\omega^2 x^2$, *verify that this rule reproduces the exact harmonic oscillator spectrum* $E_n = (n + \frac{1}{2})\hbar\omega$.

Exercise 5.52 *(WKB approximation in a slowly varying acoustic waveguide) This problem assumes familiarity with separation of variables for partial differential equations and basic properties of the Helmholtz equation.*

Consider time-harmonic acoustic pressure waves $p(x,z)$ *in a 2D duct of slowly varying depth, with x the horizontal coordinate and z the vertical coordinate. Assume the pressure satisfies the Helmholtz equation*

$$\frac{\partial^2 p}{\partial x^2} + \frac{\partial^2 p}{\partial z^2} + k^2 n^2(x,z)\,p = 0,$$

with a rigid boundary at $z = 0$ *and a free surface at* $z = H(x)$, *where* $H(x)$ *varies slowly with x and* $n(x,z)$ *is a slowly varying index of refraction.*

(a) *Using an adiabatic (slowly varying) modal approximation, assume a representation of the form*

$$p(x,z) = \phi_m(x,z)\,e^{i\beta_m(x)x},$$

where ϕ_m *is normalized in z for each x and* $\beta_m(x)$ *is the x-dependent propagation constant of mode m. Under the assumption of slow variation in x, derive a reduced equation of WKB type for* $A_m(x)$ *in*

$$p(x,z) \approx A_m(x)\,\phi_m(x,z)e^{i\int^x \beta_m(s)\,ds}.$$

(b) *Show that, to leading order,* $A_m(x)$ *satisfies a transport equation whose solution implies conservation of modal flux*

$$|A_m(x)|^2 c_{g,m}(x) \approx \text{constant},$$

where $c_{g,m}(x)$ *is the local group velocity of mode m.*

(c) Interpret this WKB result physically: explain how the slowly varying geometry $H(x)$ and index $n(x,z)$ affect the mode shape ϕ_m, the local phase

$$\int^x \beta_m(s)\,ds,$$

and the amplitude $A_m(x)$ as sound propagates along the duct.

Exercise 5.53 *(Large-argument asymptotics of Bessel functions) The Bessel function $J_\nu(x)$ of order ν satisfies Bessel's equation*

$$x^2y''(x)+xy'(x)+(x^2-\nu^2)y(x)=0.$$

(a) Show that the change of variables

$$y(x)=\sqrt{\frac{2}{\pi x}}\,u(x)$$

transforms Bessel's equation into

$$u''(x)+\left(1-\frac{\nu^2-\frac{1}{4}}{x^2}\right)u(x)=0.$$

(b) For fixed ν and $x\gg\nu$, show that

$$q(x)=1-\frac{\nu^2-\frac{1}{4}}{x^2}>0,$$

so that the solution is oscillatory . Using the WKB approximation

$$u(x)\approx\frac{1}{q(x)^{1/4}}\cos\left(\int^x\sqrt{q(s)}\,ds+\phi\right),$$

derive a leading-order approximation for $u(x)$ by expanding $q(x)^{\pm1/4}$ and $\sqrt{q(x)}$ for large x.

(c) Show that, to leading order,

$$\int^x\sqrt{q(s)}\,ds\sim x-\frac{\nu^2-\frac{1}{4}}{2x}+\cdots,$$

and hence obtain

$$J_\nu(x)\sim\sqrt{\frac{2}{\pi x}}\cos\left(x-\frac{\nu\pi}{2}-\frac{\pi}{4}\right),\qquad x\to\infty.$$

(You may assume that the phase ϕ is determined by matching with the known asymptotic behavior of $J_\nu(x)$.)

Exercise 5.54 *(Large-order asymptotics of Bessel functions) Consider Bessel's equation*

$$x^2y''(x)+xy'(x)+\left(x^2-\nu^2\right)y(x)=0,$$

and suppose that the order ν *is large while* $0<x<\nu$. *Introduce the scaled variable*

$$z=\frac{x}{\nu}, \qquad 0<z<1,$$

and write $y(x)=Y(z)$.

(a) Show that $Y(z)$ *satisfies*

$$\varepsilon^2z^2Y''(z)+\varepsilon^2zY'(z)+(z^2-1)Y(z)=0, \qquad \varepsilon=\frac{1}{\nu}.$$

Next, introduce the transformation

$$Y(z)=z^{-1/2}u(z),$$

and show that $u(z)$ *satisfies an equation of the form*

$$\varepsilon^2u''(z)+Q(z)u(z)=0.$$

Determine the function $Q(z)$.

(b) Show that

$$Q(z)<0, \qquad 0<z<1,$$

so that the problem is nonoscillatory (evanescent or classically forbidden) in this region.

Using the WKB approximation for exponentially decaying solutions, derive the leading-order asymptotic formula

$$J_\nu(\nu z)\approx C(1-z^2)^{-1/4}\exp\left(-\nu\int_z^1\frac{\sqrt{1-s^2}}{s}\,ds\right), \qquad 0<z<1, \quad \nu\to\infty,$$

where C *is a constant independent of* ν.

(c) Identify the turning point of the equation and explain why the solution is exponentially decaying for $x<\nu$ *and oscillatory for* $x>\nu$.

Interpret this transition in terms of evanescent and propagating regions for the Bessel function.

Exercise 5.55 *(Hermite polynomials and the parabolic cylinder equation) The Hermite functions*

$$\psi_n(x)=\frac{1}{\sqrt{2^nn!\sqrt{\pi}}}e^{-x^2/2}H_n(x),$$

satisfy the differential equation

$$\psi_n''(x)+(2n+1-x^2)\,\psi_n(x)=0.$$

(a) For large n, set $\lambda = 2n+1$ and rewrite the equation as

$$\psi''(x) + (\lambda - x^2)\psi(x) = 0.$$

Write this in the form $\psi'' + Q(x)\psi = 0$ and identify:

- *the classically allowed and forbidden regions,*
- *the turning points.*

(b) In the region where $\lambda - x^2 > 0$, use the leading-order WKB approximation with $p(x) = \sqrt{\lambda - x^2}$ to show that

$$\psi(x) \approx \frac{1}{(\lambda - x^2)^{1/4}} \cos\left(\int^x \sqrt{\lambda - s^2}\, ds + \phi\right), \qquad |x| < \sqrt{\lambda}.$$

(c) Use the normalization condition

$$\int_{-\infty}^{\infty} |\psi_n(x)|^2 dx = 1,$$

and approximate the integral by restricting to the classically allowed region and replacing $\cos^2$ by its average value. Estimate the amplitude and show that the oscillations of $H_n(x)$ become increasingly rapid as $n \to \infty$.

Exercise 5.56 *(Airy functions from a turning point problem) The Airy function* $\mathrm{Ai}(z)$ *is defined as the solution of Airy's equation*

$$y''(z) - zy(z) = 0$$

that remains bounded as $z \to +\infty$.

(a) Show that Airy's equation can be written in the form

$$\varepsilon^2 y''(z) + q(z)y(z) = 0, \qquad \varepsilon = 1, \quad q(z) = -z.$$

Identify the turning point and classify the regions $z > 0$ and $z < 0$ as oscillatory or nonoscillator y.

(b) Apply the WKB approximation in the region $z \gg 1$ to derive the leading asymptotic behavior

$$\mathrm{Ai}(z) \sim \frac{1}{2\sqrt{\pi}} z^{-1/4} \exp\left(-\frac{2}{3} z^{3/2}\right), \qquad z \to +\infty.$$

(You may assume that the exponentially growing solution is excluded by the boundedness requirement.)

(c) Apply the oscillatory WKB approximation in the region $z \ll -1$ to show that

$$\mathrm{Ai}(z) \sim \frac{1}{\sqrt{\pi}} |z|^{-1/4} \sin\left(\frac{2}{3}|z|^{3/2} + \frac{\pi}{4}\right), \qquad z \to -\infty.$$

(d) *Discuss how these two asymptotic forms are connected through the turning point at* $z = 0$ *and why Airy functions naturally appear in WKB problems near simple turning points.*

Exercise 5.57 *(Connection problem: WKB to Airy near a simple turning point) Consider the scaled Schrödinger-type equation*

$$\varepsilon^2 y''(x) - (x - x_0)y(x) = 0, \qquad 0 < \varepsilon \ll 1,$$

which has a simple turning point at $x = x_0$.

(a) *Introduce the stretched variable*

$$z = \frac{x - x_0}{\varepsilon^{2/3}}$$

and show that the equation transforms into Airy's equation

$$Y''(z) - zY(z) = 0.$$

(b) *Use the known asymptotic forms of* $\mathrm{Ai}(z)$ *for* $z \to \pm\infty$ *from the previous exercise to derive the connection formulas that relate the oscillatory WKB solution for* $x < x_0$ *to the decaying WKB solution for* $x > x_0$.

(c) *Explain how this analysis justifies the use of Airy functions as local models near turning points in the WKB theory of more general special functions, such as Bessel and Hermite functions.*

6

Systems of Ordinary Differential Equations

6.1 Preliminaries

A system of ordinary differential equations arises when several unknown functions evolve together and influence one another. Unlike a single differential equation, which describes the behavior of one quantity, a system captures interactions among multiple variables that change with respect to the same independent variable, typically time. Such systems naturally appear in many areas of science and engineering, including mechanics, electrical circuits, population dynamics, chemical reactions, and fluid motion.

The study of systems focuses on understanding how the collective behavior emerges from these interactions. Questions of interest include whether solutions exist and are unique, how they depend on initial conditions, and what long-term behavior they exhibit. Systems may display simple dynamics, such as steady states or periodic motion, or more complex phenomena, including instability and sensitivity to initial data.

Analyzing systems of ordinary differential equations provides a unified framework for modeling and interpreting coupled processes, making them an essential tool for describing real-world phenomena where variables do not evolve in isolation.

For motivational purpose we consider the second-order differential equation

$$x''(t) + a^2 x(t) = 0 \tag{6.1}$$

where a is some nonzero constant. The equation (6.1) is commonly used as a model describing a mechanical process which oscillates periodically. For example, it is used to describe oscillations of a pendulum, provided the angle change is small, or of a spring with a small deflection. We use the transformation $x_1(t) = x(t)$,which is the displacement of the oscillating object. (For pendulum, it is the angle made with the vertical.) Introducing the second transformation, the velocity, or angular velocity $x_2(t) = x'(t)$, and differentiating, we get $x_1'(t) = x'(t) = x_2$, and $x_2'(t) = x''(t) = -a^2 x(t) = -a^2 x_1(t)$. We therefore arrive at the two first-order differential

DOI: 10.1201/9781042009220-6

equations

$$x_1' = x_2$$
$$x_2' = -a^2 x_1.$$

Next, we consider the nth order differential equation

$$x^{(n)}(t) = g\left(t, x, x', \ldots, x^{(n-1)}\right)$$

and use the transformation

$$x_i(t) = x^{(i-1)}(t), \quad i = 1, 2, \ldots, n$$

to arrive at the following system

$$\begin{aligned} x_1' &= x_2 \\ x_2' &= x_3 \\ &\vdots \\ x_n' &= g\left(t, x_1, x_2, \ldots, x_n\right). \end{aligned}$$

The right side of the above system defines a vector function $f(t, \mathbf{x})$. To fully describe the solution, we impose the initial conditions

$$x^{(i-1)}(t_0) = x_{0i}, \quad i = 1, 2, \ldots, n$$

where x_{0i}, $i = 0, 1, \ldots, n-1$, are given constants. Next, we state and prove Gronwall's inequality, a fundamental result that plays a central role in establishing the boundedness, stability, and uniqueness of solutions.

Theorem 6.1 (Gronwall's inequality) *Let C be a nonnegative constant, and let u and v be nonnegative continuous functions on $[a, b]$ such that*

$$v(t) \leq C + \int_a^t v(s)u(s)\,ds, \qquad a \leq t \leq b. \tag{6.2}$$

Then

$$v(t) \leq C \exp\left(\int_a^t u(s)\,ds\right), \qquad a \leq t \leq b. \tag{6.3}$$

In particular, if $C = 0$, then $v(t) = 0$ for all $t \in [a, b]$.

Proof *Assume first that $C > 0$, and define*

$$h(t) = C + \int_a^t v(s)u(s)\,ds.$$

Then $h(a) = C$ and, by (6.2),

$$h'(t) = v(t)u(t) \leq h(t)u(t).$$

Hence

$$h'(t) - u(t)h(t) \leq 0.$$

Multiplying by the integrating factor

$$\exp\left(-\int_a^t u(s)\,ds\right),$$

we obtain

$$\left[h(t)\exp\left(-\int_a^t u(s)\,ds\right)\right]' \leq 0.$$

Integrating from a to t gives

$$h(t)\exp\left(-\int_a^t u(s)\,ds\right) - h(a) \leq 0.$$

Since $h(a) = C$, it follows that

$$h(t) \leq C\exp\left(\int_a^t u(s)\,ds\right).$$

Finally, because $v(t) \leq h(t)$, we have

$$v(t) \leq C\exp\left(\int_a^t u(s)\,ds\right).$$

Now suppose $C = 0$. Then (6.2) becomes

$$v(t) \leq \int_a^t v(s)u(s)\,ds.$$

For each integer $m \geq 1$, we also have

$$v(t) \leq \frac{1}{m} + \int_a^t v(s)u(s)\,ds.$$

Applying the result already proved with $C = 1/m$, we obtain

$$v(t) \leq \frac{1}{m}\exp\left(\int_a^t u(s)\,ds\right), \qquad a \leq t \leq b.$$

Letting $m \to \infty$ gives $v(t) \leq 0$. Since v is nonnegative, it follows that

$$v(t) = 0, \qquad a \leq t \leq b.$$

This completes the proof.

The next theorem shows the significance of Theorem 6.1.

Theorem 6.2 *Suppose $\phi(t)$ is continuous on $[0,\infty)$ and differentiable on $(0,\infty)$. If*

$$\lim_{t\to\infty}\phi(t)=0,$$

and

$$\int_0^\infty |\phi'|dt<\infty,$$

then all solutions of

$$x''+(1+\phi(t))x=0 \tag{6.4}$$

are bounded.

Proof *Multiply both sides of* (6.4) *by x' and integrate the resulting equation from* 0 *to t to arrive at*

$$\frac{(x'(t))^2}{2}+\frac{(x(t))^2}{2}+\int_0^t \phi(s)x(s)x'(s)ds=c_1,$$

where c_1 is a constant. Next, we perform an integration by parts on the third term from the left. To do so, we let

$$u=\phi,\quad dv=xx'ds,$$

then

$$du=\phi'ds,\quad v=\frac{xx'}{2}.$$

Consequently, we have after rearranging

$$\begin{aligned} x^2(t)\big(1+\phi(t)\big) &= 2c_2-(x'(t))^2+\int_0^t \phi'(s)x^2(s)ds \\ &\leq 2c_2+\int_0^t \phi'(s)x^2(s)ds, \end{aligned}$$

where c_2 is a constant. Since $\lim_{t\to\infty}\phi(t)=0$, we may choose t_0 large enough so that $1+\phi(t)\geq\frac{1}{2}$ for all $t\geq t_0$. This translates into

$$x^2(t)(\frac{1}{2})\leq x^2(t)\big(1+\phi(t)\big)\leq 2c_2+\int_0^t \phi'(s)x^2(s)ds.$$

Or

$$x^2(t)\leq 4c_2+2\int_0^t |\phi'(s)|x^2(s)ds,\ t\geq t_0.$$

Thus, by Theorem 6.1 we have that

$$\begin{aligned} x^2(t) &\leq 4c_2e^{2\int_0^t|\phi'(s)|ds} \\ &\leq 4c_2e^{2\int_0^\infty|\phi'(s)|ds}=M<\infty,\ \textit{for some positive constant } M \end{aligned}$$

and for all $t\geq 0$. Since $x(t)$ is continuous on $0\leq t\leq t_0$, and t_0 is arbitrary, it follows that $x(t)$ is bounded. That is, there exists a positive constant M_1 so that

$$|x(t)|\leq M_1\quad \textit{for all}\quad t\geq 0.$$

Example 6.1 *(Pendulum with friction) Fig. 6.1 shows a simple pendulum with massless string of length l and a bob of mass m acting under the external influence of gravity. Let g be the acceleration of gravity, m is the mass of the bob, l is the length of the rod, and b is the coefficient of friction (damping), and they are all positive constants. The period is the time it takes for the bob to swing from its farthest right position to its left farthest position and back. Let $s(t)$ be the distance along the arc from the lowest point of the bob at time t, with displacement to the right considered positive. Let $\theta(t)$ be the corresponding angle with respect to the vertical. The gravitational force is directed downward and has a magnitude mg. The force acting in the tangential direction is $-mg\sin(\theta)$. Then the motion is modeled by the second-order differential equation*

$$\frac{d^2s}{dt^2} = -g\sin(\theta).$$

Now $s = L\theta$ as L is the radius of the circle, and so

$$\frac{d^2\theta}{dt^2} = -\frac{g}{L}\sin(\theta).$$

We add damping to the model and assume that the damping force is proportional to the velocity. As θ is our independent variable, the damping is represented as a force proportional to angular velocity, say, $-b\frac{d\theta}{dt}$. The negative sign indicates that the damping force is in the opposite direction of the motion. Then the model takes the shape

$$mL\theta'' + mg\sin(\theta) + bL\theta' = 0. \tag{6.5}$$

Assume the initial data

$$\theta(t_0) = \theta_0, \quad \theta'(t_0) = \omega_0. \tag{6.6}$$

We put (6.5) into a system. Thus, we let $x_1(t) = \theta(t)$ and $x_2(t) = \theta'(t)$ and obtain the nonlinear two-dimensional system

$$\begin{aligned} x_1' &= x_2 \\ x_2' &= -\frac{g}{L}\sin(x_1) - \frac{b}{m}x_2 \end{aligned} \tag{6.7}$$

with corresponding initial conditions

$$x_1(t_0) = \theta_0, \; x_2(t_0) = \omega_0. \tag{6.8}$$

Let

$$f(t,x) = \begin{pmatrix} x_2 \\ \\ -\frac{g}{L}\sin(x_1) - \frac{b}{m}x_2 \end{pmatrix}$$

Define the energy function $E : [t_0,\infty) \times \mathbb{R}^2 \to [0,\infty)$ by

$$E(t) := E(t,x_1,x_2) = \frac{1}{2}x_2^2(t) + \frac{g}{L}\Big(1 - \cos(x_1(t))\Big). \tag{6.9}$$

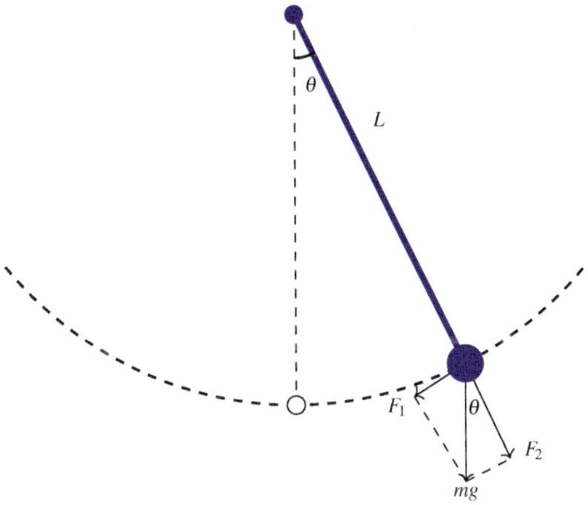

FIGURE 6.1: Pendulum with friction.

Then, along the solutions of (6.7) *we have that*

$$\begin{aligned}\frac{d}{dt}E(t) &= x_2\,x_2' + \frac{g}{L}\sin(x_1)\,x_1' \\ &= x_2\Big(-\frac{g}{L}\sin(x_1) - \frac{b}{m}x_2\Big) + \frac{g}{L}\sin(x_1)x_2 \\ &= -\frac{b}{m}x_2^2 \le 0.\end{aligned}$$

Let $t_0 \le t < \alpha$ *and integrate the above inequality from* t_0 *to* t *to get*

$$E(t) - E(t_0) \le 0,$$

or

$$E(t) \le \frac{1}{2}\omega_0^2 + \frac{g}{L}\Big(1 - \cos(\theta_0)\Big).$$

We have from (6.9) *that* $\frac{1}{2}x_2^2(t) \le E(t)$, *so that*

$$|x_2(t)| \le \sqrt{2\Big[\frac{1}{2}\omega_0^2 + \frac{g}{L}\Big(1 - \cos(\theta_0)\Big)\Big]} := M.$$

Left to bound $x_1(t)$. *From the definition of energy,*

$$E(t) = \frac{1}{2}x_2^2(t) + \frac{g}{L}(1 - \cos(x_1(t))),$$

and the estimate

$$E(t) \le E(t_0) = \frac{1}{2}\omega_0^2 + \frac{g}{L}(1 - \cos(\theta_0)) =: E_0,$$

it follows in particular that

$$\frac{g}{L}\big(1-\cos(x_1(t))\big) \le E_0 \qquad \textit{for all } t \ge t_0.$$

Hence

$$1-\cos(x_1(t)) \le \frac{L}{g}E_0 = \frac{L}{g}\left[\frac{1}{2}\omega_0^2 + \frac{g}{L}\big(1-\cos(\theta_0)\big)\right] = \frac{L}{2g}\,\omega_0^2 + 1 - \cos(\theta_0).$$

Since $-1 \le \cos(x_1(t)) \le 1$*, the inequality*

$$1-\cos(x_1(t)) \le C,$$

with some finite constant $C > 0$ *determined by* θ_0, ω_0, g, L*, implies that* $\cos(x_1(t))$ *stays in a compact subinterval of* $[-1,1]$*. In particular, define*

$$C := \min\left\{2, \frac{L}{g}E_0\right\} \in (0,2],$$

so that

$$1-\cos(x_1(t)) \le C$$

implies that

$$\cos(x_1(t)) \ge 1 - C > -1.$$

Therefore

$$x_1(t) \in \big[-\arccos(1-C), \arccos(1-C)\big] \qquad \textit{for all } t \ge t_0,$$

and hence

$$|x_1(t)| \le K := \arccos(1-C),$$

where $K > 0$ *depends only on the initial data and the parameters* g, L*. Thus, there exists a positive constant* K *such that*

$$|x_1(t)| \le K \qquad \textit{for all } t \ge t_0.$$

Thus, we have shown that

$$|x| := \sqrt{x_1^2 + x_2^2} \le D,$$

for a positive constant D*. It is worth noting that the energy function will be referred to as Lyapunov function, which we will study in more detail in Chapter 8.* □

In Example 6.1 the energy function had to be chosen just right in order to yield the right results. Generally speaking, for any second order differential equation of the form

$$x'' + f(x)x' + g(x) = 0 \tag{6.10}$$

and under more conditions on g, the energy function may take the form

$$E(t) := E(t, x_1, x_2) = \frac{1}{2}x_2^2(t) + \int_0^{x_1} g(s)ds,$$

by transforming (6.10) into a system using the transformations

$$x_i(t) = x^{(i-1)}(t), \quad i = 1, 2.$$

On the other hand, if we don't transform (6.10) into a system, we may use the energy function

$$E(t) := E(t,x) = \frac{1}{2}(x')^2 + \int_0^{x(t)} g(s)ds,$$

where $\frac{1}{2}(x')^2$ corresponds to the *kinetic energy*, and $\int_0^{x(t)} g(s)ds$ represents *potential energy*.

6.1.1 Exercises

Exercise 6.1 *(a) Consider the third-order linear ODE*

$$y^{(3)}(t) - 2y''(t) + 5y'(t) - 3y(t) = \cos t.$$

Introduce new variables

$$x_1(t) = y(t), \qquad x_2(t) = y'(t), \qquad x_3(t) = y''(t),$$

and rewrite the equation as a first-order system

$$\mathbf{x}'(t) = F(t, \mathbf{x}(t)),$$

by explicitly finding differential equations for x_1', x_2', x_3', where

$$\mathbf{x}(t) = \begin{pmatrix} x_1(t) \\ x_2(t) \\ x_3(t) \end{pmatrix}.$$

(b) Write your system from part (a) in matrix-vector form

$$\mathbf{x}'(t) = A\mathbf{x}(t) + \mathbf{b}(t),$$

for a suitable constant 3×3 matrix A and a vector $\mathbf{b}(t)$ depending only on t.

(c) Write the nonlinear fourth-order ODE

$$y^{(4)}(t) + y(t)y''(t) - \big(y'(t)\big)^2 = 0.$$

in the form $\mathbf{x}'(t) = G(\mathbf{x}(t))$.

Exercise 6.2 *Let $\phi : [0, \infty) \to \mathbb{R}$ be given by*

$$\phi(t) = \frac{\alpha}{(1+t)^p},$$

for constants $\alpha \in \mathbb{R}$ and $p > 0$, and consider

$$x''(t) + \big(1 + \phi(t)\big)x(t) = 0, \qquad t \geq 0. \tag{6.11}$$

(a) *Show that $\phi(t) \to 0$ as $t \to \infty$ for every $p > 0$. Compute $\phi'(t)$ and determine for which values of p the integral*

$$\int_0^\infty |\phi'(s)|\,ds$$

is finite.

(b) *For those p found in part (a) where the integral is finite, use Theorem 6.2 to conclude that every solution of (6.11) is bounded on $[0,\infty)$.*

(c) *Consider now the borderline case where the above integral diverges. Discuss, without full proof, whether the boundedness of all solutions can still be expected in this case. In particular, comment on whether Theorem 6.2 gives a sufficient condition, a necessary condition, or both, for boundedness of all solutions.*

(d) *(Optional) For a fixed choice of (α, p) with $p > 1$, use the energy identity and Gronwall's inequality to derive an explicit uniform bound*

$$|x(t)| \leq M(\alpha, p, x(0), x'(0)) \qquad \text{for all } t \geq 0,$$

and express M in terms of the initial data and $\int_0^\infty |\phi'(s)|\,ds$.

Exercise 6.3 *Consider the second-order differential equation*

$$x''(t) + f(x(t))x'(t) + g(x(t)) = 0, \tag{6.12}$$

where $f, g \in C^1(\mathbb{R})$. Define the energy function

$$E(t) = \frac{1}{2}\left(x'(t)\right)^2 + \int_0^{x(t)} g(s)\,ds.$$

(a) *Compute $\dfrac{d}{dt}E(t)$ along a solution $x(t)$ of (6.12) and show that*

$$\frac{d}{dt}E(t) = -f(x(t))\left(x'(t)\right)^2.$$

(b) *Suppose $f(x) \geq 0$ for all $x \in \mathbb{R}$. Prove that $E(t)$ is nonincreasing along any solution of (6.12). Interpret this fact in terms of kinetic and potential energy.*

(c) *Assume in addition that*

$$\lim_{|x| \to \infty} \int_0^x g(s)\,ds = +\infty.$$

Show that all solutions of (6.12) are bounded on their maximal interval of existence.

(d) *Rewrite* (6.12) *as a first-order system*

$$x_1'(t) = x_2(t), \qquad x_2'(t) = -f(x_1(t))x_2(t) - g(x_1(t)),$$

and express the same energy as

$$E(t, x_1, x_2) = \frac{1}{2}x_2^2 + \int_0^{x_1} g(s)\,ds.$$

Verify directly that $\frac{d}{dt}E(t, x_1(t), x_2(t)) = -f(x_1(t))x_2^2(t)$ *along solutions of the system.*

Exercise 6.4 *Consider the nonlinear oscillator*

$$x''(t) + \alpha x'(t) + x(t) + x^3(t) = 0, \tag{6.13}$$

where $\alpha > 0$ *is a constant.*

(a) *Write* (6.13) *in the form*

$$x'' + f(x)x' + g(x) = 0$$

by identifying $f(x)$ *and* $g(x)$.

(b) *Define the energy*

$$E(t) = \frac{1}{2}\left(x'(t)\right)^2 + \int_0^{x(t)} g(s)\,ds,$$

and compute this integral explicitly to obtain

$$E(t) = \frac{1}{2}\left(x'(t)\right)^2 + \frac{1}{2}x^2(t) + \frac{1}{4}x^4(t).$$

(c) *Differentiate* $E(t)$ *with respect to* t *and use* (6.13) *to show that*

$$\frac{d}{dt}E(t) = -\alpha\left(x'(t)\right)^2 \leq 0.$$

Conclude that $E(t)$ *is nonincreasing and hence bounded from above by* $E(0)$.

(d) *Show that the level sets* $\{(x, x') : E(t) = \text{constant}\}$ *are closed and bounded subsets of the phase plane, and use this to prove that every solution of* (6.13) *is bounded for all* $t \geq 0$.

(e) *Discuss the physical interpretation of* (6.13) *briefly as a unit-mass particle moving in the potential*

$$V(x) = \frac{1}{2}x^2 + \frac{1}{4}x^4,$$

subject to a linear damping force proportional to its velocity. Explain how the sign of $\frac{d}{dt}E(t)$ *reflects the dissipative nature of the system.*

Exercise 6.5 *(Generalized Gronwall's inequality) Show that if f,u,v are nonnegative continuous functions on $[a,b]$ such that*

$$u(t) \leq f(t) + \int_a^t v(s)u(s)ds, \; a \leq t \leq b.$$

then

$$u(t) \leq f(t) + \int_0^t v(s)f(s)e^{\int_s^t v(r)dr}ds, \; a \leq t \leq b.$$

Exercise 6.6 *Find all nonnegative continuous functions $f : [0,1] \to \mathbb{R}$ such that*

$$f(t) \leq \int_0^t f(s)ds, \; 0 \leq t \leq 1.$$

Exercise 6.7 *Let K_1, K_2 be positive constants and f be a nonnegative continuous function on $[a,b]$ such that if*

$$f(t) \leq K_1 + K_2 \int_a^t f(s)ds, \; a \leq t \leq b$$

then f satisfies

$$f(t) \leq K_1 e^{K_2(b-a)}$$

Exercise 6.8 *(Another form of Gronwall's inequality) Let K_1, K_2 be positive constants and let C_0 be a nonnegative constant. Suppose that f is a nonnegative continuous function on $[0,h]$ satisfying*

$$f(t) \leq C_0 + \int_0^t \left(K_1 f(s) + K_2\right) ds, \qquad 0 \leq t \leq h.$$

Show that

$$f(t) \leq C_0 e^{K_1 t} + \frac{K_2}{K_1}\left(e^{K_1 t} - 1\right), \qquad 0 \leq t \leq h.$$

Exercise 6.9 *Consider the system*

$$x' = -x^3, \qquad y' = -cy + x^{2\lambda} y \cos^2(y),$$

with initial conditions

$$x(0) = x_0, \qquad y(0) = y_0.$$

Use Gronwall's inequality to show that:

(a) If $c = 0$ and $\lambda > 1$, then $y(t)$ is bounded for all $t \geq 0$.

(b) If $c > 0$ and $\lambda = 1$, then $y(t) \to 0$ as $t \to \infty$.

6.2 Fundamental Matrix

The concept of a fundamental matrix plays a central role in the theory of linear systems of ordinary differential equations. It provides a compact and systematic way to describe the entire family of solutions of a homogeneous system, encoding how initial data evolve in time under the dynamics of the system. By organizing a set of linearly independent solutions into a single matrix-valued function, the fundamental matrix reveals the intrinsic structure of the system, including properties such as linear independence, stability, and invariance under changes of initial conditions. Moreover, it serves as a foundational tool for analyzing nonhomogeneous systems, formulating variation-of-parameters methods, and constructing solution operators, thereby linking abstract theory with practical solution techniques.

Consider the linear system of ordinary differential equations of the form

$$\begin{aligned} x_1' &= a_{11}(t)x_1(t) + a_{12}(t)x_2(t) + \ldots + a_{1n}(t)x_n(t) + g_1(t) \\ x_2' &= a_{21}(t)x_1(t) + a_{22}(t)x_2(t) + \ldots + a_{2n}(t)x_n(t) + g_2(t) \\ &\ldots \\ x_n' &= a_{n1}(t)x_1(t) + a_{n2}(t)x_2(t) + \ldots + a_{nn}(t)x_n(t) + g_n(t) \end{aligned}$$

where the functions a_{ij}, g_i, $1 \le i \le n, 1 \le j \le n$, are continuous real-valued functions on an interval I. Using vector and matrix notations, the above system is equivalent to the vector equation

$$x'(t) = A(t)x(t) + g(t), \tag{6.14}$$

where

$$x := \begin{pmatrix} x_1 \\ x_2 \\ \vdots \\ x_n \end{pmatrix}, \quad x' := \begin{pmatrix} x_1' \\ x_2' \\ \vdots \\ x_n' \end{pmatrix}$$

and

$$A(t) := \begin{pmatrix} a_{11} & a_{12} & \cdots & a_{1n} \\ a_{21} & a_{22} & \cdots & a_{2n} \\ \vdots & \vdots & \ddots & \vdots \\ a_{n1} & a_{n2} & \cdots & a_{nn} \end{pmatrix}, \quad g(t) := \begin{pmatrix} g_1(t) \\ g_2(t) \\ \vdots \\ g_n(t)I. \end{pmatrix}$$

for $t \in I$. Note that the matrix functions A and g are continuous on an interval I if and only if all their entries are continuous on I.

Definition 6.1 *We say the $n \times 1$ vector y is a solution of 6.14 on I provided y is continuously differentiable on I and*

$$y'(t) = A(t)y(t) + g(t),$$

for all $t \in I$.

We denote the space of n-tuple functions $x : I \to \mathbb{R}^n$ that are continuously differentiable on an interval I by $C^1(I, \mathbb{R}^n)$.

Next, we define matrix norm. In many areas of applications, it is useful to quantify the size or magnitude of a matrix in a consistent way. This leads to the notion of a *matrix norm*, which generalizes the concept of a vector norm and provides a tool for analyzing stability, convergence, and error estimates in matrix computations. We now state the axiomatic definition of a matrix norm.

Definition 6.2 *A matrix norm is a function that assigns to each square matrix A a nonnegative real number $||A||$ such that the following properties hold:*

1. $||A|| \geq 0$ *for any square matrix*
2. $||A|| = 0$ *if and only if* $A = 0$ *(0 matrix)*
3. $||kA|| = |k|||A||$, *for any constant k*
4. $||A+B|| \leq ||A|| + ||B||$, *where B is a square matrix*
5. $||AB|| \leq ||A||||B||$.

Example 6.2 *Let A be a matrix with entries a_{ij}, $1 \leq i \leq n, 1 \leq j \leq n$; that is $A = (a_{ij})$, then we define the following possible norms on the matrix A.*

1. *The* 1*-norm:*

$$||A||_1 = \max_{1 \leq j \leq n} \Big(\sum_{i=1}^{n} |a_{ij}| \Big),$$

(the maximum absolute column sum).

2. *The infinity norm:*

$$||A||_\infty = \max_{1 \leq i \leq n} \Big(\sum_{j=1}^{n} |a_{ij}| \Big),$$

(sum the absolute values along each row and then take the biggest value).

3. *The Euclidean norm:*

$$||A||_E = \sqrt{\sum_{i=1}^{n} \sum_{j=1}^{n} (a_{ij})^2},$$

(square root of the sum of each entry).

□

We note that if I is the identity matrix, then $||I||_1 = ||I||_\infty = 1$.
Solving the nonhomogeneous system

$$x'(t) = A(t)x(t) + g(t), \tag{6.15}$$

requires solving the homogenous system

$$x'(t) = A(t)x(t), \tag{6.16}$$

where $A(t)$ is an $n \times n$ matrix of coefficients $a_{ij}(t)$ that are assumed to be continuous on an interval I. Recall that a solution $x(t)$ of (6.16) is an n-tuple of C^1 functions $x_i : I \to \mathbb{R}$. We adopt the notation that

$$x(t) = \begin{pmatrix} x_1(t) \\ x_2(t) \\ \vdots \\ x_n(t) \end{pmatrix}.$$

Any solution x is a C^1 vector-valued function $x : I \to \mathbb{R}^n$, and the collection of all such functions is denoted by $C^1(I, \mathbb{R}^n)$. Hence, if $\mathscr{S}$ denotes the *solution space* of (6.16), then $\mathscr{S} \subset C^1(I, \mathbb{R}^n)$.

Definition 6.3 *Let $f_1, f_2, \ldots, f_n$ be a set of $n \times 1$ vector functions defined on an interval I. We say the set $\{f_1, f_2, \ldots, f_n\}$ is linearly dependent on I if there exists constants $c_1, c_2, \ldots, c_n$ not all zero, such that*

$$c_1 f_1 + c_2 f_2 + \ldots + c_n f_n = 0$$

for every $t \in I$. If the set of functions is not linearly dependent on the interval I, it is said to be linearly independent.

Definition 6.4 *(Fundamental set of solutions). A set of n solutions of the linear differential system* (6.16) *all defined on the same open interval I, is called a fundamental set of solutions on I if the solutions are linearly independent functions on I.*

We have the following corollary.

Corollary 3 *Let $A(t)$ be an $n \times n$ matrix of coefficients $a_{ij}(t)$ that are continuous on an interval I. If $\{\phi_1(t), \phi_2(t), \ldots, \phi_n(t)\}$ form a fundamental set of solutions on I, then the general solution of* (6.16) *is given by*

$$x(t) = c_1\phi_1(t) + c_2\phi_2(t) + \ldots + c_n\phi_n(t),$$

for constants $c_i, i = 1, 2, \ldots, n$.

Definition 6.5 *(Fundamental matrix) A matrix solution Φ is called a fundamental matrix solution (or in short, fundamental matrix) of* (6.16) *on an interval I if its columns form a fundamental set of solutions. If in addition, $\Phi(t_0) = I$, a fundamental matrix solution is called the principal fundamental matrix solution (or in short, principal matrix).*

Example 6.3 *We claim the matrix*

$$\Phi(t) = \begin{pmatrix} e^{-t} & te^{-t} \\ -e^{-t} & (1-t)e^{-t} \end{pmatrix}$$

is the fundamental matrix for the system

$$x'(t) = \begin{pmatrix} 0 & 1 \\ -1 & -2 \end{pmatrix} \begin{pmatrix} x_1 \\ x_2 \end{pmatrix}.$$

To see this,

$$\Phi'(t) = \begin{pmatrix} -e^{-t} & e^{-t} - te^{-t} \\ e^{-t} & -e^{-t} - (1-t)e^{-t} \end{pmatrix} = \begin{pmatrix} -e^{-t} & (1-t)e^{-t} \\ e^{-t} & -2e^{-t} + te^{-t} \end{pmatrix}.$$

On the other hand

$$\begin{pmatrix} 0 & 1 \\ -1 & -2 \end{pmatrix} \Phi = \begin{pmatrix} 0 & 1 \\ -1 & -2 \end{pmatrix} \begin{pmatrix} e^{-t} & te^{-t} \\ -e^{-t} & (1-t)e^{-t} \end{pmatrix} = \begin{pmatrix} -e^{-t} & (1-t)e^{-t} \\ e^{-t} & te^{-t} - 2e^{-t} \end{pmatrix}.$$

This proves the claim. □

We now characterize precisely when a matrix-valued function forms a fundamental matrix for a linear system. The following theorem shows that this occurs exactly when the matrix solves the system and is invertible at the initial time.

Theorem 6.3 *The matrix Φ is a fundamental matrix of $x' = Ax$ at t_0 if and only if*

(a) Φ is a solution of $x' = Ax$
and

(b) $\det\Phi(t_0) \neq 0$.

Proof *($\Rightarrow$) Suppose Φ is a fundamental matrix of $x' = Ax$ at t_0. By definition, its columns form a fundamental set of solutions of $x' = Ax$, so each column solves $x' = Ax$. Hence, Φ itself satisfies*

$$\Phi'(t) = A\Phi(t),$$

which is (a). Moreover, a fundamental set of solutions is linearly independent at t_0, which is equivalent to

$$\det\Phi(t_0) \neq 0,$$

giving (b).

($\Leftarrow$) Conversely, assume (a) and (b). Condition (a) says the columns of Φ are solutions of $x' = Ax$, and (b) implies these columns are linearly independent at t_0. Therefore, they form a fundamental set of solutions, so Φ is a fundamental matrix of $x' = Ax$ at t_0 by definition.

For the next two theorems, we suppress the argument t.

Theorem 6.4 *If Φ is a fundamental matrix of* (6.16)*, then ΦC, where C is an arbitrary $n \times n$ nonsingular constant matrix, is a fundamental matrix of* (6.16)*. Conversely, If Ψ is another fundamental matrix to* (6.16)*, then there exists a constant nonsingular $n \times n$ matrix such that $\Psi = \Phi C$ for all $t \in I$.*

Proof *Let Φ be a fundamental matrix of* (6.16) *on I. Then*

$$(\Phi C)' = \Phi' C = (A(t)\Phi)C = A(t)(\Phi C).$$

So, ΦC solves (6.16)*. Next*

$$\det(\Phi C) = \det\Phi \det C \neq 0$$

hence, ΦC is a fundamental matrix of (6.16)*. Conversely, let Φ and Ψ be two fundamental matrix solutions of* (6.16)*. Since $\Phi\Phi^{-1} = I$, we have that $\Phi'\Phi^{-1} + \Phi(\Phi^{-1})' = 0$, or $(\Phi^{-1})' = -\Phi^{-1}\Phi'\Phi^{-1}$. Now we differentiate $\Phi^{-1}\Psi$, and obtain*

$$\begin{aligned}
(\Phi^{-1}\Psi)' &= (\Phi^{-1})'\Psi + \Phi^{-1}\Psi' \\
&= -\Phi^{-1}\Phi'\Phi^{-1}\Psi + \Phi^{-1}\Psi' \\
&= \big(-\Phi^{-1}A(t)\Phi\Phi^{-1} + \Phi^{-1}A(t)\big)\Psi \\
&= \big(-\Phi^{-1}A(t) + \Phi^{-1}A(t)\big)\Psi \\
&= 0.
\end{aligned}$$

Therefore, integrating the above expression gives $\Phi^{-1}\Psi = C$, for some $n \times n$ constant matrix. Or, $\Psi = \Phi C$. Furthermore, as Ψ and Φ are fundamental matrix solutions, we have $\det\Phi \neq 0$, *and* $\det\Psi \neq 0$, *and hence* $\det C \neq 0$. *This completes the proof.*

We remark that it is not true in general that $C\Phi$ is a fundamental matrix of (6.16).

Theorem 6.5 *If Φ is a fundamental matrix of* (6.16) *on an interval I, then Φc solves* (6.16) *on I for every $n \times 1$ vector c.*

Proof *Let Φ be a fundamental matrix of* (6.16) *on I. Then*

$$(\Phi c)' = \Phi' c = (A(t)\Phi)c = A(t)(\Phi c).$$

So, Φc solves (6.16).

Theorem 6.6 *If $\Phi(t)$ is a fundamental matrix of* (6.16) *on an interval I, then $\Phi(t)c$, with $c = \Phi^{-1}(t_0)x_0$, is a solution of* (6.16) *with $x(t_0) = x_0$.*

Proof *First, $\Phi^{-1}(t_0)$ exists by the definition of fundamental matrix. Hence, by Theorem 6.5 $x(t) = \Phi(t)\Phi^{-1}(t_0)x_0$ solves* (6.16)*. Moreover, $x(t_0) = \Phi(t_0)\Phi^{-1}(t_0)x_0 = Ix_0 = x_0$. This completes the proof.*

6.3 $x' = Ax$

Now we consider homogeneous systems with constant matrices. In particular, we look at

$$x'(t) = Ax(t) \tag{6.17}$$

where A is an $n \times n$ constant matrix. The fundamental solution of (6.17) can be constructed from a knowledge of the eigenvalues and eigenvectors of A. We begin with the following definition.

Definition 6.6 *Le A be an $n \times n$ constant matrix, in short "matrix." A number λ is said to be an eigenvalue of A if there exists a nonzero vector v of the linear system*

$$Av = \lambda v. \tag{6.18}$$

The solution vector v is said to be an eigenvector corresponding to the eigenvalue λ. We may refer to λ and v as an eigenpair.

Now, assume a solution x of (6.17) in the form

$$x(t) = ve^{\lambda t},$$

for $x \in \mathbb{R}^n$ and $\lambda \in \mathbb{R}$. Then we have

$$\lambda ve^{\lambda t} = Ave^{\lambda t}.$$

Or

$$(A - \lambda I)v = 0, \ \text{ if and only if } \det(A - \lambda I) = 0.$$

Thus, we have the following straight forward result.

Theorem 6.7 *If λ_0, v_0 is an eigenpair of A, then*

$$x(t) = e^{\lambda_0 t} v_0 = v_0 e^{\lambda_0 t}$$

is a solution of $x' = Ax$.

Proof *Let $x(t) = e^{\lambda_0 t} v_0$, then*

$$x'(t) = \lambda_0 e^{\lambda_0 t} v_0 = \lambda_0 v_0 e^{\lambda_0 t} = Av_0 e^{\lambda_0 t} = Ae^{\lambda_0 t} v_0 = Ax,$$

as desired. This completes the proof.

Theorem 6.8 *(Independent eigenvectors) Let $v_1, v_2, \ldots, v_p$ be the corresponding eigenvectors to the distinct eigenvalues $\lambda_1, \lambda_2, \ldots, \lambda_p$ of a matrix A. Then $v_1, v_2, \ldots, v_p$ are linearly independent.*

Proof *Suppose $v_1, v_2, \ldots, v_j$ are linearly independent for a positive integer j. If $j < p$, then v_{j+1} can be written as a linear combination of the vectors, $v_1, v_2, \ldots, v_j$. That is there are constants $c_1, c_2, \ldots, c_j$ such that*

$$v_{j+1} = c_1 v_1 + c_2 v_2 + \ldots + c_j v_j.$$

Multiply from the left by the matrix A and apply the fact that $Av_i = \lambda_i v_i$ for $i = 1, 2, \ldots j$ to arrive at

$$\begin{aligned} Av_{j+1} &= \lambda_{j+1} v_{j+1} \\ &= \lambda_{j+1}(c_1 v_1 + c_2 v_2 + \ldots + c_j v_j) \\ &= c_1 \lambda_{j+1} v_1 + c_2 \lambda_{j+1} v_2 + \ldots + c_j \lambda_{j+1} v_j. \end{aligned}$$

On the other hand

$$\begin{aligned} Av_{j+1} &= A(c_1 v_1 + c_2 v_2 + \ldots + c_j v_j) \\ &= c_1 A v_1 + c_2 A v_2 + \ldots + c_j A v_j \\ &= c_1 \lambda_1 v_1 + c_2 \lambda_2 v_2 + \ldots + c_j \lambda_j v_j. \end{aligned}$$

Subtracting the two equations gives

$$c_1(\lambda_{j+1} - \lambda_1) v_1 + c_2(\lambda_{j+1} - \lambda_2) v_2 + \ldots + c_j(\lambda_{j+1} - \lambda_j) v_j = 0.$$

Since $v_1, v_2, \ldots, v_j$ are linearly independent, we must have that

$$c_1(\lambda_{j+1} - \lambda_1) = 0,\ c_2(\lambda_{j+1} - \lambda_2) = 0, \ldots, c_j(\lambda_{j+1} - \lambda_j) = 0.$$

But since

$$\lambda_{j+1} - \lambda_j \neq 0, \text{ for all } j = 1, 2, \ldots, n$$

this could only be possible if

$$c_1 = c_2, \ldots, c_j = 0.$$

This implies that the vector

$$v_{j+1} = 0, \text{ (zero vector)}$$

which is a contradiction, since v_{j+1} is the eigenvector corresponding to λ_{j+1}. This completes the proof.

As a result of Definition 6.4, Corollary 3, and Theorem 6.8, we have the following theorem concerning systems with constant matrices such as (6.17).

Theorem 6.9 *(Distinct eigenvalues) Let $\lambda_1, \lambda_2, \ldots, \lambda_n$ be n distinct real eigenvalues of the matrix A of (6.17) and let $K_1, K_2, \ldots, K_n$ be the corresponding eigenvectors. Then the general solution of (6.17) on the interval $I = (-\infty, \infty)$ is given by*

$$x(t) = c_1 K_1 e^{\lambda_1 t} + c_2 K_2 e^{\lambda_2 t}, \ldots, c_n K_n e^{\lambda_n t}$$

for constants $c_i, i = 1, 2, \ldots, n$.

Remark 20 *If we let*

$$\phi_i = K_i e^{\lambda_i t}, \quad i = 1, 2, \ldots, n$$

then the fundamental matrix of (6.17) *is formed by taking its columns to be*

$$\Phi = \left[\phi_1(t), \phi_2(t), \ldots, \phi_n(t)\right].$$

Example 6.4 *(Distinct eigenvalues) Consider the linear system*

$$x' = \begin{pmatrix} -4 & 1 & 1 \\ 1 & 5 & -1 \\ 0 & 1 & -3 \end{pmatrix} \begin{pmatrix} x_1 \\ x_2 \\ x_3 \end{pmatrix}.$$

Then the eigenpairs are given by

$$\lambda_1 = -3, K_1 = \begin{pmatrix} 1 \\ 0 \\ 1 \end{pmatrix}, \quad \lambda_2 = -4, K_2 = \begin{pmatrix} 10 \\ -1 \\ 1 \end{pmatrix}, \quad \lambda_3 = 5, K_3 = \begin{pmatrix} 1 \\ 8 \\ 1 \end{pmatrix}$$

and hence the general solution can be written as

$$x(t) = c_1 \begin{pmatrix} 1 \\ 0 \\ 1 \end{pmatrix} e^{-3t} + c_2 \begin{pmatrix} 10 \\ -1 \\ 1 \end{pmatrix} e^{-4t} + c_3 \begin{pmatrix} 1 \\ 8 \\ 1 \end{pmatrix} e^{5t}.$$

Moreover, the fundamental matrix $\Phi(t)$ *is given by*

$$\Phi(t) = \begin{pmatrix} e^{-3t} & 10e^{-4t} & e^{5t} \\ 0 & -e^{-4t} & 8e^{5t} \\ e^{-3t} & e^{-4t} & e^{5t} \end{pmatrix}. \tag{6.19}$$

Now, suppose we impose that every solution satisfies the initial condition

$$x(t_0) := x_0 = \begin{pmatrix} x_{01} \\ x_{02} \\ x_{03} \end{pmatrix},$$

then by Theorem 6.6 the unique solution is

$$x(t) = \Phi(t)\Phi^{-1}(t_0)x_0,$$

where $\Phi(t)$ *is given by* (6.19). □

Next, we discuss repeated roots. For example, the system

$$x' = \begin{pmatrix} 3 & -18 \\ 2 & -9 \end{pmatrix} \begin{pmatrix} x_1 \\ x_2 \end{pmatrix} \tag{6.20}$$

has the repeated eigenvalue $\lambda_1 = \lambda_2 = -3$. If $K_1 = \begin{pmatrix} k_1 \\ k_2 \end{pmatrix}$ is the corresponding eigenvector, then we have the two equations $6k_1 - 18k_2 = 0,\ 2k_1 - 6k_2 = 0$, which are both equivalent to $k_1 = 3k_2$. As a consequence, we have the single eigenvector $K_1 = \begin{pmatrix} 3 \\ 1 \end{pmatrix}$ and so one solution is given by

$$\phi_1 = \begin{pmatrix} 3 \\ 1 \end{pmatrix} e^{-3t}.$$

But since we are interested in finding the general solution, we need to examine the question of finding another solution.
In general, if m is a positive integer and $(\lambda - \lambda_1)^m$ is a factor of the characteristic equation $\det(A - \lambda I) = 0$, while $(\lambda - \lambda_1)^{m+1}$ is not a factor, then λ_1 is said to be an eigenvalue of multiplicity m. Below, we discuss two such scenarios:

(a) For some $n \times n$ matrices A, it may be possible to find m linearly independent eigenvectors $K_1, K_2, \ldots, K_m$ corresponding to an eigenvalue λ_1 of multiplicity $m \le n$. In this case, the general solution of the system contains the linear combination

$$c_1 K_1 e^{\lambda_1 t} + c_2 K_2 e^{\lambda_2 t} + \ldots + c_m K_m e^{\lambda_n t}.$$

(b) If there is one eigenvector corresponding to an eigenvalue of multiplicity m, then m linearly independent solutions of the form

$$\begin{aligned} \phi_1 &= K_{11} e^{\lambda_1 t} \\ \phi_2 &= K_{21} t e^{\lambda_1 t} + K_{22} e^{\lambda_1 t} \\ &\vdots \\ \phi_m &= K_{m1} \frac{t^{m-1}}{(m-1)!} e^{\lambda_1 t} + K_{m2} \frac{t^{m-2}}{(m-2)!} e^{\lambda_1 t} + \cdots K_{mm} e^{\lambda_1 t}, \end{aligned}$$

where K_{ij} are column vectors that can always be found, and they are known as *generalized eigenvectors*. For an illustration of case (b), we suppose λ_1 is an eigenvalue of multiplicity two with only one corresponding eigenvector. To find the second eigenvector, in general, we assume a second solution of

$$x' = Ax$$

of the form

$$\phi_2(t) = Kte^{\lambda_1 t} + Pe^{\lambda_1 t}, \tag{6.21}$$

where

$$P = \begin{pmatrix} p_1 \\ p_2 \\ \vdots \\ p_n \end{pmatrix} \text{ and } K = \begin{pmatrix} k_1 \\ k_2 \\ \vdots \\ k_n \end{pmatrix}$$

are to be found. Differentiate $\phi_2(t)$ and substitute back into $x' = Ax$ to get

$$(AK - \lambda_1 K)te^{\lambda_1 t} + (AP - \lambda_1 P - K)e^{\lambda_1 t} = 0.$$

Since the above equation must hold for all t, we must have

$$(A - \lambda_1 I)K = 0 \tag{6.22}$$

and

$$(A - \lambda_1 I)P = K. \tag{6.23}$$

Equation (6.22) implies K must be an eigenvector of A associated with the eigenvalue λ_1, and hence we have one solution $\phi_1(t) = Ke^{\lambda_1 t}$. To find the second solution given by (6.21) we must solve for the vector P in (6.23). Now, we go back to finding a second solution for (6.19). We already know K, which is given by K_1. Let $P = \begin{pmatrix} p_1 \\ p_2 \end{pmatrix}$. Then from equation (6.23), we have $(A+3I)P = K$, which implies that $6p_1 - 18p_2 = 3$, or $2p_1 - 6p_2 = 1$. Since these two equations are equivalent, we may choose $p_1 = 1$ and find $p_2 = 1/6$. However, for simplicity, we shall choose $p_1 = 1/2$ so that $p_2 = 0$. Using (6.21), we find that

$$\phi_2(t) = \begin{pmatrix} 3 \\ 1 \end{pmatrix} te^{-3t} + \begin{pmatrix} 1/2 \\ 0 \end{pmatrix} e^{-3t}.$$

Thus, the general solution is

$$x(t) = c_1\phi_1(t) + c_2\phi_2(t) = c_1 \begin{pmatrix} 3 \\ 1 \end{pmatrix} e^{-3t} + c_2\left(\begin{pmatrix} 3 \\ 1 \end{pmatrix} te^{-3t} + \begin{pmatrix} 1/2 \\ 0 \end{pmatrix} e^{-3t}\right).$$

Moreover, the fundamental matrix $\Phi(t)$ is given by

$$\Phi(t) = \begin{pmatrix} 3e^{-3t} & 3te^{-3t} + \frac{1}{2}e^{-3t} \\ e^{-3t} & te^{-3t} \end{pmatrix}.$$

6.3.1 Exponential Matrix e^{At}

In the previous section, we looked at the concept of fundamental matrix $\Phi(t)$ for the linear system $x'(t) = A(t)x(t)$ and applied it to only constant matrices. In this section, we study, in detail, such constant matrices; namely, we consider

$$x'(t) = Ax(t), \quad x(t_0) = x_0, \tag{6.24}$$

where A is an $n \times n$ constant matrix and $x \in \mathbb{R}^n$. We note that

$$||Ax - Ay|| \leq K||x - y||$$

where $K > 0$ is constant and $||\cdot||$ is the matrix norm of A defined by

$$||A|| = \sup_{x \in \mathbb{R}^n - \{0\}} \frac{|Ax|}{|x|}.$$

We may define Piccard's iteration by

$$x_{n+1}(t) = x_0 + \int_{t_0}^{t} Ax_n(s)ds, \; x_0(t) = x_0 \tag{6.25}$$

with the understanding that

$$x_i = (\phi_1^i, \phi_2^i, \ldots, \phi_n^i)^T, \; i = 1, 2, \ldots, n$$

where $\phi_j^i \in \mathbb{R}^n, \; j = 1, 2, \ldots, n$. Setting $n = 0$ in (6.25) we get

$$x_1 = x_0 + \int_{t_0}^{t} Ax_0 ds = x_0 + (t - t_0)Ax_0,$$

Similarly, for $n = 1$, we have

$$\begin{aligned} x_2 &= x_0 + \int_{t_0}^{t} Ax_1(s)ds = x_0 + \int_{t_0}^{t} A(x_0 + (s - t_0))Ax_0 ds \\ &= x_0 + (t - t_0)Ax_0 + \frac{(t - t_0)^2}{2} A^2 x_0. \end{aligned}$$

Using an induction argument, we arrive at

$$x_n = x_0 + (t - t_0)Ax_0 + \frac{(t - t_0)^2}{2} A^2 x_0 + \ldots + \frac{(t - t_0)^n}{n!} A^n x_0 = \sum_{k=0}^{n} \frac{(t - t_0)^k}{k!} A^k x_0.$$

Letting $n \to \infty$, we have

$$x(t) = e^{A(t - t_0)} x_0, \tag{6.26}$$

where

$$e^{A(t - t_0)} = \sum_{n=0}^{\infty} \frac{(t - t_0)^n}{n!} A^n. \tag{6.27}$$

It readily follows from the ratio test that the infinite series in (6.27) converges absolutely for all $t - t_0 \in \mathbb{R}$. Thus $e^{A(t-t_0)}$ exists for all time. We note that the infinite series given by (6.27) can be differentiated term by term and hence

$$\begin{aligned}
\frac{d}{dt}e^{A(t-t_0)} &= \sum_{n=0}^{\infty} n\frac{(t-t_0)^{n-1}}{n!}A^n \\
&= \sum_{n=1}^{\infty} \frac{(t-t_0)^{n-1}}{(n-1)!}A^n \\
&= A\sum_{n=1}^{\infty} \frac{(t-t_0)^{n-1}}{(n-1)!}A^{n-1} \\
&= A\sum_{k=0}^{\infty} \frac{(t-t_0)^{k}}{k!}A^{k} \\
&= Ae^{A(t-t_0)}.
\end{aligned}$$

This shows that if

$$x(t) = e^{A(t-t_0)}x_0,$$

then $x(t)$ satisfies $x' = Ax$. Moreover,

$$x(t) = e^{A(t-t_0)}x_0 = \Big[I + (t-t_0)A + \frac{(t-t_0)^2}{2}A^2 + \ldots + \frac{(t-t_0)^n}{n!}A^n + \ldots\Big]x_0,$$

and hence

$$x(t_0) = Ix_0 = x_0.$$

We have the following definition.

Definition 6.7 *Let A be an $n \times n$ constant matrix. The matrix exponential is defined by*

$$e^{A(t-t_0)} = \sum_{n=0}^{\infty} \frac{(t-t_0)^n}{n!}A^n.$$

It is the unique matrix-valued solution of the initial value problem

$$x'(t) = Ax(t), \qquad x(t_0) = I,$$

where I is the identity matrix.

Thus, we have already proved that

$$x(t) = e^{A(t-t_0)}x_0 \tag{6.28}$$

is the unique solution of

$$x'(t) = Ax(t), \;\; x(t_0) = x_0,$$

for all $t \in \mathbb{R}$.

Example 6.5 *Consider the system*

$$x' = \begin{pmatrix} 3 & 0 \\ 0 & -1 \end{pmatrix} \begin{pmatrix} x_1 \\ x_2 \end{pmatrix}, \quad x(0) = \begin{pmatrix} 2 \\ -2 \end{pmatrix}$$

then the matrix A is diagonal and given by

$$A = \begin{pmatrix} 3 & 0 \\ 0 & -1 \end{pmatrix}.$$

Since A is a diagonal matrix, by induction, one can easily show that

$$A^n = \begin{pmatrix} 3^n & 0 \\ 0 & (-1)^n \end{pmatrix}, \quad n = 1, 2, \ldots.$$

Hence

$$I + tA + \frac{(tA)^2}{2} + \ldots + \frac{(tA)^n}{n!} = \begin{pmatrix} 1 + \frac{3t}{1!} + \frac{(3t)^2}{2!} + \ldots + \frac{(3t)^n}{n!} & 0 \\ 0 & 1 + \frac{(-t)}{1!} + \frac{(-t)^2}{2!} + \ldots + \frac{(-t)^n}{n!} \end{pmatrix}.$$

We deduce from the exponential function properties that the exponential matrix e^{At} is given by

$$e^{At} = \begin{pmatrix} e^{3t} & 0 \\ 0 & e^{-t} \end{pmatrix}$$

and the solution is

$$x(t) = e^{At} x_0 = \begin{pmatrix} e^{3t} & 0 \\ 0 & e^{-t} \end{pmatrix} \begin{pmatrix} 2 \\ -2 \end{pmatrix}.$$

□

In the next theorem, we list some properties of the exponential matrix. For simplicity, we start at the initial time zero. That is, we take $t_0 = 0$. We are concerned with

$$x' = Ax, \tag{6.29}$$

where A is an $n \times n$ constant matrix and $x \in \mathbb{R}^n$.

Theorem 6.10 *Assume A and B are $n \times n$ constant matrices, then:*

(1) $\frac{d}{dt} e^{At} = Ae^{At}, \quad t \in \mathbb{R}$

(2) $\det(e^{At}) \neq 0$, $t \in \mathbb{R}$, *and* e^{At} *is a fundamental matrix solution for* (6.29).

(3) $e^{At} = I + At + \frac{1}{2}(At)^2 + \frac{1}{6}(At)^3 + \ldots + \frac{1}{k!}(At)^k + \ldots$

(4) $e^{At}e^{As} = e^{A(t+s)}, \quad t, s \in \mathbb{R}$

(5) $\left(e^{At}\right)^{-1} = e^{-At}$

(6) If $AB = BA$, *then* $e^{At}B = Be^{At}$

(7) If $AB = BA$, *then* $e^{At}e^{Bt} = e^{(A+B)t}, \quad t \in \mathbb{R}$

(8) If P *is a nonsingular matrix , then* $e^{PBP^{-1}} = Pe^{B}P^{-1}$.

Proof *The proof of* (1) *is already done. For* (2), *we already know that* e^{At} *solves* (6.29). *Moreover, at* $t = 0$, $e^{At} = I$ *and hence* $\det(e^{A0}) = \det(I) \neq 0$. *By the uniqueness of solution, this must hold for all* $t \in \mathbb{R}$. *Now, since* e^{At} *solves* (6.29) *we have* e^{At} *is a fundamental matrix for* (6.29). *The proof of* (3) *is already done. For* (4), *we let* $L(t) = e^{At}e^{As} - e^{A(t+s)}$, *where* $s \in \mathbb{R}$ *is fixed and* $t \in \mathbb{R}$. *Then*

$$L'(t) = Ae^{At}e^{As} - Ae^{A(t+s)} = A\left(e^{At}e^{As} - e^{A(t+s)}\right),$$

or $L'(t) = AL(t)$, *from which we conclude that* $L(t)$ *solves* (6.29). *On the other hand,* $L(0) = e^{As} - e^{As} = 0$, *and hence by the uniqueness of solutions, we must have* $L(t) = 0$ *for all* $t \in \mathbb{R}$. *This completes the proof of* (4).
By the proof of (4), *we have that*

$$e^{At}e^{-At} = e^{At-At} = e^{A(t+(-t))} = I.$$

This implies that

$$\left(e^{At}\right)^{-1} = e^{-At}, \; t \in \mathbb{R}.$$

Next, we discuss an important topic that relies on *Jordan canonical form* and helps us compute the matrix exponential e^{At}. The following theorem is well known in linear algebra; for a proof, see Brauer and Nohel, [5].

Theorem 6.11 *(Jordan Canonical Form) For a general* $n \times n$ *matrix* A, *we can construct a matrix* C *such that*

$$C^{-1}AC = \begin{pmatrix} J_1 & 0 & \cdots & 0 \\ 0 & J_2 & \ddots & \vdots \\ \vdots & \ddots & \ddots & \vdots \\ 0 & \cdots & 0 & j_k \end{pmatrix}, \tag{6.30}$$

where

$$J_k = \lambda_k I + N_k,$$

with

$$N_k = \begin{pmatrix} 0 & 1 & 0 & 0 & \cdots & 0 & 0 \\ 0 & 0 & 1 & 0 & \cdots & 0 & 0 \\ \cdot & \cdot & \cdot & \cdot & \cdots & \cdot & \cdot \\ 0 & 0 & 0 & 0 & \cdots & 0 & 1 \\ 0 & 0 & 0 & 0 & \cdots & 0 & 0 \end{pmatrix}.$$

Remark 21 *(The simplest case) If A has n real and distinct eigenvalues* $\lambda_1, \lambda_2, \ldots, \lambda_n$*, then the matrix* $C^{-1}AC$ *will be a diagonal matrix with entries* λ_i*. That is*

$$C^{-1}AC = \begin{pmatrix} \lambda_1 & 0 & \cdots & 0 \\ 0 & \lambda_2 & \ddots & \vdots \\ \vdots & \ddots & \ddots & \vdots \\ 0 & \cdots & 0 & \lambda_n \end{pmatrix}.$$

In this case, the matrix C is formed from the corresponding eigenvectors of the matrix A.

If (6.30) holds, then let

$$B = C^{-1}AC,$$

and hence

$$e^{tA} = e^{tCBC^{-1}} = Ce^{tB}C^{-1}.$$

In Section 6.2, we encountered different scenarios concerning the type and nature of the eigenvalues when computing eigenvectors. The next result, called *Putzer's algorithm*, allows us to find the fundamental matrix, or in particular the matrix exponential e^{At}, when the matrix A is constant, regardless of the order or multiplicities of the eigenvalues. Putzer's algorithm relies on the following theorem, known as the Cayley–Hamilton Theorem.

Theorem 6.12 *(Cayley-Hamilton) Every* $n \times n$ *constant matrix satisfies its characteristic equation*

$$p_A(\lambda) := \det(A - \lambda I) = 0.$$

In other words, if

$$p_A(\lambda) = (-1)^n\left(\lambda^n + c_1\lambda^{n-1} + \ldots + c_{n-1}\lambda + c_n\right),$$

then

$$p_A(A) = (-1)^n\left(A^n + c_1A^{n-1} + \ldots + c_{n-1}A + c_nI\right) = 0.$$

Theorem 6.13 (Putzer algorithm) *Suppose A is an $n \times n$ constant matrix with eigenvalues $\lambda_1, \lambda_2, \ldots, \lambda_n$, listed with algebraic multiplicity so that*

$$\det(\lambda I - A) = \prod_{k=1}^{n} (\lambda - \lambda_k).$$

Let $\mu_1(t), \mu_2(t), \ldots, \mu_n(t)$ be the scalar functions defined by the initial value problems

$$\mu_1'(t) = \lambda_1 \mu_1(t), \quad \mu_1(0) = 1; \quad \mu_j'(t) = \lambda_j \mu_j(t) + \mu_{j-1}(t), \quad \mu_j(0) = 0,\ j = 2, 3, \ldots, n. \tag{6.31}$$

Define matrices $P_0, P_1, \ldots, P_n$ recursively by

$$P_0 = I, \qquad P_j = \prod_{k=1}^{j} (A - \lambda_k I), \quad j = 1, 2, \ldots, n. \tag{6.32}$$

Then

$$e^{At} = \sum_{j=0}^{n-1} \mu_{j+1}(t) P_j. \tag{6.33}$$

Proof *Define*

$$\Phi(t) = \sum_{j=0}^{n-1} \mu_{j+1}(t) P_j.$$

Evaluating at $t = 0$ and using (6.31) gives

$$\begin{aligned}
\Phi(0) &= \sum_{j=0}^{n-1} \mu_{j+1}(0) P_j \\
&= \mu_1(0) P_0 + \mu_2(0) P_1 + \cdots + \mu_n(0) P_{n-1} \\
&= \mu_1(0) P_0 + 0 + \cdots + 0 \\
&= I,
\end{aligned}$$

since $\mu_1(0) = 1$ and $P_0 = I$.

Next we differentiate $\Phi(t)$ with respect to t. From (6.31) we obtain

$$\begin{aligned}
\Phi'(t) &= \sum_{j=0}^{n-1} \mu_{j+1}'(t) P_j \qquad (6.34) \\
&= \lambda_1 \mu_1(t) P_0 + \sum_{j=1}^{n-1} \mu_{j+1}'(t) P_j \\
&= \lambda_1 \mu_1(t) P_0 + \sum_{j=1}^{n-1} \big(\lambda_{j+1} \mu_{j+1}(t) + \mu_j(t)\big) P_j.
\end{aligned}$$

On the other hand,

$$A\Phi(t) = \sum_{j=0}^{n-1} \mu_{j+1}(t) AP_j.$$

Using (6.32) *we have*

$$P_j = \prod_{k=1}^{j} (A - \lambda_k I) = (A - \lambda_{j+1} I) \prod_{k=1}^{j} (A - \lambda_k I)$$

for $j = 0, 1, \ldots, n-1$, *which implies*

$$P_{j+1} = (A - \lambda_{j+1} I) P_j = AP_j - \lambda_{j+1} P_j,$$

and hence

$$AP_j = P_{j+1} + \lambda_{j+1} P_j, \qquad j = 0, 1, \ldots, n-1.$$

Substituting this into the expression for $A\Phi(t)$ *yields*

$$\begin{aligned} A\Phi(t) &= \sum_{j=0}^{n-1} \mu_{j+1}(t) \big(P_{j+1} + \lambda_{j+1} P_j\big) \\ &= \lambda_1 \mu_1(t) P_0 + (\lambda_2 \mu_2(t) + \mu_1(t)) P_1 + \cdots + (\lambda_n \mu_n(t) + \mu_{n-1}(t)) P_{n-1} + \mu_n(t) P_n. \end{aligned}$$

By the Cayley–Hamilton theorem, the characteristic polynomial

$$p(\lambda) = \prod_{k=1}^{n} (\lambda - \lambda_k)$$

satisfies $p(A) = 0$, *and thus*

$$P_n = \prod_{k=1}^{n} (A - \lambda_k I) = 0.$$

Therefore, the last term $\mu_n(t) P_n$ *vanishes, and the remaining sum coincides with the right-hand side of* (6.34). *Hence*

$$\Phi'(t) = A\Phi(t), \qquad \Phi(0) = I.$$

By the uniqueness of the solution of this matrix initial value problem, we conclude that $\Phi(t) = e^{At}$, *which establishes* (6.33).

Example 6.6 *Consider the linear system*

$$x' = \begin{pmatrix} 2 & 2 & -1 \\ 2 & -1 & 2 \\ -1 & 2 & 2 \end{pmatrix} \begin{pmatrix} x_1 \\ x_2 \\ x_3 \end{pmatrix}$$

satisfying the initial condition

$$x(0) := x_0 = \begin{pmatrix} 1 \\ -1 \\ 2 \end{pmatrix}.$$

The matrix $A = \begin{pmatrix} 2 & 2 & -1 \\ 2 & -1 & 2 \\ -1 & 2 & 2 \end{pmatrix}$ *has the eigenvalues* $3, 3, -3$. *The eigenvalues may be ordered in any way. We chose to let* $\lambda_1 = 3, \lambda_2 = 3,$ *and* $\lambda_3 = -3$. *Using* (6.32) *with* $j = 1, 2$, *we see that*

$$P_0 = I,\ P_1 = A - 3I,\ P_2 = (A - 3I)^2.$$

Similarly, by (6.31)

$$\mu_1' = 3\mu_1,\ \mu_1(0) = 1;\ \mu_2' = 3\mu_2 + \mu_1(t),\ \mu_2(0) = 0,\ \mu_3' = -3\mu_3 + \mu_2(t),\ \mu_3(0) = 0.$$

Solving the above initial value problems gives

$$\mu_1(t) = e^{3t},\ \ \mu_2(t) = te^{3t},\ \text{and}\ \mu_3(t) = \frac{te^{3t}}{6} - \frac{e^{3t}}{36} + \frac{e^{-3t}}{36}.$$

We obtain by using (6.33) *with* $n = 3$

$$e^{At} = \sum_{j=0}^{2} \mu_{j+1}(t) P_j = \mu_1 P_0 + \mu_2 P_1 + \mu_3 P_2$$

and the unique solution is given by

$$x(t) = e^{At} x_0.$$

□

Example 6.7 *Assume we have three masses,* $m_1, m_2,$ *and* m_3, *connected by four springs with spring constants* $k_1, k_2, k_3,$ *and* k_4, *respectively. As shown in Fig. 6.2, the first and last springs are attached to fixed walls, and the masses slide along a frictionless horizontal surface. The displacements of the masses from their equilibrium positions are denoted by* $x, y,$ *and* z, *each taken positive to the right.*

The motion of the masses depends strongly on the initial conditions. For example, one may initially displace one or more of the masses by pulling them to the right or left and then releasing them. We now derive the equations of motion.

The springs obey Hooke's law: the restoring force is proportional to the elongation or compression of the spring. The elongations of the first three springs are x, $y - x$,

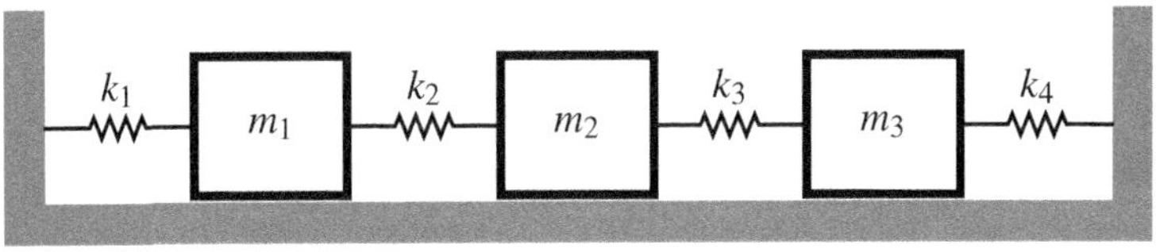

FIGURE 6.2: Three masses connected by four springs.

and $z - y$, respectively. Thus, using Newton's second law, we obtain the system of second-order differential equations

$$\begin{aligned} m_1x''(t) &= -k_1x(t) + k_2[y(t) - x(t)], \\ m_2y''(t) &= -k_2[y(t) - x(t)] + k_3[z(t) - y(t)], \\ m_3z''(t) &= -k_3[z(t) - y(t)] - k_4z(t). \end{aligned} \tag{6.35}$$

System (6.35) *can be written as a second-order vector-matrix system*

$$M\mathbf{x}''(t) = A\mathbf{x}(t),$$

where the displacement vector $\mathbf{x}$, *the mass matrix* M, *and the stiffness matrix* A *are given by*

$$\mathbf{x} = \begin{pmatrix} x \\ y \\ z \end{pmatrix}, \quad M = \begin{pmatrix} m_1 & 0 & 0 \\ 0 & m_2 & 0 \\ 0 & 0 & m_3 \end{pmatrix}, \quad A = \begin{pmatrix} -k_1 - k_2 & k_2 & 0 \\ k_2 & -k_2 - k_3 & k_3 \\ 0 & k_3 & -k_3 - k_4 \end{pmatrix}.$$

Now let

$$m_1 = m_2 = m_3 = 1, \qquad k_1 = 2, \qquad k_2 = 1, \qquad k_3 = 1, \qquad k_4 = 2.$$

Then the system becomes

$$\mathbf{x}''(t) = A\mathbf{x}(t).$$

Using Putzer's algorithm, we find that the solution is

$$\begin{aligned} \begin{pmatrix} x \\ y \\ z \end{pmatrix} &= (a_1 \cos t + b_1 \sin t) \begin{pmatrix} 1 \\ 2 \\ 1 \end{pmatrix} \\ &\quad + (a_2 \cos(\sqrt{3}t) + b_2 \sin(\sqrt{3}t)) \begin{pmatrix} 1 \\ 0 \\ -1 \end{pmatrix} \\ &\quad + (a_3 \cos(2t) + b_3 \sin(2t)) \begin{pmatrix} 1 \\ -1 \\ 1 \end{pmatrix}, \end{aligned}$$

where $a_1, a_2, a_3, b_1, b_2,$ and b_3 are arbitrary constants. □

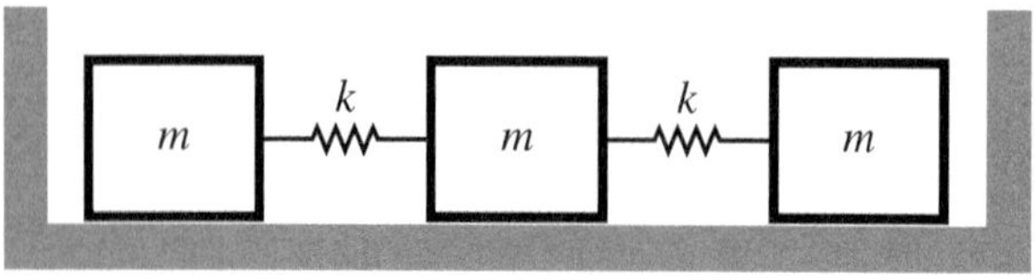

FIGURE 6.3: Three identical boxcars connected by identical springs.

Example 6.8 *(Boxcars) A special case of a mass-spring system is boxcars. In Fig. 6.3 we assume three identical boxcars on a level track connected by two identical springs. Except for the springs on fixed ends, this problem is the same as in Example 6.7. With this in mind, we have that* $k_1 = k_4 = 0$, *and* $k_2 = k_3 = k$. *In addition,* $m_1 = m_2 = m_3 = m$. *Consequently, we arrive at the system*

$$M\mathbf{x}''(t) = A\mathbf{x}(t),$$

where the displacement $\mathbf{x}$, *mass matrix* M *and stiffness matrix* A *are given by*

$$\mathbf{x} = \begin{pmatrix} x \\ y \\ z \end{pmatrix}, \quad M = \begin{pmatrix} m & 0 & 0 \\ 0 & m & 0 \\ 0 & 0 & m \end{pmatrix}, \quad A = \begin{pmatrix} -k & k & 0 \\ k & -2k & k \\ 0 & k & -k \end{pmatrix}.$$

For illustrative purposes, we take $\frac{k}{m} = 1$. *Then we have the second order system*

$$\mathbf{x}'' = \begin{pmatrix} -1 & 1 & 0 \\ 1 & -2 & 1 \\ 0 & 1 & -1 \end{pmatrix} \mathbf{x},$$

which has the vector solution

$$\begin{aligned} \begin{pmatrix} x \\ y \\ z \end{pmatrix} &= (a_1 + b_1 t) \begin{pmatrix} 1 \\ 1 \\ 1 \end{pmatrix} \\ &+ (a_2 \cos(t) + b_2 \sin(t)) \begin{pmatrix} 1 \\ 0 \\ -1 \end{pmatrix} \\ &+ (a_3 \cos(\sqrt{3}t) + b_3 \sin(\sqrt{3}t)) \begin{pmatrix} 1 \\ -2 \\ 1 \end{pmatrix} \end{aligned}$$

where a_1, a_2, a_3, b_1, b_2, *and* b_3 *are arbitrary constants.* □

6.4 $x' = A(t)x + g(t)$

Now we consider the nonhomogeneous system

$$x' = A(t)x + g(t) \tag{6.36}$$

where $A(t)$ is an $n \times n$ matrix and $g(t)$ is an $n \times 1$ vector that are continuous on some interval Ia. It is known that system (6.36) has a unique solution on I, if an initial condition is specified for $t_0 \in I$. We are interested in finding the *general solution* of the nonhomogeneous system given by (6.36). We associate with the nonhomogeneous system the complementary homogeneous system

$$x' = A(t)x. \tag{6.37}$$

Knowing the solution of (6.37) is enough to find the general solution of (6.36), and such a solution will be called the *variation-of-parameters formula*

Definition 6.8 *Let x_h be the homogeneous solution of* (6.37) *and x_p be the particular solution of* (6.36) *on an interval I. Then the general solution $x : I \to \mathbb{R}^n$ of* (6.36) *on the interval I is given by*

$$x(t) = x_h(t) + x_p(t).$$

We have seen that if Φ is the fundamental matrix of the homogeneous system (6.37), then $x_h = \Phi(t)k$ for some constant $n \times 1$ vector k.

Theorem 6.14 *Let $A(t)$ be an $n \times n$ matrix and $g(t)$ be an $n \times 1$ vector that are continuous on some interval I. Suppose Φ is the fundamental matrix of the homogeneous system* (6.37). *Then $x(t)$ is a solution of*

$$x' = A(t)x + g(t), \;\; x(t_0) = x_0, \; t \geq t_0 \tag{6.38}$$

on I if and only if x satisfies

$$x(t) = \Phi(t)\Phi^{-1}(t_0)x_0 + \int_{t_0}^{t} \Phi(t)\Phi^{-1}(s)g(s)ds, \tag{6.39}$$

where $t_0 \in I$ and $x_0 \in \mathbb{R}^n$.

Proof *First we note that the solution exists and is unique. Let Φ be the fundamental matrix of the homogeneous system* (6.37). *Suppose x satisfies* (6.39). *Taking the derivative of t in* (6.39) *gives*

$$\begin{aligned} x'(t) &= \Phi'(t)\Phi^{-1}(t_0)x_0 + \Phi'(t)\int_{t_0}^{t} \Phi^{-1}(s)g(s)ds + \Phi(t)\Phi^{-1}(t)g(t) \\ &= A(t)\Phi(t)\Phi^{-1}(t_0)x_0 + A(t)\Phi(t)\int_{t_0}^{t} \Phi^{-1}(s)g(s)ds + g(t) \end{aligned}$$

$$\begin{aligned} &= A(t)\left[\Phi(t)\Phi^{-1}(t_0)x_0 + \Phi(t)\int_{t_0}^{t}\Phi^{-1}(s)g(s)ds\right] + g(t) \\ &= A(t)x(t) + g(t). \end{aligned}$$

Finally, replacing t by t_0 *in* (6.39) *gives* $x(t_0) = Ix_0 = x_0$. *Now we assume* $x(t)$ *is the unique solution of* (6.38), *and show it satisfies* (6.39). *As we have mentioned earlier,* $x_h = \Phi(t)k$ *for some constant* $n \times 1$ *vector* k. *To derive the particular solution* x_p, *we assume a particular solution of the form*

$$x_p(t) = \Phi(t)u(t),$$

for some continuous function u *on* I, *to be found. Differentiate* x_p *and plug back into* (6.38).

$$x_p' = \Phi'(t)u(t) + \Phi(t)u'(t).$$

Or,

$$\Phi'(t)u(t) + \Phi(t)u'(t) = A(t)\Phi(t)u(t) + g(t).$$

Using the fact that

$$\Phi'(t) = A(t)\Phi(t),$$

the above expression reduces to

$$\Phi(t)u'(t) = g(t).$$

Multiply the above expression by $\Phi^{-1}(t)$ *from the left and get* $u'(t) = \Phi^{-1}(t)g(t)$, *or* $u(t) = \int_{t_0}^{t}\Phi^{-1}(s)g(s)ds$. *Finally using* $x_p(t) = \Phi(t)u(t)$ *gives,*

$$x_p(t) = \Phi(t)\int_{t_0}^{t}\Phi^{-1}(s)g(s)ds.$$

It follows that the general solution is

$$x(t) = x_h(t) + x_p(t) = \Phi(t)k + \int_{t_0}^{t}\Phi(t)\Phi^{-1}(s)g(s)ds. \tag{6.40}$$

Using the initial condition in (6.40) *we get*

$$x_0 = \Phi(t_0)k + 0,$$

or

$$k = \Phi^{-1}(t_0)x_0.$$

A substitution of k into (6.40) *gives* (6.39). *This completes the proof.*

Equation (6.39) is called *variation of parameters formula*.

Remark 22 *In the case that A is an $n \times n$ constant matrix, the exponential matrix is equivalent to*

$$e^{A(t-t_0)} = \Phi(t)\Phi^{-1}(t_0),$$

which is the principal matrix. In this case, (6.39) *reduces to or takes the form*

$$x(t) = e^{A(t-t_0)}x_0 + \int_{t_0}^{t} e^{A(t-s)}g(s)ds. \tag{6.41}$$

Example 6.9 *Solve*

$$x'(t) = \begin{pmatrix} 4 & 2 \\ 3 & -1 \end{pmatrix} x + \begin{pmatrix} -15te^{-2t} \\ -4te^{-2t} \end{pmatrix}, \quad x(0) = \begin{pmatrix} 7 \\ 3 \end{pmatrix}.$$

Using Putzer's algorithm we find that

$$e^{At} = \frac{1}{7}\begin{pmatrix} e^{-2t} + 6e^{5t} & -2e^{-2t} + 2e^{5t} \\ -3e^{-2t} + 3e^{5t} & 6e^{-2t} + e^{5t} \end{pmatrix}.$$

Note that $e^{A0} = I$. Using formula (6.41), *we have*

$$\begin{aligned} x(t) &= \frac{1}{7}\begin{pmatrix} e^{-2t} + 6e^{5t} & -2e^{-2t} + 2e^{5t} \\ -3e^{-2t} + 3e^{5t} & 6e^{-2t} + e^{5t} \end{pmatrix}\begin{pmatrix} 7 \\ 3 \end{pmatrix} \\ &+ \int_{t_0}^{t} \frac{1}{7}\begin{pmatrix} e^{-2(t-s)} + 6e^{5(t-s)} & -2e^{-2(t-s)} + 2e^{5(t-s)} \\ -3e^{-2(t-s)} + 3e^{5(t-s)} & 6e^{-2(t-s)} + e^{5(t-s)} \end{pmatrix}\begin{pmatrix} -15se^{-2s} \\ -4se^{-2s} \end{pmatrix} ds \\ &= \frac{1}{14}\begin{pmatrix} (6 + 28t - 7t^2)e^{-2t} + 92e^{5t} \\ (-4 + 14t + 21t^2)e^{-2t} + 46e^{5t} \end{pmatrix}. \end{aligned}$$

□

6.5 Exercises

Exercise 6.10 *Show that $||A||_1$ is a norm. That is, verify all the items in Definition 6.2.*

Exercise 6.11 *Show that if A is an $n \times n$ invertible matrix, then*

$$||A|| > 0, \text{ and } ||A^{-1}|| \geq \frac{1}{||A||}.$$

Exercise 6.12 *Use Theorem 6.9 to solve the linear system*

$$x' = \begin{pmatrix} 5 & 2 & 0 \\ 2 & 5 & 0 \\ -3 & 4 & 6 \end{pmatrix} \begin{pmatrix} x_1 \\ x_2 \\ x_3 \end{pmatrix}.$$

satisfying the initial condition

$$x(t_0) := x_0 = \begin{pmatrix} 1 \\ 0 \\ 2 \end{pmatrix},$$

for all $t \geq t_0$.

Exercise 6.13 *Use Theorem 6.9 to solve the linear system with initial condition*

$$x' = \begin{pmatrix} 1 & -12 & -14 \\ 1 & 2 & -3 \\ 1 & 1 & -2 \end{pmatrix} \begin{pmatrix} x_1 \\ x_2 \\ x_3 \end{pmatrix}, \quad x(0) := x_0 = \begin{pmatrix} 4 \\ 6 \\ -7 \end{pmatrix},$$

Exercise 6.14 *(a) Let*

$$x'(t) = Ax(t),$$

where A is a constant $n \times n$ matrix. Suppose A has only one eigenvector K associated with an eigenvalue λ_1 of multiplicity three. We can find a second solution of the form (6.21) where the eigenvector P satisfies (6.23). Show that if we assume a third solution, $\phi_3(t)$ of the form

$$\phi_3(t) = K\frac{t^2}{2}e^{\lambda_1 t} + Pte^{\lambda_1 t} + Qe^{\lambda_1 t},$$

then the vector Q is given by the equation

$$(A - \lambda_1 I)Q = P.$$

(b) Find the fundamental matrix for the linear system

$$x' = \begin{pmatrix} 1 & 0 & 0 \\ 2 & 2 & -1 \\ 0 & 1 & 0 \end{pmatrix} \begin{pmatrix} x_1 \\ x_2 \\ x_3 \end{pmatrix}.$$

Exercise 6.15 *(a) Let*

$$x'(t) = Ax(t),$$

where A is an $n \times n$ constant matrix with real entries. Let K be an eigenvector corresponding to the complex eigenvalue $\lambda_1 = \alpha + i\beta$, α and β are real. Show that the two linearly independent corresponding solutions $\phi_1(t)$ and $\phi_2(t)$ are given by

$$\phi_1(t) = \Big(\Re(K)\cos(\beta t) - Im(K)\sin(\beta t)\Big)e^{\alpha t},$$

and

$$\phi_2(t) = \Big(Im(K)\cos(\beta t) + \Re(K)\sin(\beta t)\Big)e^{\alpha t}.$$

For example if $K = \begin{pmatrix} 1 \\ 2i \end{pmatrix}$, *then*

$$K = \begin{pmatrix} 1 \\ 0 \end{pmatrix} + \begin{pmatrix} 0 \\ 2i \end{pmatrix} = \Re(K) + Im(K).$$

(b) Find the fundamental matrix for the linear system

$$x' = \begin{pmatrix} 1 & 2 \\ -1/2 & 1 \end{pmatrix} \begin{pmatrix} x_1 \\ x_2 \end{pmatrix}.$$

Exercise 6.16 *Let Φ be a fundamental matrix of $x' = Ax$ and C is nonsingular with $AC = CA$. Prove that $C\Phi$ is now a fundamental matrix of $x' = Ax$. (See Theorem 6.4)*

Exercise 6.17 *Prove parts* (5), (6) *and* (7) *of Theorem 6.10.*

Exercise 6.18 *Finish the calculation in Example 6.6 and show that*

$$e^{At} = \cosh(3t)I + \frac{1}{3}\sinh(3t)A.$$

Exercise 6.19 *Use Theorem 6.13 to compute e^{At} for:*

(a)

$$A = \begin{pmatrix} 2 & 0 & 0 \\ 1 & 2 & 0 \\ 1 & 0 & 3 \end{pmatrix}.$$

(b)

$$A = \begin{pmatrix} 5 & -4 & 0 \\ 1 & 0 & 2 \\ 0 & 2 & 5 \end{pmatrix}.$$

Exercise 6.20 *Use Theorem 6.13 to compute e^{At} for*

$$x' = \begin{pmatrix} -1 & -2 & 2 \\ 3 & 5 & -3 \\ 3 & 4 & -2 \end{pmatrix} \begin{pmatrix} x_1 \\ x_2 \\ x_3 \end{pmatrix}$$

satisfying the initial condition

$$x(0) := x_0 = \begin{pmatrix} 1 \\ -1 \\ 1 \end{pmatrix}.$$

Exercise 6.21 (Putzer algorithm for a 2×2 matrix) *Let*

$$A = \begin{pmatrix} 2 & 1 \\ 0 & 3 \end{pmatrix}.$$

Compute e^{At}.

Exercise 6.22 (Repeated eigenvalue case) *Consider the matrix*

$$A = \begin{pmatrix} 1 & 1 \\ 0 & 1 \end{pmatrix},$$

which has a repeated eigenvalue.

(a) Compute the eigenvalues of A and list them with multiplicity.

(b) Using Theorem 6.13, take $\lambda_1 = \lambda_2 = 1$ and define

$$P_0 = I, \qquad P_1 = A - \lambda_1 I.$$

(c) Write down and solve the system

$$\mu_1'(t) = \lambda_1 \mu_1(t), \quad \mu_1(0) = 1; \qquad \mu_2'(t) = \lambda_2 \mu_2(t) + \mu_1(t), \quad \mu_2(0) = 0.$$

(d) Show that

$$e^{At} = \mu_1(t) P_0 + \mu_2(t) P_1$$

reduces to the familiar formula

$$e^{At} = \begin{pmatrix} e^t & te^t \\ 0 & e^t \end{pmatrix}.$$

(e) Explain briefly why Putzer's algorithm works even when A is not diagonalizable.

Exercise 6.23 *Let*

$$A = \begin{pmatrix} 0 & 1 & 0 \\ 0 & 0 & 1 \\ -2 & -3 & -1 \end{pmatrix}.$$

(a) Compute the characteristic polynomial of A and find its eigenvalues $\lambda_1, \lambda_2, \lambda_3$, counting multiplicities.

(b) Choose an ordering of the eigenvalues and define

$$P_0 = I, \qquad P_1 = (A - \lambda_1 I), \qquad P_2 = (A - \lambda_2 I)(A - \lambda_1 I).$$

(c) Write down the coupled system of scalar ODEs

$$\mu_1'(t) = \lambda_1 \mu_1(t), \quad \mu_1(0) = 1,$$
$$\mu_2'(t) = \lambda_2 \mu_2(t) + \mu_1(t), \quad \mu_2(0) = 0,$$
$$\mu_3'(t) = \lambda_3 \mu_3(t) + \mu_2(t), \quad \mu_3(0) = 0.$$

Solve successively for $\mu_1(t)$, then $\mu_2(t)$, and finally $\mu_3(t)$.

(d) Use Theorem 6.13 to write

$$e^{At} = \mu_1(t)P_0 + \mu_2(t)P_1 + \mu_3(t)P_2,$$

and simplify the expression as far as possible.

(e) Discuss (without full computation) how the structure of the eigenvalues (real vs. complex, distinct vs. repeated) affects the qualitative behavior of solutions to the system $\mathbf{x}' = A\mathbf{x}$.

Exercise 6.24 *(Comparison with diagonalization) Let*

$$A = \begin{pmatrix} 4 & 1 \\ 0 & 2 \end{pmatrix}.$$

(a) Compute e^{At} using diagonalization: find an invertible matrix S and a diagonal matrix D such that $A = SDS^{-1}$ and then use $e^{At} = Se^{Dt}S^{-1}$.

(b) Compute e^{At} again using Putzer's algorithm. Explicitly determine the functions μ_1, μ_2 and the matrices P_0, P_1.

(c) Show that the two expressions obtained in parts (a) and (b) coincide.

(d) Comment on the relative advantages of Putzer's algorithm versus diagonalization in the following situations:

- *A is diagonalizable with distinct eigenvalues*
- *A has repeated eigenvalues but is still diagonalizable.*
- *A is not diagonalizable.*

7

Stability of Linear Systems

This chapter is devoted to the study of stability properties of linear differential systems, a central topic in the qualitative theory of differential equations and dynamical systems. We begin with a general introduction to stability theory and the fundamental concepts that describe the long-term behavior of solutions, followed by precise definitions and illustrative examples. The stability of non-autonomous linear systems of the form $x'(t) = A(t)x(t)$ is then examined, emphasizing the role of fundamental matrices and growth estimates. We next focus on autonomous linear systems $x'(t) = Ax$, where stability can be characterized in terms of the spectral properties of the coefficient matrix. The chapter concludes with a discussion of autonomous systems in the plane, providing a geometric interpretation of stability through phase portraits and classification of equilibrium points. Throughout the chapter, exercises are included to reinforce the theoretical results and develop practical computational skills.

7.1 Introduction to Stability Theory

The qualitative behavior of solutions of differential equations is often more important than their explicit closed-form expressions. In particular, when studying dynamical systems arising in physics, engineering, and applied mathematics, one is frequently concerned with the long-time behavior of trajectories near equilibrium points or reference solutions. The notion of *stability* in the sense of Lyapunov provides a rigorous framework to address such questions.

In this chapter, we develop several complementary approaches to stability analysis for systems of ordinary differential equations. We begin with linear systems $x' = A(t)x$ and use the fundamental matrix associated with it to develop theorems regarding stability and asymptotic stability. In the time-invariant case, the matrix exponential e^{At} and the spectral properties of A provide powerful criteria for stability and asymptotic behavior. We then formalize several notions of stability for equilibrium points, including stability, asymptotic stability, and exponential stability. For linear systems, these properties can be completely characterized in terms of the fundamental matrix or, in the autonomous case, the eigenvalues of the system matrix. For nonlinear systems, however, direct spectral analysis is generally not sufficient,

DOI: 10.1201/9781042009220-7

and one must resort to indirect methods based on linearization or to more general nonlinear tools.

A particularly illuminating case is that of planar systems

$$x' = P(x,y), \qquad y' = Q(x,y),$$

with P and Q linear functions of x and y. In this setting, the phase portrait near an equilibrium can be completely classified in terms of the eigenvalues of the associated 2×2 matrix. Negative real parts correspond to sinks, where trajectories are attracted to the equilibrium, positive real parts correspond to sources (or repellers), and eigenvalues of opposite sign yield saddles with both attracting and repelling directions. Complex eigenvalues with nonzero real part give rise to spiral sinks or spiral sources, while purely imaginary eigenvalues produce centers with closed orbits. These canonical phase portraits provide a geometric realization of the algebraic stability criteria and serve as a guide for the analysis of more general nonlinear planar systems.

7.2 Definitions and Examples

Consider the scalar differential equation

$$x' = ax, \quad t \geq 0, \tag{7.1}$$

where a is a constant. The general solution of (7.1) is

$$x(t) = Ce^{at}, \quad C \in \mathbb{R}.$$

In particular, $x(t) \equiv 0$ is a solution, since it satisfies the differential equation identically. If $a > 0$, every nonzero solution grows without bound as $t \to \infty$; if $a < 0$, every solution decays exponentially to 0 as $t \to \infty$. Thus, the long–time behavior of solutions is completely determined by the sign of a. In the case $a < 0$, the zero solution attracts all nearby solutions and is therefore asymptotically stable, a notion we will develop rigorously in this chapter.

When $a = 0$, all solutions are constant. In this case, any solution that starts close to the zero solution stays close for all $t \geq 0$, so $x \equiv 0$ is stable, but not asymptotically stable, since nearby solutions do not converge to zero. We recall the following. Consider the system of ordinary differential equations

$$x' = f(t,x) \tag{7.2}$$

where $f \in C\big([0,\infty) \times D, \mathbb{R}^n\big), D \subset \mathbb{R}^n$ that is open. We say a vector $x^* \in \mathbb{R}^n$ is an *equilibrium*, or *constant solution*, or *equilibrium solution* of (7.2) if

$$f(t,x^*) = 0, \;\; 0 \leq t < w,$$

for a positive constant w. In practice, it would be easier to talk about the *zero solution*, or $x = 0$. To do so, we translate every nonzero equilibrium point to zero by the change of variables

$$\tilde{x}(t) = x(t) - x^*,$$

and have from (7.2) that

$$\tilde{x}'(t) = (x(t) - x^*)' = x'(t) = f(t, \tilde{x}(t) + x^*).$$

In this case, $\tilde{x}(t) \equiv 0$ is an equilibrium point of (7.2), which we might call "*the zero solution*" of (7.2). Thus, in most cases, we may require $f(t,0) = 0,\ 0 \le t < w$, when we talk about the zero solution.

In the above example, we found the solution explicitly. This is no longer true for the nonlinear differential equation

$$x'(t) = \frac{x(x-1)(x+2)}{1+x^2}, \quad t \ge 0, \quad x(0) = x_0. \tag{7.3}$$

The right-hand side of (7.3) is nonlinear, and an explicit closed-form solution cannot be obtained. However, we can still analyze the qualitative behavior of solutions. First observe that

$$x = -2, \quad x = 0, \quad x = 1$$

are constant (equilibrium) solutions of (7.3). Since the right-hand side of (7.3) is continuous for all $x \in \mathbb{R}$, solutions exist for all $t \ge 0$. Moreover, by uniqueness, any solution passing through a point (t_0, x_0) is uniquely determined and different from any other solution curves cannot intersect.

For the following discussion, we refer the reader to Fig. 7.1.

(a) If $x < -2$, then $x'(t) < 0$ by (7.3), so solutions decrease without bound as t increases and move away from the equilibrium $x = -2$.

(b) If $-2 < x < 0$, then $x'(t) > 0$, so solutions increase toward the equilibrium $x = 0$ and away from $x = -2$.

(c) If $0 < x < 1$, then $x'(t) < 0$, so solutions decrease toward $x = 0$ and away from $x = 1$. Note that in regions (b) and (c), solutions cannot cross the equilibrium $x = 0$ because of uniqueness.

(d) If $x > 1$, then $x'(t) > 0$, so solutions increase without bound and move away from the equilibrium $x = 1$.

Combining these observations with Fig. 7.1, we conclude that the equilibrium $x = 0$ is an attractor (asymptotically stable), while the equilibria $x = -2$ and $x = 1$ are repellers (unstable); we will return to these stability notions later in the chapter.

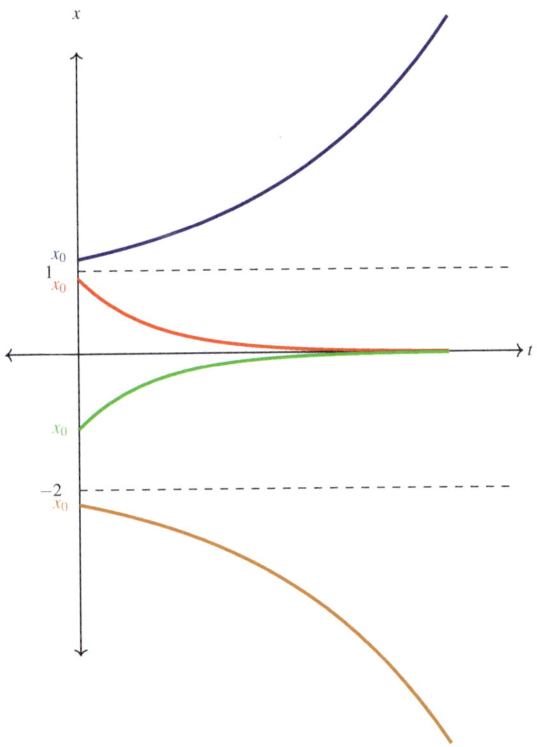

FIGURE 7.1: Diagram for the constant solutions.

Example 7.1 *(Lotka–Volterra predator–prey model) Let $x(t)$ and $y(t)$ be the number of prey and predators at time t, respectively. To keep the model simple, we will make the following assumptions:*

- *The predator species is dependent on a single prey species as its only food supply,*
- *The prey species has an unlimited food supply and*
- *There is no threat to the prey other than the specific predator.*

In the absence of predation, the prey population would grow at a natural rate

$$\frac{dx}{dt} = ax(t), \ a > 0.$$

On the other hand, in the absence of prey, the predator population would decline at a natural rate

$$\frac{dy}{dt} = -cy(t), \ c > 0.$$

The effects of predators eating prey are an interaction rate of decline $(-bxy, \ b > 0)$ in the prey population x, and an interaction rate of growth $(dxy, \ d > 0)$ of the predator *population y. Hence, one obtains the predator–prey model*

$$\frac{dx}{dt} = ax - bxy$$

$$\frac{dy}{dt} = -cy + dxy. \tag{7.4}$$

The Lotka–Volterra model consists of a system of linked differential equations that cannot be separated from each other and that cannot be solved in closed form. Thus, we search for the equilibrium solutions by setting the right sides of both equations in (7.4) equal to zero. That is

$$x(a - by) = 0, \;\; y(-c + dx) = 0.$$

From the first equation, we obtain $x = 0, y = a/b$, *and from the second equation, we get* $y = 0, x = c/d$. *Now, if* $x = 0$, *we have from the second equation that* $-cy = 0$ *and hence* $y = 0$. *Thus, we have the equilibrium solution*

$$x_1^* = (0,0).$$

Similarly, if $y = a/b$, *then* $-\frac{c}{b}a + dx\frac{a}{b} = 0$, *which implies that* $x = c/d$. *So the other equilibrium solution is*

$$x_2^* = (c/d, a/b).$$

The first equilibrium solution x_1^* *effectively represents the extinction of both species. If both populations are at* 0, *then they will continue to be so indefinitely. The second equilibrium solution* x_2^* *represents a fixed point at which both populations sustain their current, nonzero numbers, and do so indefinitely (see Fig. 7.2). The levels of the population at which this equilibrium is achieved depend on the chosen values of the parameters* a, b, c, *and* d. *We will further discuss the stability of both equilibrium solutions,* x_1^* *and* x_2^* *and hope that* x_1^* *is unstable.* □

Next, we consider (7.2) with the initial condition $x(t_0) = x_0$, and assume that $f(t,0) = 0$.

Definition 7.1 *The zero solution* $(x = 0)$ *of (7.2);*

(a) is stable (S) if for each $\varepsilon > 0$ *and* $t_0 \geq 0$, *there is a* $\delta = \delta(t_0, \varepsilon) > 0$ *such that* $|x(t_0)| < \delta$ *implies* $|x(t, t_0, x_0)| < \varepsilon$,

(b) is uniformly stable (US) if δ *is independent of* t_0,

(c) is unstable if it is not stable,

(d) is asymptotically stable (AS) if it is stable and $\lim_{t\to\infty} |x(t, t_0, x_0)| = 0$.

(e) is uniformly asymptotically stable (UAS) if it is (US) and there exists a $\gamma > 0$ *with the property that for each* $\mu > 0$ *there exists* $T = T(\mu) > 0$ *such that* $|x(t_0)| < \gamma$, $t \geq t_0 + T$ *implies* $|x(t, t_0, x_0)| < \mu$.

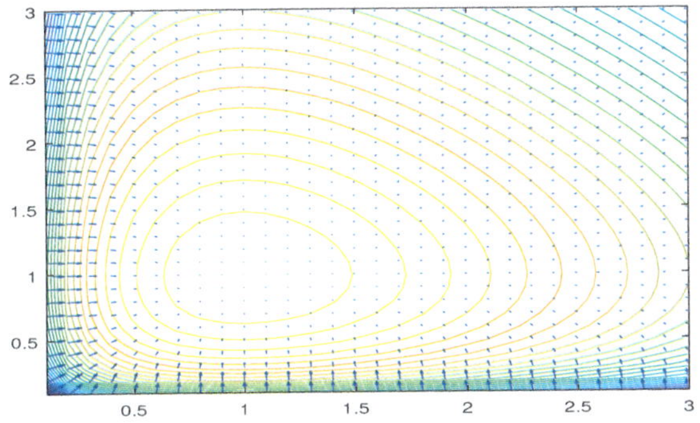

FIGURE 7.2: Prey-Predator dynamics as described by the level curves. The arrows describe the velocity and direction of solutions. We chose $a = b = c = d = 1$.

There are various stability definitions that can be found in many of the referenced books. As far as this book is concerned, the stability definitions that we have listed are the most important ones.

Next, we explore the equivalence between (UAS) and (AS) when the system under consideration is autonomous of the form

$$x'(t) = f(x), \quad x(t_0) = x_0.$$

We show that for the above autonomous system, the usual definition of asymptotic stability is equivalent to uniform asymptotic stability (UAS) in the sense of item (e). This relies on the time-translation invariance of autonomous systems. Let $\phi(t,x_0)$ denote the solution at time t with initial condition $x(0) = x_0$, for the autonomous system $x' = f(x)$. Then the flow satisfies

$$\psi(t,\phi(s,x_0)) = \phi(t+s,x_0), \qquad \text{for all } t,s \text{ for which both sides are defined.}$$

For a general (possibly time-varying) system

$$x'(t) = f(t,x), \quad x(t_0) = x_0, \tag{7.5}$$

we write $x(t,t_0,x_0)$ for the solution.

Definition 7.2 *The equilibrium* $x = 0$ *of* (7.5) *is* asymptotically stable *if*

it is stable: for every $\varepsilon > 0$*, there exists* $\delta(\varepsilon) > 0$ *such that*

$$|x_0| < \delta(\varepsilon) \quad \textit{implies} \quad |\phi(t,x_0)| < \varepsilon \quad \forall t \geq 0,$$

- *and there exists* $\rho > 0$ *such that*

$$|x_0| < \rho \quad \textit{implies} \quad \lim_{t\to\infty} \phi(t,x_0) = 0.$$

For (7.5) *with solution* $x(t,t_0,x_0)$, *the equilibrium* 0 *is* uniformly asymptotically stable *if:*

it is uniformly stable (US): for each $\varepsilon > 0$, *there exists* $\delta(\varepsilon) > 0$ *such that for all* $t_0 \geq 0$,

$$|x_0| < \delta(\varepsilon) \quad \textit{implies} \quad |x(t,t_0,x_0)| < \varepsilon \quad \forall t \geq t_0,$$

- *and there exists* $\gamma > 0$ *such that for each* $\mu > 0$ *there exists* $T(\mu) > 0$ *with*

$$|x(t_0)| < \gamma, \ t \geq t_0 + T(\mu) \quad \textit{implies} \quad |x(t,t_0,x_0)| < \mu.$$

For an autonomous system, $x(t,t_0,x_0)$ can be written as

$$x(t,t_0,x_0) = \phi(t - t_0, x_0),$$

so the dependence on t_0 enters only through the time difference $t - t_0$. This is the crucial property.

Claim. For the autonomous system $x' = f(x)$, (AS) is equivalent to (UAS).

(AS $\Rightarrow$ UAS).

(i) Uniform stability. Let 0 be asymptotically stable. Then, in particular it is stable, so for every $\varepsilon > 0$ there exists $\delta(\varepsilon) > 0$ such that

$$|x_0| < \delta(\varepsilon) \quad \text{implies} \quad |\phi(t,x_0)| < \varepsilon \quad \forall t \geq 0.$$

For the autonomous system, the solution with initial time t_0 satisfies

$$x(t,t_0,x_0) = \phi(t - t_0, x_0).$$

Thus, if $|x_0| < \delta(\varepsilon)$ and $t \geq t_0$, then $t - t_0 \geq 0$ and

$$|x(t,t_0,x_0)| = |\phi(t - t_0, x_0)| < \varepsilon.$$

The same $\delta(\varepsilon)$ works for all $t_0 \geq 0$, hence the equilibrium is uniformly stable (US).

(ii) Uniform attractivity. By asymptotic stability, there exists $\rho > 0$ such that

$$|x_0| < \rho \quad \text{implies} \quad \lim_{t \to \infty} \phi(t,x_0) = 0.$$

Fix $0 < \mu < \rho$. For each x_0 with $|x_0| < \rho$, the convergence $\phi(t,x_0) \to 0$ implies that there exists $T_{x_0}(\mu)$ such that

$$|\phi(t,x_0)| < \mu \quad \text{for all } t \geq T_{x_0}(\mu).$$

To obtain a *uniform* time $T(\mu)$ valid for all initial conditions with $|x_0| < \rho$, one uses continuity of the flow and compactness of the closed ball $\overline{B_\rho}$. The standard argument (see, e.g., Lyapunov stability texts) shows that there exists $T(\mu) > 0$ such that

$$|x_0| < \rho \quad \text{implies} \quad |\phi(t, x_0)| < \mu \quad \forall t \geq T(\mu),$$

i.e., convergence to 0 is uniform on the ball of radius ρ.

Now, for the system written with initial time t_0,

$$x(t, t_0, x_0) = \phi(t - t_0, x_0).$$

Fix $\gamma := \rho$. For any $t_0 \geq 0$ and any x_0 with $|x_0| < \gamma$, if $t \geq t_0 + T(\mu)$ then $t - t_0 \geq T(\mu)$, hence

$$|x(t, t_0, x_0)| = |\phi(t - t_0, x_0)| < \mu.$$

Thus, the uniform attractivity condition in (e) holds with the same γ and $T(\mu)$, independently of t_0. Together with uniform stability from (i), this proves that the equilibrium is uniformly asymptotically stable in the sense of (e).

(UAS $\Rightarrow$ AS).

Assume now that the equilibrium is uniformly asymptotically stable in the sense of (e) for the autonomous system $x' = f(x)$. In particular:

- uniform stability (US) implies stability in the usual sense, by taking $t_0 = 0$ and using the same $\delta(\varepsilon)$.
- The uniform attractivity condition with $t_0 = 0$ says: there exists $\gamma > 0$ such that for each $\mu > 0$ there exists $T(\mu) > 0$ with

  $$|x_0| < \gamma,\ t \geq T(\mu) \quad \text{implies} \quad |x(t, 0, x_0)| < \mu.$$

  Equivalently,

  $$|x_0| < \gamma \quad \text{implies} \quad \lim_{t \to \infty} x(t, 0, x_0) = 0.$$

Hence, the equilibrium is stable, and all solutions starting sufficiently close to 0 converge to 0 as $t \to \infty$, which is precisely asymptotic stability in the usual sense (AS).

Conclusion. For an autonomous system $x' = f(x)$, the time-translation invariance of the flow implies that the dependence on the initial time t_0 in the definition of uniform asymptotic stability is, in fact, redundant. Therefore, for such systems, asymptotic stability (AS) is equivalent to uniform asymptotic stability (UAS) as defined in item (e).

We furnish a few examples to illustrate the above definitions.

Example 7.2 *In this example, we show the zero solution is (US) and (AS) but not (UAS). Thus, we consider the initial value problem given by*

$$(1-t)x'(t) = x,\ x(t_0) = x_0,\ t \geq t_0 > 1.$$

For simplicity, we may write our problem as

$$x' = -\frac{x}{t-1},\ x(t_0) = x_0,\ t \geq t_0 > 1. \tag{7.6}$$

Separating the variables in (7.6) and integrating from t_0 to t and then applying the initial data, we arrive at the solution

$$x(t,t_0,x_0) = \frac{t_0-1}{t-1}x_0,\ t \geq t_0 > 1. \tag{7.7}$$

For any $\varepsilon > 0$ set $\delta = \varepsilon$, so that for $|x_0| < \delta$ we have that $t \geq t_0 > 1$, and

$$|x(t,t_0,x_0)| = \left|\frac{t_0-1}{t-1}x_0\right| = \left|\frac{t_0-1}{t-1}\right||x_0| \leq |x_0| < \delta = \varepsilon.$$

This shows the zero solution is (US). Moreover,

$$\lim_{t\to\infty}|x(t,t_0,x_0)| = \lim_{t\to\infty}\left|\frac{t_0-1}{t-1}\right||x_0| = 0,$$

and so the zero solution is (AS). On the other hand, if we evaluate the solution $x(t,t_0,x_0) = \frac{t_0-1}{t-1}x_0$ at $t = t_0+T$, we get $x(t_0+T,t_0,x_0) = \frac{t_0-1}{t_0+T-1}x_0 \to 1$ as $t_0 \to \infty$. This implies that $|x(t,t_0,x_0)| \not< \varepsilon$ for $t \geq t_0+T$. We conclude that the zero solution is not (UAS). Example 7.2 shows that (US) plus (AS) does not necessarily imply (UAS). □

Example 7.3 *The differential equation*

$$x' = 0,\ x(t_0) = x_0$$

has the constant solution $x(t) = x_0$, and hence the zero solution is (US) but not (AS). On the other hand

$$x' = x,\ x(t_0) = x_0$$

has the solution $x(t) = x_0e^{t-t_0}$, and its zero solution is unstable. □

Example 7.4 *The differential equation*

$$x' = -x,\ x(t_0) = x_0$$

has the solution

$$x(t,t_0,x_0) = x_0e^{-(t-t_0)}, t \geq t_0 \geq 0$$

and its zero solution is (US) and (UAS). To see this, for any $\varepsilon > 0$, we set $\delta = \varepsilon$. Then for $|x_0| < \delta$, we have that $x(t,t_0,x_0)| = |x_0e^{-(t-t_0)}| \leq |x_0| \leq \varepsilon = \delta$. So the zero *solution is (US). For the (UAS), we must show that there exists a $\gamma > 0$ with the property that for each $\mu > 0$, there exists $T = T(\mu) > 0$ such that $|x(t_0)| < \gamma$, $t \geq t_0 + T$ implies $|x(t,t_0,x_0)| < \mu$. Set $\delta = 1$ for the uniform stability and let $\gamma = \delta = 1$. Let $T(\mu) > -Ln(\mu)$, then for $t \geq t_0 + T(\mu)$,*

$$|x(t,t_0,x_0)| = |x_0 e^{-(t-t_0)}| \leq e^{-T(\mu)} \leq e^{Ln(\mu)} = \mu.$$

This shows the zero solution is (UAS). □

7.3 Stability of $x' = A(t)x$

Now we are in a position to discuss the stability of the zero solution of the linear system

$$x'(t) = A(t)x(t), \;\; x(t_0) = x_0 \tag{7.8}$$

where $A(t)$ is an $n \times n$ matrix with continuous entries on $[0,\infty)$, using the concept of the fundamental matrix. We begin with the following theorem.

Theorem 7.1 *Let $\Phi(t)$ be a fundamental matrix of* (7.8). *Then the zero solution of* (7.8) *is*

(a) stable if and only if there exists a constant $M > 0$ such that

$$|\Phi(t)| \leq M, \qquad t \geq 0,$$

(b) asymptotically stable if and only if

$$|\Phi(t)| \to 0 \quad \text{as } t \to \infty.$$

Proof *(a)*

($\Leftarrow$) Let $\Phi(t)$ be a fundamental matrix of (7.8). Then the unique solution is

$$x(t) = \Phi(t)\Phi^{-1}(t_0)x_0.$$

Assume that $|\Phi(t)| \leq M$ for all $t \geq 0$. For any $\varepsilon > 0$, choose

$$\delta = \frac{\varepsilon}{|\Phi^{-1}(t_0)|M}.$$

If $|x_0| < \delta$, then

$$|x(t)| = |\Phi(t)\Phi^{-1}(t_0)x_0| \leq |\Phi(t)||\Phi^{-1}(t_0)||x_0| \leq M|\Phi^{-1}(t_0)|\delta = \varepsilon.$$

Thus, the zero solution is stable.

($\Rightarrow$) *Assume the zero solution is stable. Set* $\varepsilon = 1$. *Then there exists* $\delta(1,t_0) > 0$ *such that*

$$|x(t)| = |\Phi(t)\Phi^{-1}(t_0)x_0| < 1, \quad \text{for } t \geq t_0 \text{ whenever } |x_0| < \delta(1,t_0).$$

Hence,

$$|\Phi(t)\Phi^{-1}(t_0)| \leq \frac{1}{\delta(1,t_0)}.$$

Using $\Phi(t) = \Phi(t)\Phi^{-1}(t_0)\Phi(t_0)$, *we obtain*

$$|\Phi(t)| \leq |\Phi(t)\Phi^{-1}(t_0)|\,|\Phi(t_0)| \leq \frac{|\Phi(t_0)|}{\delta(1,t_0)} := M.$$

This proves (a).

(b) *From part (a), stability is already established. Now,*

$$|x(t)| = |\Phi(t)\Phi^{-1}(t_0)x_0| \to 0 \quad \text{as } t \to \infty$$

if and only if

$$|\Phi(t)| \to 0 \quad \text{as } t \to \infty.$$

This completes the proof.

Theorem 7.2 *Let* $\Phi(t)$ *be a fundamental matrix of* (7.8). *Then the zero solution of* (7.8) *is uniformly stable if and only if there exists a constant* $M > 0$ *such that*

$$|\Phi(t)\Phi^{-1}(s)| \leq M, \qquad t \geq s \geq 0. \tag{7.9}$$

Proof ($\Leftarrow$) *Let* $\Phi(t)$ *be a fundamental matrix. Then*

$$x(t) = \Phi(t)\Phi^{-1}(s)x(s).$$

Assume (7.9) *holds. Then*

$$|x(t)| = |\Phi(t)\Phi^{-1}(s)x(s)| \leq |\Phi(t)\Phi^{-1}(s)|\,|x(s)| \leq M|x(s)|.$$

Given $\varepsilon > 0$, *choose* $\delta = \varepsilon/M$. *If* $|x(s)| < \delta$, *then*

$$|x(t)| \leq M\delta = \varepsilon,$$

which proves uniform stability.

($\Rightarrow$) *Assume the zero solution is uniformly stable. Set* $\varepsilon = 1$. *Then there exists* $\delta(1) > 0$ *such that*

$$|x(t)| = |\Phi(t)\Phi^{-1}(s)x(s)| < 1, \quad \text{for all } t \geq s \geq 0 \text{ whenever } |x(s)| < \delta(1).$$

Thus,

$$|\Phi(t)\Phi^{-1}(s)| \leq \frac{1}{\delta(1)} := M,$$

which proves (7.9).

The next theorem provides necessary and sufficient conditions for the (UAS).

Theorem 7.3 *Let $\Phi(t)$ be a fundamental matrix of (7.8). Then the zero solution of (7.8) is uniformly asymptotically stable if and only if there exist positive constants M and β such that*

$$\|\Phi(t)\Phi^{-1}(s)\| \leq Me^{-\beta(t-s)}, \qquad t \geq s \geq 0. \tag{7.10}$$

Proof *($\Leftarrow$) Assume that (7.10) holds. Let $x(t)$ be the solution of (7.8) with $x(s) = x_0$. Then*

$$x(t) = \Phi(t)\Phi^{-1}(s)x_0, \qquad t \geq s \geq 0.$$

From (7.10) we obtain

$$\|x(t)\| \leq \|\Phi(t)\Phi^{-1}(s)\| \, \|x_0\| \leq Me^{-\beta(t-s)}\|x_0\|, \qquad t \geq s \geq 0.$$

Hence, the zero solution is uniformly stable (take, for example, $\gamma = 1$ in the definition). Moreover, given any $\mu > 0$, choose

$$T(\mu) = \frac{1}{\beta}\ln\frac{M}{\mu},$$

so that $Me^{-\beta T(\mu)} \leq \mu$. If $\|x_0\| < 1$ and $t \geq s + T(\mu)$, then

$$\|x(t)\| \leq Me^{-\beta(t-s)}\|x_0\| \leq Me^{-\beta T(\mu)} \leq \mu.$$

Thus, the zero solution is uniformly asymptotically stable.

($\Rightarrow$) Now assume that the zero solution of (7.8) is uniformly asymptotically stable. Then there exists $\gamma > 0$ such that for each $\mu \in (0, \gamma)$ there exists $T(\mu) > 0$ with the property that

$$\|x_0\| < \gamma, \quad t \geq t_0 + T(\mu) \quad \Longrightarrow \quad \|x(t, t_0, x_0)\| < \mu,$$

for all $t_0 \geq 0$. Let $\mu_0 \in (0, \gamma)$ be fixed and set

$$\alpha = \frac{\mu_0}{\gamma} \in (0, 1).$$

Consider the solution with initial data $x(t) = x_0$ at time t, where $\|x_0\| = \gamma$. By uniform asymptotic stability, for all $t \geq 0$, we have

$$\|\Phi(t + T(\mu_0))\Phi^{-1}(t)x_0\| = \|x(t + T(\mu_0), t, x_0)\| < \mu_0.$$

Therefore,

$$\|\Phi(t + T(\mu_0))\Phi^{-1}(t)\| \leq \frac{\mu_0}{\gamma} = \alpha < 1, \qquad t \geq 0.$$

Next, fix $t_0 \geq 0$ and $t \geq t_0$. There exists a nonnegative integer k such that

$$t_0 + kT(\mu_0) \leq t < t_0 + (k+1)T(\mu_0).$$

Define $t_j = t_0 + jT(\mu_0)$ *for* $j = 0, 1, \ldots, k$. *Then*

$$\Phi(t)\Phi^{-1}(t_0) = \Phi(t)\Phi^{-1}(t_k)\,\Phi(t_k)\Phi^{-1}(t_{k-1})\cdots\Phi(t_1)\Phi^{-1}(t_0).$$

Taking norms and using the previous inequality, we obtain

$$||\Phi(t)\Phi^{-1}(t_0)|| \le ||\Phi(t)\Phi^{-1}(t_k)|| \prod_{j=0}^{k-1} ||\Phi(t_{j+1})\Phi^{-1}(t_j)|| \le ||\Phi(t)\Phi^{-1}(t_k)||\,\alpha^k.$$

Since $\Phi(t)\Phi^{-1}(s)$ *is continuous on the compact set*

$$\{(t,s) : 0 \le s \le t \le s + T(\mu_0)\},$$

there exists a constant $M_0 \ge 1$ *such that*

$$||\Phi(t)\Phi^{-1}(s)|| \le M_0, \qquad 0 \le s \le t \le s + T(\mu_0).$$

In particular, $||\Phi(t)\Phi^{-1}(t_k)|| \le M_0$ *for all* t *with* $t_k \le t \le t_k + T(\mu_0)$. *Thus*

$$||\Phi(t)\Phi^{-1}(t_0)|| \le M_0\alpha^k, \qquad t \ge t_0 \ge 0.$$

Finally, we relate k *to* $t - t_0$. *From*

$$t < t_0 + (k+1)T(\mu_0)$$

we get

$$k + 1 > \frac{t - t_0}{T(\mu_0)}.$$

Let

$$\beta = -\frac{1}{T(\mu_0)} \ln \alpha > 0.$$

Then

$$\alpha^{k+1} \le e^{-(k+1)\beta T(\mu_0)} \le e^{-\beta(t-t_0)}.$$

Therefore,

$$||\Phi(t)\Phi^{-1}(t_0)|| \le \frac{M_0}{\alpha}\,\alpha^{k+1} \le \frac{M_0}{\alpha}\,e^{-\beta(t-t_0)}.$$

Setting $M = M_0/\alpha$ *completes the proof of* (7.10).

Example 7.5 *The nonlinear system*

$$x'(t) = \begin{pmatrix} 0 & 1 \\ \frac{-2}{t^2} & \frac{2}{t} \end{pmatrix} \begin{pmatrix} x_1 \\ x_2 \end{pmatrix}, \; t > 0,$$

has the fundamental matrix

$$\Phi(t) = \begin{pmatrix} t^2 & t \\ 2t & 1 \end{pmatrix}, \; t > 0.$$

Thus, by Theorem 7.1 its zero solution is unstable, since there is no constant M *such that* $||\Phi(t)|| \le M$.

□

Example 7.6 *The system*

$$x'(t) = \begin{pmatrix} 0 & 1 \\ -2 & -2 \end{pmatrix} \begin{pmatrix} x_1 \\ x_2 \end{pmatrix}, \ t > 0,$$

has the fundamental matrix

$$\Phi(t) = e^{-t} \begin{pmatrix} \cos(t) & \sin(t) \\ -(\cos(t) + \sin(t)) & \cos(t) - \sin(t) \end{pmatrix}, \quad t > 0.$$

After some calculations, one can verify that for a positive constant K,

$$||\Phi(t)\Phi^{-1}(s)|| \leq Ke^{-(t-s)}, t \geq s \geq 0,$$

and by Theorem 7.3 the zero solution is (UAS). □

7.3.1 Exercises

Exercise 7.1 *Discuss the flow of solutions around the equilibrium solutions as we did in Fig. 4.1 for the following systems:*

(a)

$$x'(t) = (1 - x^2)xe^x, \ x(0) = x_0 \neq 0.$$

(b)

$$x'(t) = x^3 + x^2 - 6x, \ x(0) = x_0 \neq 0.$$

Exercise 7.2 *Discuss the stability as we did in Example 7.2 for the following (IVPs):*

(a)

$$(t^2 + 9)x'(t) + tx = 0, \ x(0) = x_0 \neq 0.$$

(b)

$$(1 + t)x'(t) = -x, \ x(t_0) = x_0, \ t \geq t_0 > -1.$$

Exercise 7.3 *Show that the zero solution of*

$$x'(t) = -x^3, \ x(t_0) = x_0 \neq 0$$

is (UAS).

Exercise 7.4 *Consider the scalar non-autonomous linear differential equation*

$$x'(t) = a(t)x(t), \qquad t \geq t_0 \geq 0, \tag{7.11}$$

where $a\colon [0, \infty) \to \mathbb{R}$ is continuous. The solution with initial condition $x(t_0) = x_0$ is

$$x(t) = x_0 \exp\left(\int_{t_0}^{t} a(s)\, ds \right).$$

Using this explicit representation, analyze the stability of the zero solution.

(a) *Show that the zero solution of* (7.11) *is (Lyapunov) stable if and only if there exists a constant* $M > 0$ *such that*

$$\exp\left(\int_{t_0}^{t} a(s)\,ds\right) \leq M \qquad \text{for all } t \geq t_0 \geq 0.$$

(b) *Show that the zero solution is asymptotically stable if and only if it is stable and*

$$\int_{t_0}^{t} a(s)\,ds \longrightarrow -\infty \qquad \text{as } t \to \infty \quad \text{for every fixed } t_0 \geq 0.$$

(Equivalently, $x(t,t_0,x_0) \to 0$ *as* $t \to \infty$ *for all* t_0 *and all* x_0 *sufficiently small.)*

(c) *Prove that the zero solution is uniformly stable if and only if there exists a constant* $M > 0$ *such that*

$$\int_{t_1}^{t_2} a(s)\,ds \leq M \qquad \text{for all } 0 \leq t_1 < t_2 < \infty.$$

(Hint: relate this condition to a uniform bound on the ratio $\dfrac{|x(t_2)|}{|x(t_1)|}$ *for solutions of* (7.11).*)*

(d) *Show that the zero solution is uniformly asymptotically stable if and only if there exist constants* $M > 0$ *and* $\alpha > 0$ *such that*

$$\int_{t_0}^{t} a(s)\,ds \leq -\alpha(t-t_0) + M \qquad \text{for all } t \geq t_0 \geq 0.$$

Conclude that in this case

$$|x(t,t_0,x_0)| \leq C e^{-\alpha(t-t_0)} |x_0| \qquad \text{for all } t \geq t_0,$$

for some constant $C \geq 1$ *independent of* t_0 *and* x_0*, and interpret this as a form of uniform exponential decay.*

Exercise 7.5 *Find the fundamental matrix and study the stability of the zero solution of the given system for* $t > 0$.

(a)

$$x'(t) = \begin{pmatrix} 0 & 1 \\ -\frac{1}{2t^2} & -\frac{5}{2t} \end{pmatrix} \begin{pmatrix} x_1 \\ x_2 \end{pmatrix},$$

(b)

$$x'(t) = \begin{pmatrix} 0 & 1 \\ -\frac{8}{t^2} & -\frac{10}{t} \end{pmatrix} \begin{pmatrix} x_1 \\ x_2 \end{pmatrix}.$$

Exercise 7.6 *Let A be an $n \times n$ constant matrix. Suppose $B(t)$ is an $n \times n$ variable matrix with continuous entries for all $t \geq 0$.*

(a) Show that if every eigenvalue of the matrix A has a nonpositive real part and that the eigenvalues with zero real parts are simple, and

$$\int_0^\infty ||B(t)dt \leq E,$$

for a positive constant E, then the zero solution of

$$x' = Ax + B(t)x$$

is stable, where $x = (x_1, x_2, \dots, x_n)^T$.

(b) Show that the zero solution of

$$x' = \begin{pmatrix} -1+te^{-t} & 2+e^{-t} \\ 2+\frac{1}{(1+t)^2} & -4+(1+t)^{-3/2} \end{pmatrix} \begin{pmatrix} x_1 \\ x_2 \end{pmatrix}$$

is stable.

Exercise 7.7 *Consider again the scalar linear equation*

$$x'(t) = a(t)x(t), \qquad t \geq t_0 \geq 0,$$

with solution

$$x(t) = x_0 \exp\left(\int_{t_0}^t a(s)\, ds \right).$$

In this exercise, the coefficient $a(t)$ is given explicitly, and you are asked to analyze the stability of the zero solution using the criteria from Exercise 7.4.

(a) Let

$$a(t) = -\frac{1}{1+t}, \qquad t \geq 0.$$

Compute

$$\int_{t_0}^t a(s)\, ds \quad \text{and} \quad \exp\left(\int_{t_0}^t a(s)\, ds \right),$$

and use part (a) of Exercise 7.4 to determine whether the zero solution is stable. Is it asymptotically stable? Is it uniformly stable? Is it uniformly asymptotically stable?

(b) Now let

$$a(t) = -1 - \frac{1}{1+t}, \qquad t \geq 0.$$

Compute explicitly

$$\int_{t_0}^t a(s)\, ds \quad \text{and} \quad \exp\left(\int_{t_0}^t a(s)\, ds \right),$$

and verify directly the following properties:

- *There exists* $M > 0$ *such that*

$$\int_{t_1}^{t_2} a(s)\,ds \le M \qquad \text{for all } 0 \le t_1 < t_2 < \infty.$$

Conclude that the zero solution is uniformly stable.

- *There exist constants* $M > 0$ *and* $\alpha > 0$ *such that*

$$\int_{t_0}^{t} a(s)\,ds \le -\alpha(t - t_0) + M \qquad \text{for all } t \ge t_0 \ge 0.$$

Deduce from part (d) of Exercise 7.4 that the zero solution is uniformly asymptotically stable, and write an explicit exponential decay estimate for $|x(t)|$ *in terms of* $|x_0|$.

(c) Consider finally

$$a(t) = \frac{\sin t}{1+t}, \qquad t \ge 0.$$

Show that

$$\sup_{0 \le t_1 < t_2 < \infty} \int_{t_1}^{t_2} a(s)\,ds < \infty,$$

and use this to conclude that the zero solution is uniformly stable . Does the zero solution converge to zero as $t \to \infty$ *for all initial data? Justify your answer by examining the behavior of* $\int_{t_0}^{t} a(s)\,ds$ *as* $t \to \infty$.

7.4 Stability of $x' = Ax$

In this section, we investigate the stability of linear autonomous systems governed by a constant matrix. The long-term behavior of solutions is completely determined by the associated exponential matrix, whose spectral properties provide a precise classification of stability, asymptotic stability, and instability. By analyzing the eigenvalues of this exponential matrix, we obtain simple and powerful criteria that link algebraic information about the system matrix to the qualitative dynamics of the underlying differential equation.

In this section, we give explicit criteria that satisfy those conditions that were imposed, but only for the linear system with constant coefficients given by

$$x' = Ax \tag{7.12}$$

where A is an $n \times n$ matrix with constant entries. We begin with the following lemma.

Lemma 3 *Let*

$$r(t) = t^m e^{-\alpha t},\ t \geq 0$$

where m is a nonnegative integer, and α is a positive constant. Then there exists a positive constant D depending on m and α such that $r(t) \leq D$.

Proof *If $m = 0$ then $r(t) \leq 1 := D$. On the other hand, if $m > 0$, then by repeated application of L'Hôpital's rule we have $\lim_{t\to\infty} t^m e^{-\alpha t} = 0$. Moreover, since $r(0) = 0$ and $r(t) \geq 0$, we have that $r(t) \leq D$, for some positive constant D. This completes the proof.*

Definition 7.3 *Let A be a square matrix. We denote the spectrum of A by*

$$\sigma(A) = \{\lambda : \det(A - \lambda I) = 0\}.$$

Thus, the notation

$$Re\big(\sigma(A)\big) < \zeta,$$

for some real number ζ it is to indicate that all the eigenvalues of A have real parts less than ζ.

The next Theorem is concerned with eigenvalues of multiplicity m.

Theorem 7.4 *Let λ be an eigenvalue of A such that* $\mathrm{Re}(\lambda) < \zeta$. *Then for any integer $m \geq 0$, there exists a positive constant D such that*

$$t^m e^{\lambda t} \leq De^{\zeta t}, \qquad \textit{for all } t \geq 0.$$

Proof *Write $\lambda = \alpha + i\beta$, where $\alpha, \beta \in \mathbb{R}$. Since* $\mathrm{Re}(\lambda) = \alpha < \zeta$, *we have $\alpha - \zeta < 0$.*

By Lemma 3, there exists a constant $D > 0$ such that

$$t^m e^{(\alpha-\zeta)t} \leq D, \qquad \textit{for all } t \geq 0.$$

Multiplying both sides by $e^{\zeta t}$ gives

$$t^m e^{\alpha t} \leq De^{\zeta t}, \qquad \textit{for all } t \geq 0$$

If $\beta = 0$, then $\lambda = \alpha$ and the result follows immediately. If $\beta \neq 0$, then

$$|t^m e^{\lambda t}| = t^m |e^{(\alpha+i\beta)t}| = t^m e^{\alpha t} \leq De^{\zeta t},$$

which completes the proof.

Theorem 7.5 *Suppose that*

$$\max_{\lambda\in\sigma(A)} \mathrm{Re}(\lambda) < \zeta.$$

Then any solution $x(t)$ of (7.12) *satisfies*

$$\|x(t)\| \leq De^{\zeta t}, \qquad \textit{for all } t \geq 0,$$

for some positive constant D depending on the initial condition.

Proof *Let $\lambda_1, \lambda_2, \ldots, \lambda_k$ be the distinct eigenvalues of A. A fundamental set of solutions of* (7.12) *consists of vector-valued functions of the form*

$$x_j(t) = t^m e^{\lambda_i t} v,$$

where v is a (generalized) eigenvector corresponding to λ_i, and m is a nonnegative integer determined by the size of the Jordan block.

By Theorem 7.4, for each such solution, there exists a constant $V_j > 0$ such that

$$\|x_j(t)\| \leq V_j e^{\zeta t}, \qquad t \geq 0.$$

Any solution $x(t)$ can be written as

$$x(t) = c_1 x_1(t) + c_2 x_2(t) + \cdots + c_n x_n(t).$$

Hence, for $t \geq 0$,

$$\|x(t)\| \leq |c_1| \|x_1(t)\| + \cdots + |c_n| \|x_n(t)\| \leq (|c_1| V_1 + \cdots + |c_n| V_n) e^{\zeta t}.$$

Setting $D = |c_1| V_1 + \cdots + |c_n| V_n$ completes the proof.

The next theorem plays an important role in the study of stability. Its proof relies on Putzer's algorithm (Theorem 6.13).

Theorem 7.6 *Let λ_i, $i = 1, 2, \ldots, k$, be the distinct eigenvalues of A, where λ_i has algebraic multiplicity n_i and*

$$n_1 + \cdots + n_k = n.$$

Repeat each eigenvalue according to its algebraic multiplicity to form the list

$$\lambda_1, \lambda_2, \ldots, \lambda_n.$$

Assume there exists a constant ζ such that

$$\operatorname{Re} \lambda_i < \zeta, \qquad i = 1, 2, \ldots, k.$$

Then there exists a constant $D > 0$ such that

$$\|e^{At}\| \leq D e^{\zeta t}, \qquad t \geq 0.$$

Proof *From Theorem 6.13, we know that*

$$e^{At} = \sum_{j=0}^{n-1} \mu_{j+1}(t) P_j, \tag{7.13}$$

where the scalar functions $\mu_1(t), \ldots, \mu_n(t)$ satisfy

$$\mu_1'(t) = \lambda_1 \mu_1(t), \quad \mu_1(0) = 1; \qquad \mu_j'(t) = \lambda_j \mu_j(t) + \mu_{j-1}(t), \quad \mu_j(0) = 0,\ j = 2, \ldots, n,$$

and the matrices $P_0, \ldots, P_n$ *are defined by*

$$P_0 = I, \qquad P_j = \prod_{k=1}^{j}(A - \lambda_k I), \quad j = 1, 2, \ldots, n.$$

By solving the scalar equations, we obtain

$$\mu_1(t) = e^{\lambda_1 t}, \qquad \mu_{j+1}(t) = e^{\lambda_{j+1} t}\int_0^t e^{-\lambda_{j+1} s}\mu_j(s)\,ds,$$

for $j = 1, 2, \ldots, n-1$. *Without loss of generality, we order the eigenvalues* $\lambda_j = \rho_j + i\sigma_j$ *so that their real parts are nondecreasing:*

$$\rho_1 \le \rho_2 \le \cdots \le \rho_n.$$

We claim that

$$|\mu_j(t)| \le \frac{t^{j-1}}{(j-1)!}e^{\rho_j t} \quad \text{for } t > 0,\ j = 1, 2, \ldots, n. \tag{7.14}$$

For $j = 1$ *this is immediate, since* $\mu_1(t) = e^{\lambda_1 t}$ *and* $|\mu_1(t)| \le e^{\rho_1 t}$. *Assume* (7.14) *holds for some* j. *Then*

$$\mu_{j+1}(t) = e^{\lambda_{j+1} t}\int_0^t e^{-\lambda_{j+1} s}\mu_j(s)\,ds,$$

and hence

$$|\mu_{j+1}(t)| \le e^{\rho_{j+1} t}\int_0^t e^{-\rho_{j+1} s}|\mu_j(s)|\,ds \le e^{\rho_{j+1} t}\int_0^t e^{-\rho_{j+1} s}\frac{s^{j-1}}{(j-1)!}e^{\rho_j s}\,ds.$$

Since $\rho_j \le \rho_{j+1}$, *we have* $e^{(\rho_j - \rho_{j+1})s} \le 1$, *so*

$$|\mu_{j+1}(t)| \le e^{\rho_{j+1} t}\frac{1}{(j-1)!}\int_0^t s^{j-1}\,ds = \frac{t^j}{j!}e^{\rho_{j+1} t}.$$

This proves (7.14) *by induction. In particular, since* ρ_n *is the largest real part, we obtain*

$$|\mu_{j+1}(t)| \le \frac{t^j}{j!}e^{\rho_n t}, \qquad j = 0, 1, \ldots, n-1.$$

Now fix any number $\rho^* > 0$. *For each integer* $j \ge 0$ *there exists a constant* $h_j > 0$ *such that*

$$t^j \le h_j e^{\rho^* t}, \qquad t \ge 0.$$

Hence

$$|\mu_{j+1}(t)| \le \frac{h_j}{j!}e^{(\rho_n + \rho^*)t}, \qquad t \ge 0.$$

Let

$$C = \max_{0\le j\le n-1} \frac{h_j}{j!}.$$

Then

$$|\mu_{j+1}(t)| \le Ce^{(\rho_n+\rho^*)t}, \qquad j = 0, 1, \ldots, n-1,\ t \ge 0.$$

Since $\rho_n < \zeta$ *by hypothesis, we can choose* $\rho^* > 0$ *such that* $\rho_n + \rho^* = \zeta$*. Using* (7.13) *and an induced matrix norm, we obtain*

$$\|e^{At}\| \le \sum_{j=0}^{n-1} |\mu_{j+1}(t)|\,\|P_j\| \le C\Big(\sum_{j=0}^{n-1} \|P_j\|\Big)e^{\zeta t} =: De^{\zeta t},$$

for all $t \ge 0$*, where* $D > 0$ *is a constant. This completes the proof.*

As a direct consequence of Theorem 7.6 we have the following lemma.

Lemma 4 *Suppose all the eigenvalues of* A *have negative real parts. Then there are positive constants* h*,* K*, and* ρ *such that*

$$\mu_{j+1}(t) \le he^{-\rho t} \text{ and } \|e^{At}\| \le Ke^{-\rho t},\ t \ge 0. \tag{7.15}$$

Theorem 7.7 *(i) Assume every eigenvalue of the matrix* A *has a nonpositive real part and that the eigenvalues with real part zero are simple. Then there exists a positive constant* D *such that*

$$\|e^{At}\| \le D, \qquad t \ge 0.$$

Moreover, if x *is a solution of* (7.12)*, then it is bounded, and the zero solution is stable.*

(ii) If $\mathrm{Re}(\sigma(A)) < 0$*, then the zero solution is asymptotically stable.*

(iii) If there exists an eigenvalue of A *with a positive real part, then the zero solution is unstable.*

Proof *(i) Suppose* $\lambda = i\beta$ *is a simple purely imaginary eigenvalue. Then it contributes a bounded function of the form* $e^{i\beta t}$ *to a fundamental set of solutions. Hence, we obtain a fundamental set of solutions in which every element is bounded on* $[0,\infty)$*. Since the matrix exponential* e^{At} *is formed from such a fundamental set, an argument similar to that in the proof of Theorem 7.6 shows that there exists a positive constant* D *such that*

$$\|e^{At}\| \le D, \qquad t \ge 0.$$

Now let x *be a solution of* (7.12) *with* $x(t_0) = x_0$*. Then*

$$x(t) = e^{A(t-t_0)}x_0, \qquad t \ge t_0.$$

Let $\varepsilon > 0$ *and set* $\delta = \varepsilon/D$. *If* $|x_0| < \delta$, *then*

$$|x(t)| = \|e^{A(t-t_0)}\|\,|x_0| \leq D|x_0| < D\delta = \varepsilon,$$

for all $t \geq t_0$. *This shows that solutions are bounded and that the zero solution is stable. This completes the proof of (i).*

(ii) Since $\text{Re}(\sigma(A)) < 0$, *the zero solution is stable by part (i). Choose a real number* $\zeta < 0$ *such that*

$$\text{Re}(\sigma(A)) < \zeta.$$

By Theorem 7.6, there exists a positive constant D such that

$$\|e^{At}\| \leq De^{\zeta t}, \qquad t \geq 0.$$

Hence, for any initial condition $x(0) = x_0$, *the corresponding solution of* (7.12) *satisfies*

$$|x(t,0,x_0)| = |e^{At}x_0| \leq D|x_0|e^{\zeta t} \to 0 \quad \text{as } t \to \infty.$$

Thus, the zero solution is asymptotically stable. This completes the proof of (ii).

(iii) Suppose that A has an eigenvalue λ *with* $\text{Re}(\lambda) > 0$. *Then the corresponding component of the solution contains a term of the form* $e^{\lambda t}$, *which grows exponentially as* $t \to \infty$. *Hence, there exist solutions that become unbounded, and therefore, the zero solution is unstable. This completes the proof.*

Remark 23 *Consider the two matrices*

$$A = \begin{pmatrix} 0 & 0 \\ 0 & 0 \end{pmatrix} \qquad \text{and} \qquad B = \begin{pmatrix} 0 & 1 \\ 0 & 0 \end{pmatrix}.$$

Each matrix has the eigenvalue 0 *with algebraic multiplicity* 2. *However, the stability properties of the corresponding systems are different.*

For the system $x' = Ax$, *every solution is constant, so the zero solution is stable.*

For the system $x' = Bx$, *we have*

$$x_1' = x_2, \qquad x_2' = 0.$$

Hence

$$x_2(t) = c_1,$$

and therefore

$$x_1(t) = c_1 t + c_2.$$

Thus, whenever $c_1 \neq 0$, *the solution grows linearly and becomes unbounded as* $t \to \infty$. *Hence, the zero solution is not stable.*

Therefore, when eigenvalues with zero real part have multiplicity greater than one, one must examine the Jordan structure carefully before drawing conclusions about stability.

Theorem 7.8 *If all the eigenvalues of A are distinct with*

$$Re\big(\sigma(A)\big) < -\zeta, \quad \zeta > 0$$

then the zero solution of (7.12) *is (UAS).*

Proof *By Lemma 4, we have that*

$$||e^{At}|| \leq Ke^{-\zeta t},\, t \geq 0$$

and hence

$$||e^{A(t-s)}|| \leq Ke^{-\zeta(t-s)},\, t \geq s \geq 0.$$

Now if x is a solution of (7.12) *with* $x(s) = x_0$, *then*

$$x(t) = e^{A(t-s)}x_0, \ for\ t \geq s$$

and the results follow from Theorem 7.3, since

$$e^{A(t-s)} = \Phi(t)\Phi^{-1}(s).$$

This completes the proof.

7.4.1 Exercises

Exercise 7.8 *Let*

$$A_1 = \begin{pmatrix} -2 & 0 \\ 0 & -1 \end{pmatrix}, \qquad A_2 = \begin{pmatrix} 0 & 1 \\ 0 & 0 \end{pmatrix}.$$

(a) *Find the eigenvalues of A_1 and A_2 and their algebraic multiplicities.*

(b) *Using Theorem 7.7, determine the stability type of the zero solution of $x' = A_1x$ and of $x' = A_2x$ (stable, asymptotically stable, unstable).*

(c) *Explain how Remark 23 applies to A_2 and why the multiplicity of the eigenvalue with zero real part is crucial for the stability conclusion.*

Exercise 7.9 *Consider the matrix*

$$A = \begin{pmatrix} -1 & 1 \\ -4 & -1 \end{pmatrix}.$$

Discuss the stability of the zero solution of $x' = Ax$.

Exercise 7.10 *Let A be a 3×3 real matrix with distinct eigenvalues*

$$\lambda_1 = -3, \qquad \lambda_2 = -1, \qquad \lambda_3 = -2,$$

and suppose A is diagonalizable over $\mathbb{C}$.

(a) *Using Definition 7.3, write $\sigma(A)$ and $Re(\sigma(A))$ and find a real number $\zeta < 0$ such that $Re(\sigma(A)) < \zeta$.*

(b) Apply Theorem 7.6 to show that there exists a constant $D > 0$ such that

$$\|e^{At}\| \leq De^{\zeta t}, \qquad t \geq 0.$$

(c) Use Theorem 7.8 to conclude that the zero solution of $x' = Ax$ is uniformly asymptotically stable.

(d) Explain briefly why the distinctness of the eigenvalues is assumed in Theorem 7.8, and how this relates to the construction of the fundamental matrix via exponentials of the form $t^m e^{\lambda t}$.

Exercise 7.11 *Let A be a real 2×2 matrix with eigenvalues*

$$\lambda_1 = 0, \qquad \lambda_2 = -1,$$

and suppose λ_1 is simple (algebraic multiplicity one).

(a) Give an example of such a matrix A and compute its eigenvectors.

(b) Using Theorem 7.7(i), discuss whether the zero solution of $x' = Ax$ is stable, and whether solutions are bounded for all $t \geq 0$.

(c) Construct an initial condition x_0 for your example in part (a) such that the corresponding solution does not converge to zero as $t \to \infty$. Explain why this does not contradict Theorem 7.7(ii).

(d) Compare your example with the matrices in Remark 23, emphasizing the role of simple versus multiple eigenvalues with zero real part.

Exercise 7.12 *Discus the stability of the system $x' = Ax$, where*

$$A = \begin{pmatrix} -3 & 0 & 0 \\ 0 & -1 & 0 \\ 0 & 0 & -2 \end{pmatrix}.$$

7.5 Autonomous Systems in the Plane

In most applications, the corresponding modeling systems can be written as two-dimensional autonomous systems in the form

$$x' = P(x,y),\ y' = Q(x,y) \tag{7.16}$$

where P and Q are continuous on some subset of $\mathbb{R}^2$. Just as a reminder that the system is called *autonomous* since P and Q do not depend explicitly on t. System (7.16) can be written in the form $z' = f(z)$, where

$$z = \begin{pmatrix} x \\ y \end{pmatrix} \text{ and } f(z) = \begin{pmatrix} P(z) \\ Q(z) \end{pmatrix}.$$

Thus the second order differential equation

$$x'' = g(x, x')$$

can be put in the form of (7.16) as the system

$$x' = y, \;\; y' = g(x, y).$$

We begin with the following lemma. We denote the solution of (7.16) by the pair $(x(t), y(t))$, $t \in (a, b)$.

Lemma 5 *Let $(x(t), y(t))$ be a solution of (7.16). Then, for any constant k, the pair $(x(t+k), y(t+k))$ is also a solution of (7.16).*

Proof *Let $x_1 = x(t+k)$ and $x_2 = y(t+k)$. Then by the chain rule we have*

$$x_1' = x'(t+k) = P(x(t+k), y(t+k)) = P(x_1, x_2),$$

$$x_2' = y'(t+k) = Q(x(t+k), y(t+k)) = Q(x_1, x_2),$$

which implies that the pair $(x_1(t), x_2(t))$ is a solution of (7.16) for $t \in (a-k, b-k)$. This completes the proof.

The results of Lemma 5 do not hold for non-autonomous systems. To see this, we consider the system

$$x' = ty, \; y' = y.$$

It readily follows that

$$x(t) = (t-1)e^t, \quad y(t) = e^t$$

is a solution of the system. However, for any nonzero constant k, we have $x(t+k) = (t+k-1)e^{t+k}$ and $x'(t+k) = (t+k)e^{t+k} \neq te^{t+k} = ty(t+k)$.

Definition 7.4 *The solution $(x(t), y(t))$ of (7.16) as t varies describes parametrically a curve in the plane that we call a trajectory (or orbit).*

Theorem 7.9 *Different trajectories of (7.16) cannot intersect.*

Proof *Let C_1 and C_2 be two trajectories represented by $(x_1(t), y_1(t))$ and $(x_2(t), y_2(t))$, respectively, with a common point (x_0, y_0). Then there are two different times (otherwise uniqueness is violated), say t_1 and t_2, such that*

$$(x_0, y_0) = (x_1(t_1), y_1(t_1)) = (x_2(t_2), y_2(t_2)).$$

By Lemma 5,

$$x(t) = x_1(t + t_1 - t_2), \; y(t) = y_1(t + t_1 - t_2)$$

is a solution. Since $(x(t_2), y(t_2)) = (x_0, y_0)$, we have that $x(t)$ and $y(t)$ must agree respectively with $x_2(t)$ and $y_2(t)$ due to uniqueness. Thus, C_1 and C_2 must coincide. This completes the proof.

Remark 24 *Be aware that a trajectory is a curve that is represented parametrically by one or more solutions. Thus,* $x(t), y(t)$*, and* $x(t+k), y(t+k)$ *for* $k \neq 0$ *represent distinct solutions but the same trajectory.*

Example 7.7 *The damped harmonic motion represented by the second order differential equation*

$$u'' + 2u' + 2u = 0,$$

can be easily transformed into a planar autonomous system

$$x' = y,$$

$$y' = -2x - 2y.$$

The eigenvalues of the systems are found to be $\lambda = -1 \pm i$*. Hence, the solution is*

$$x = e^{-t}\left[(c_1 \cos(t) + c_2 \sin(t)\right).$$

$$y = e^{-t}\left[(c_2 - c_1)\cos(t) - (c_1 + c_2)\sin(t)\right],$$

To describe the trajectories, we find that

$$x^2 + y^2 = (c_1^2 + c_2^2)e^{-2t},$$

which describes a family of spirals. □

We begin a detailed study of the linear plane autonomous system

$$X' = AX \tag{7.17}$$

where

$$A = \begin{pmatrix} a & b \\ c & d \end{pmatrix}$$

is a nonsingular matrix and

$$X - \begin{pmatrix} x \\ y \end{pmatrix}.$$

The characteristic equation for this system is

$$\lambda^2 - (a+d)\lambda + ad - bc = 0.$$

By introducing

$$p = a + d = \text{tr}(A), \qquad q = ad - bc = \det(A),$$

the characteristic equation becomes

$$\lambda^2 - p\lambda + q = 0.$$

If we let λ_1 and λ_2 be the roots of this equation, then

$$\lambda_{1,2} = \frac{1}{2}\left[p \pm \sqrt{p^2 - 4q}\right]. \tag{7.18}$$

Hence, the stability of the zero solution depends on the discriminant

$$\Delta = p^2 - 4q = (\mathrm{tr}(A))^2 - 4\det(A).$$

Theorem 7.10 *Consider the linear system* (7.17) *with characteristic equation*

$$\lambda^2 - p\lambda + q = 0, \qquad p = \mathrm{tr}(A),\ q = \det(A).$$

Let λ_1, λ_2 be the eigenvalues of A. Then the zero solution of (7.17) *is classified as follows.*

(a) The zero solution is ***asymptotically stable*** *if and only if*

$$p < 0 \quad \text{and} \quad q > 0.$$

(b) The zero solution is ***stable but not asymptotically stable*** *if and only if*

$$p = 0 \quad \text{and} \quad q > 0 \quad (\text{equivalently } \Delta < 0,\ p = 0).$$

(c) The zero solution is ***unstable*** *if*

$$p > 0 \quad \text{or} \quad q < 0.$$

Remark 25 *In the case where $p = 0$ and $q = 0$, further analysis is required to determine the stability of the zero solution.*

Note that according to Theorem 7.10 the zero solution $(0,0)$ of the system in Example 7.7 is asymptotically stable.

Next, we consider the behavior of the trajectories of system (7.17) for various cases of the eigenvalues. In what to follow, we use the matrix transformation

$$Y = BX,\ \det(B) \neq 0$$

so that the essential behavior near the critical point $(0,0)$ remains unchanged.
1. *Real and distinct roots.* If we let

$$B = \begin{pmatrix} c & \lambda_1 - a \\ c & \lambda_2 - a \end{pmatrix}$$

then the system (7.17) transforms into the system

$$x' = \lambda_1 x,$$

$$y' = \lambda_2 y,$$

where, for simplicity x and y are again used as the new coordinates. For some arbitrary constants c_1 and c_2, the solutions of the system are

$$x(t) = c_1 e^{\lambda_1 t}, \;\; y(t) = c_2 e^{\lambda_2 t}. \tag{7.19}$$

Eliminating the time t in (7.19) gives

$$y = c x^{\frac{\lambda_2}{\lambda_1}}, \tag{7.20}$$

where c is an arbitrary constant. When λ_1, λ_2 have the same sign, then equation (7.20) represents parabolic curves that are tangent to the origin as shown in Fig. 7.3(a). The origin is called a *proper node*. We have the following three sub cases.
1.(a) If both λ_1, λ_2 are negative then the origin is asymptotically stable. In this case, we have a *stable node.*
1.(b) If λ_1, λ_2 are positive then the origin is unstable. In this case, we have an *unstable node.*
1.(c) If λ_1, λ_2 have opposite signs, then (7.20) represents hyperbolic curves as shown in Fig. 7.3(b). The origin is unstable and is called a *saddle point* and is unstable.

2. *Real and equal roots.* In this case $\Delta = (a-d)^2 + 4bc = 0$, which implies that

$$\lambda_1 = \lambda_2 = \frac{(a+d)}{2} := \lambda.$$

2.(a) If b or $c = 0$ and $a = d$, then the system (7.17) reduces to

$$x' = \lambda x,$$

$$y' = \lambda y.$$

As before, solving the system and eliminating t leads to the trajectories, that are lines through the origin given by

$$y = \frac{c_2}{c_1} x.$$

Thus, if $\lambda < 0$, the origin is asymptotically stable and is a *proper node* , as shown in Fig. 7.3(c). If $\lambda > 0$, then the origin is unstable.
2.(b) This is a more complicated case. If we let

$$B = \begin{pmatrix} \frac{a-d}{2b} & 1 \\ \frac{1}{b} & 0 \end{pmatrix}, \; b \neq 0,$$

then the system (7.17) transforms into the system

$$x' = \lambda x,$$

$$y' = x + \lambda y.$$

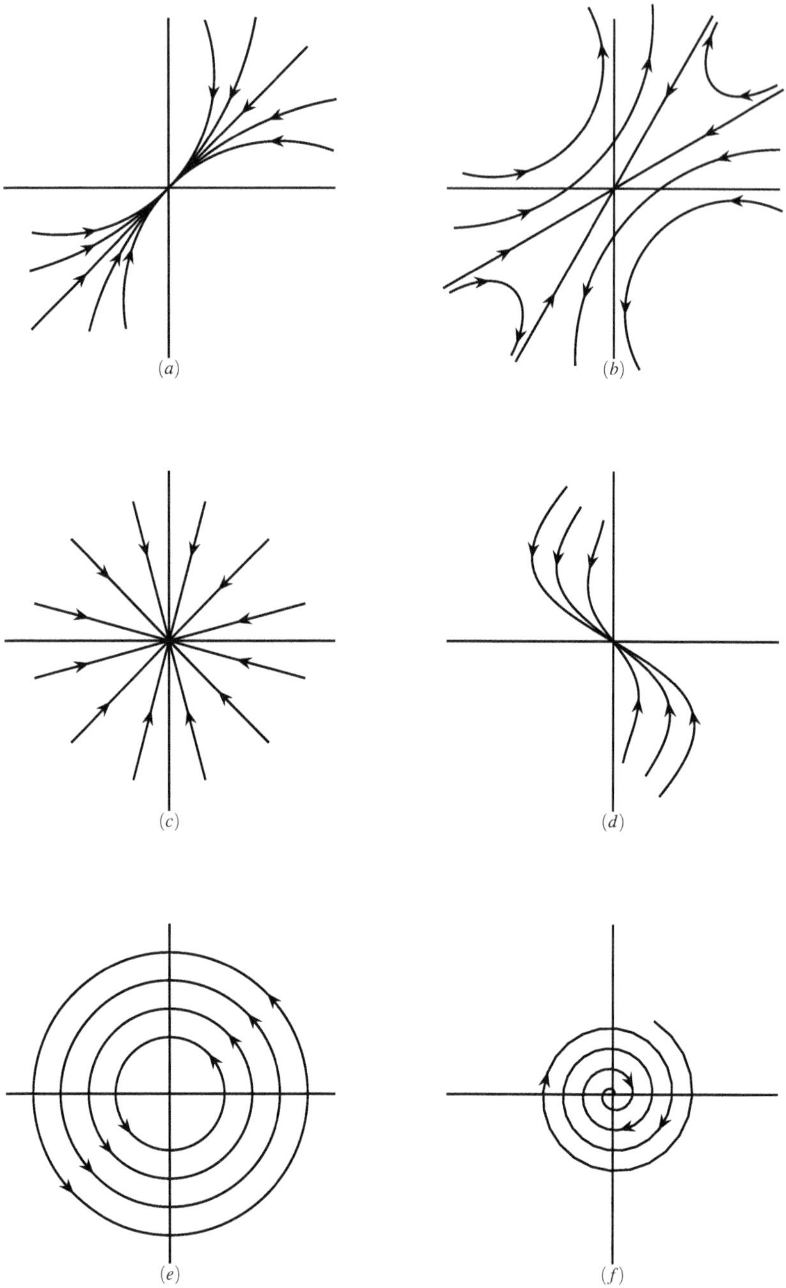

FIGURE 7.3: A phase portrait gallery.

For some arbitrary constants c_1 and c_2, the solutions of the system are

$$x(t) = c_1 e^{\lambda t}, \;\; y(t) = (c_1 t + c_2) e^{\lambda t}. \tag{7.21}$$

The trajectories are shown in Fig. 7.3(d), and the origin is an *improper node*; it is asymptotically stable when $\lambda < 0$ and unstable when $\lambda > 0$.

3. *Complex roots.* In this case, we let

$$\lambda_1 = \alpha + i\beta, \text{ and } \lambda_2 = \alpha - i\beta$$

and chose the matrix

$$B = \begin{pmatrix} c & \alpha - a \\ 0 & \beta \end{pmatrix}$$

so that the system (7.17) transforms into

$$x' = \alpha x - \beta y, \qquad y' = \beta x + \alpha y.$$

3.(a) For $\alpha = 0$, the above system becomes

$$x' = -\beta y, \qquad y' = \beta x$$

with solutions

$$x(t) = c_1 \cos(\beta t) + c_2 \sin(\beta t), \;\; y(t) = c_1 \sin(\beta t) + c_2 \cos(\beta t)$$

for arbitrary constants c_1 and c_2. Hence, it easy to verify that the trajectories are circles given by

$$x^2 + y^2 = c_1^2 + c_2^2$$

and are shown in Fig. 7.3(e) In this case, the origin is stable but not asymptotically stable.

3.(b) For $\alpha \neq 0$, the above system has the solutions

$$x(t) = e^{\alpha t}(c_1 \cos(\beta t) + c_2 \sin(\beta t)), \;\; y(t) = e^{\alpha t}(c_1 \sin(\beta t) + c_2 \cos(\beta t))$$

for arbitrary constants c_1 and c_2. The trajectories are family of spiral given by

$$x^2 + y^2 = (c_1^2 + c_2^2) e^{2\alpha t}$$

and are shown in Fig. 7.3(f). In this case, the origin is a *focal point* that is asymptotically stable when $\alpha < 0$ and unstable when $\alpha > 0$.

The solutions of a planar system of linear differential equations can be classified according to the trace and the determinant of the coefficient matrix. This diagram show schematically the different types of solutions.

7.5.1 Exercises

Exercise 7.13 *Consider the autonomous planar system*

$$\begin{cases} x' = -x + 2y, \\ y' = -2x - y. \end{cases}$$

(a) Write the system in matrix form $X' = AX$ *and compute* $\mathrm{tr}(A)$ *and* $\det(A)$*.*

(b) Compute the discriminant $\Delta = (\mathrm{tr}(A))^2 - 4\det(A)$*.*

(c) Determine the eigenvalues of A *and classify the equilibrium at* $(0,0)$ *using Theorem 7.10 (sink, source, saddle, center, etc.).*

(d) Describe qualitatively the phase portrait near $(0,0)$ *(spiral, node, saddle) and indicate whether the origin is stable, asymptotically stable, or unstable.*

Exercise 7.14 *Transform the second-order ODE*

$$u'' + 3u' + 2u = 0$$

into an autonomous system in the plane and analyze its phase portrait.

(a) Introduce $x = u$ *and* $y = u'$ *and derive the corresponding system*

$$x' = P(x,y), \qquad y' = Q(x,y).$$

(b) Write the system in matrix form $X' = AX$ *and compute* $\mathrm{tr}(A)$*,* $\det(A)$ *and the discriminant* Δ*.*

(c) Find the eigenvalues of A *and classify the equilibrium at* $(0,0)$ *via Theorem 7.10.*

(d) Determine whether the trajectories are straight lines, parabolas, hyperbolas, circles, or spirals, and whether the origin is a stable node, unstable node, saddle, center, or spiral point.

Exercise 7.15 *Consider the linear autonomous system*

$$\begin{cases} x' = 4x + y, \\ y' = -2x + y. \end{cases}$$

(a) Compute the associated matrix A*, its trace* $p = \mathrm{tr}(A)$*, and determinant* $q = \det(A)$*.*

(b) Compute the discriminant $\Delta = p^2 - 4q$ *and determine the eigenvalues using*

$$\lambda_{1,2} = \frac{1}{2}\left(p \pm \sqrt{\Delta}\right).$$

(c) Use Theorem 7.10 to decide whether the origin is asymptotically stable, stable, or unstable.

(d) Classify the type of critical point (proper node, improper node, saddle, spiral source, spiral sink, etc.) and sketch a qualitative phase portrait indicating the main families of trajectories.

Exercise 7.16 *Consider the planar system with complex eigenvalues*

$$\begin{cases} x' = -x - 4y, \\ y' = x - y. \end{cases}$$

(a) Write the system as $X' = AX$ *and compute* $\operatorname{tr}(A)$, $\det(A)$, *and the discriminant* Δ.

(b) Verify that the eigenvalues are complex and compute their real and imaginary parts.

(c) Using the discussion of case 3 in the lecture (complex roots), determine whether the origin is a center, spiral sink, or spiral source.

(d) Show that trajectories satisfy an equation of the form

$$x^2 + y^2 = Ce^{2\alpha t}$$

for some constant C *and* α *equal to the real part of the eigenvalues. Interpret this in terms of spirals moving toward or away from the origin.*

Exercise 7.17 *This exercise illustrates Lemma 5 and Theorem 7.9. Consider the autonomous system*

$$\begin{cases} x' = y, \\ y' = -x. \end{cases}$$

(a) Verify directly that

$$x(t) = \cos t, \qquad y(t) = -\sin t$$

is a solution and describes the corresponding trajectory in the phase plane.

(b) Using Lemma 5, show that for any constant k, the pair

$$x_k(t) = \cos(t+k), \qquad y_k(t) = -\sin(t+k)$$

is also a solution and represents the same trajectory.

(c) Explain why this example illustrates the fact that different solutions (shifted in time) can generate the same trajectory, as emphasized in the remark following Theorem 7.9.

(d) Argue why, for this system, different trajectories (distinct circles) cannot intersect in the phase plane, and relate this to the uniqueness of solutions.

Exercise 7.18 *Describe the type of stability of the origin of each system and sketch the trajectories.*

(a) $x' = x, \ y' = 2x + 2y.$

(b) $x' = -x, \ y' = x - y.$

(c) $x' = -x + 2y, \; y' = x - y.$
(d) $x' = -x + y, \; y' = 2x.$
(e) $x' = 2x - 8y, \; y' = x - 2y.$
(f) $x' = -3x + 2y, \; y' = -2x.$

Exercise 7.19 *Consider the damped harmonic motion represented by the second order differential equation*

$$mx'' + ax' + kx = 0,$$

where m, a, and k are positive constants. By changing the equation into a system, discuss the nature and stability of the origin in the following cases: $a = 0$; $a^2 - 4mk = 0$; $a^2 - 4mk < 0$; $a^2 - 4mk > 0$. Interpret the results physically.

8

Stability and Bifurcations in Nonlinear Systems

In many real-world applications, the governing differential equations are inherently nonlinear, and linear models provide at best a local approximation. Nonlinear systems exhibit a much wider variety of dynamical phenomena than their linear counterparts, including multiple equilibria, limit cycles, and even chaotic behavior. Understanding the stability and qualitative behavior of such systems is essential in physics, engineering, biology, economics, and control theory, where one must often assess the robustness of equilibria or design mechanisms to stabilize desired operating points.

This chapter develops a systematic framework for the analysis of nonlinear systems. We introduce stability indicators that allow one to classify equilibrium points and invariant sets according to their long-term behavior. Particular emphasis is placed on *bifurcation theory*, which studies how qualitative changes in dynamics occur when parameters are varied. In low-dimensional settings, bifurcation diagrams and the geometry of planar curves provide an intuitive description of transitions such as saddle node, transcritical, pitchfork, and Hopf bifurcations.

A central tool throughout is *stability via linearization*: by expanding the vector field around an equilibrium and analyzing the linearized system, one can often infer local stability properties of the full nonlinear system. This linear stability analysis is closely related to the spectrum of the Jacobian matrix and provides simple algebraic criteria for classifying equilibria, while also serving as a first step toward detecting bifurcations. However, linearization is inconclusive in certain critical cases, and more powerful nonlinear methods are required.

To address these limitations, we develop *Lyapunov's direct method* for nonlinear systems. Lyapunov functions play the role of generalized energy functionals whose monotonicity along trajectories yields rigorous stability and instability results without solving the system explicitly. This approach leads to local and global criteria for stability, asymptotic stability, and exponential stability, and it is particularly useful near bifurcation points where linear analysis alone does not capture the full dynamics. Together, linearization, bifurcation analysis, and Lyapunov methods provide a coherent toolkit for understanding and controlling nonlinear dynamical systems.

DOI: 10.1201/9781042009220-8

8.1 Bifurcations in Scalar Systems

In this section, we limit our discussion to the scalar differential equation

$$x' = f(x), \quad t \geq 0 \tag{8.1}$$

where $f : \mathbb{R} \to \mathbb{R}$ is continuously differentiable. We assume x^* is an equilibrium solution of (8.1). We always assume the equilibria are isolated. That is, if x^* is an equilibrium, then there is an open interval containing x^* but no other equilibria. Fig. 8.1 shows a generic plot of $f(x)$ for $x^* = a, b, c$. The equilibrium solution a is an *attractor* (stable) since arrows to its left and right point toward it. The equilibrium solution b is a *repeller* (unstable) since the arrow to its left and the arrow to its right point away from it. Finally, the equilibrium solution c is *semi-stable* since the arrow to its left points toward it and the one to its right points away from it. In the case of the equilibrium solution a, we may say it is *locally asymptotically stable* with respect to small perturbation or deviation from the equilibrium solution. To better illustrate the notion consider the scenario of a population of foxes in a protected refuge that is in a locally asymptotically stable state a. Suppose a weather event caused a small number of the foxes to die. Local asymptotic stability means that the system will eventually return to the original state a. However, if the number of dead foxes is large, then the perturbation is not small, and in this event, there is no guarantee that the population of foxes will return to the original state a. In the case that population return to state a for all perturbations, no matter how large, then the state a is called *globally asymptotically stable.*

Now we examine the dynamics of (8.1) in the neighborhood of its equilibrium solution x^*. If (8.1) is in an equilibrium state x^* and $\eta(t)$ represents a small perturbation

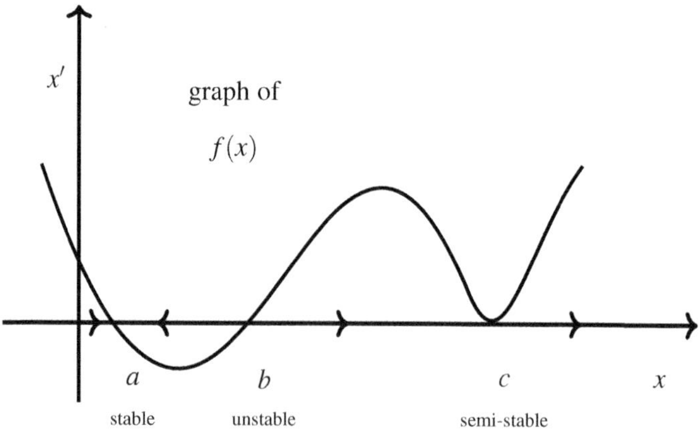

FIGURE 8.1: Generic phase line diagram.

from that state, then $x(t) = x^* + \eta(t)$. It follows that $x'(t) = \eta'(t)$. Thus, using Taylor series expansion about x^* we have

$$\begin{aligned} x'(t) = \eta'(t) = f(x^* + \eta) &= f(x^*) + \frac{f'(x^*)}{1!}(x^* + \eta - x^*) + \frac{f''(x^*)}{2!}(x^* + \eta - x^*)^2 + \dots \\ &= f'(x^*)\eta + \frac{f''(x^*)}{2!}\eta^2 + \dots \\ &= f'(x^*)\eta + O(\eta^2) + \dots \end{aligned}$$

If $f'(x^*) \neq 0$, then the term $|f'(x^*)\eta| \gg |\frac{f''(x^*)}{2!}\eta^2|$. Thus, we may neglect the term $O(\eta^2)$ and higher-order terms for that same reason. Then, we arrive at the linearization of the system about the equilibrium state x^* given by

$$\eta'(t) = s\eta, \tag{8.2}$$

where $s = f'(x^*)$ is the slope of $f(x)$ at x^*. Equation (8.2) has the solution

$$\eta(t) = ce^{st} \tag{8.3}$$

for some constant c. It is clear from (8.3) that if $s > 0$, then solutions grow exponentially and the equilibrium solution x^* is unstable. On the other hand, if $s < 0$, then the solutions decay exponentially to zero, and the equilibrium solution x^* is asymptotically stable and hence it is stable. If $s = 0$, then nothing can be said about the stability. Note that different stability may be obtained for different systems when $s = 0$. To see this, we consider the scalar differential equations

$$x' = x^2, \;\; x' = -x^2, \;\; x' = x^3, \;\; \text{and} \;\; x' = -x^3.$$

All of them share the equilibrium solution $x^* = 0$ and $s = f'(0) = 0$. Consider the first equation with $f(x) = x^2$. Then $x'(x) > 0$ for $x < 0$, and for $x > 0$. Hence, trajectories approach 0 from the left but move away on the right. Therefore, $x^* = 0$ is semi-stable. In similar arguments one finds that $x^* = 0$ is semi-stable for the second equation, unstable for the third equation, and (asymptotically) stable for the fourth equation.

Example 8.1 *If $x(t)$ represents a population at time t, $r > 0$ is the growth rate, and $K > 0$ is the carrying capacity, then the logistics model of the population growth is given by*

$$x' = rx(1 - \frac{x}{K}). \tag{8.4}$$

Setting the right-hand side of (8.4) *equal to zero, we arrive at the two equilibrium solutions*

$$x^* = 0, \; K.$$

Let $f(x) = rx(1 - \frac{x}{K})$, then $f'(x) = r - \frac{2r}{K}x$. It follows from (8.2) *that*

$$s = f'(0) = r, \;\; \text{and} \;\; s = f'(K) = -r.$$

We conclude that $x^* = 0$ is unstable and $x^* = K$ is stable. □

Although the dynamic of scalar differential equations is simple since either solutions converge to a constant or become unbounded as $t \to \infty$, they can have some interesting behaviors if solutions depend on some embedded parameters. To be specific, when the parameters are varied and pass some critical values, the systems may experience some abrupt changes, or undergo some qualitative changes. These qualitative changes are called *bifurcations*, and the parameter values at which the bifurcations occur are called *bifurcation points*. We are interested in scalar differential equations of the form

$$x' = f(x,\mu), \tag{8.5}$$

where f is continuous in both arguments and $\mu \in \mathbb{R}$ is a parameter.

Thus, the system is said to undergo a bifurcation as μ crosses some critical value μ_0 if, in some way, the nature of the solutions changes qualitatively between the region $\mu < \mu_0$ and $\mu > \mu_0$.

The three most important one-dimensional equilibrium bifurcations are described locally by the following differential equations:

$$x' = \mu - x^2, \quad \text{saddle node;}$$

$$x' = \mu x - x^2, \quad \text{transcritical;}$$

$$x' = \mu x - x^3, \quad \text{pitchfork.}$$

We will study each of these in more detail below. We begin by considering *saddle node bifurcation* which is basic mechanism for creation and destruction of fixed points. As the parameter varies, the two fixed points move toward each other, collide, and mutually annihilate. That is, the saddle-node bifurcation results in fixed points being created or destroyed.

Example 8.2 *Consider the scalar differential equation with parameter* μ

$$x' = \mu - x^2. \tag{8.6}$$

Clearly there are no equilibrium solutions for $\mu < 0$ *and* $x^* = \pm\sqrt{\mu}$ *for* $\mu > 0$. *It is clear that* $s = f'(\sqrt{\mu}) = -2\sqrt{\mu} < 0$, *and* $s = f'(-\sqrt{\mu}) = 2\sqrt{\mu} > 0$, *which imply that the dashed branch* $x = \sqrt{\mu}$ *is stable and the branch* $x = -\sqrt{\mu}$ *is unstable. The Bifurcation diagram is shown in Fig. 8.2, and since there is an exchange of stability at* $\mu = 0$, *we have a bifurcation at* $\mu = 0$. *This type of bifurcation is known by the names saddle node bifurcation, fold bifurcation, turning point, and blue-sky bifurcation. Note that at the bifurcation point* $\mu = 0$, $s = f'(0) = 0$ *and hence the linearization term disappears when equilibrium solutions coalesce.* □

Remark 26 *In* (8.6) *of Example 8.2, if we change the term* $-x^2$ *to* x^2 *then*

$$x' = \mu + x^2.$$

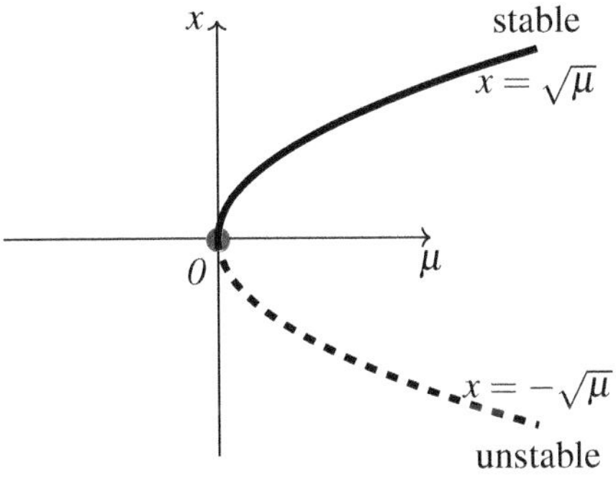

FIGURE 8.2: Bifurcation diagram determined by $x' = \mu - x^2$.

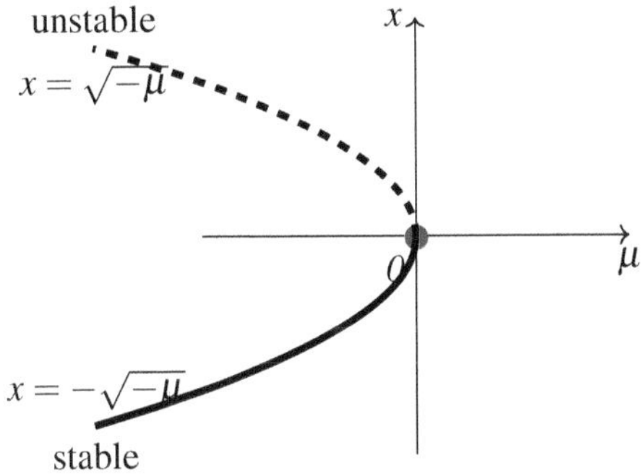

FIGURE 8.3: Bifurcation diagram determined by $x' = \mu + x^2$.

There are no equilibrium solutions for $\mu > 0$, and the curve $y = \mu + x^2$ will not intersect the x-axis. As $\mu \searrow 0$, then the graph $y = \mu + x^2$ intersects the origin tangentially and we have one equilibrium solution. As μ becomes negative, two equilibrium solutions appear at $x = \pm\sqrt{-\mu}$. Then, using the stability indicator, we see that $x = -\sqrt{-\mu}$ is stable, and $x = \sqrt{-\mu}$ is unstable. We conclude there is a saddle node bifurcation at $(0,0)$ as indicated in the bifurcation diagram shown in Fig. 8.3.

Next, we discuss a class of bifurcation that occur as stability changes when going from one equilibrium solution to another. This type of bifurcation is called *transcritical bifurcation*

Example 8.3 *To illustrate the idea, we consider the population growth model with carrying capacity , μ*

$$x' = x(\mu - x).$$

The system has the equilibrium solutions $x^* = 0, \mu$. Using $s = f'(x^*)$, it follows that $x^* = 0$ is stable of $\mu < 0$, and unstable if $\mu > 0$. Similarly, $x^* = \mu$ is stable for $\mu > 0$,

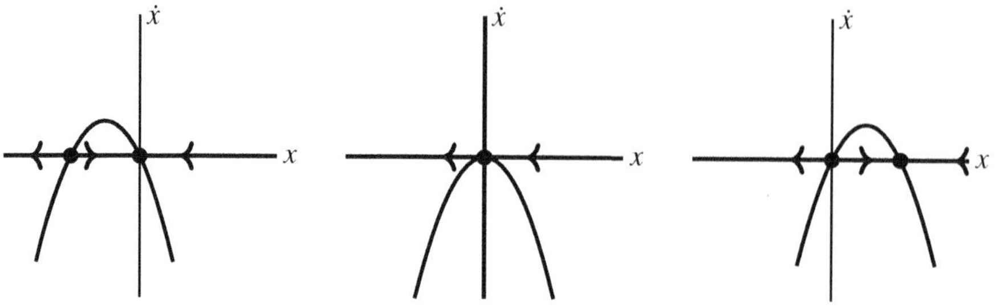

(a) $\mu < 0$ (b) $\mu = 0$ (c) $\mu > 0$

FIGURE 8.4: Phase portraits for different values of μ.

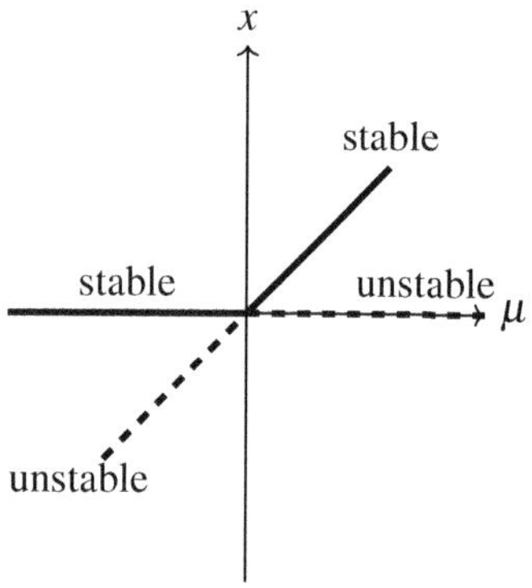

FIGURE 8.5: Bifurcation diagram that shows $(\mu, x) = (0,0)$ is a transcritical bifurcation point.

and unstable for $\mu < 0$. *Fig. 8.4 Shows the Phase portraits for different values of* μ. *Also, Fig. 8.5 shows the bifurcation diagram. The stable branches are shown with the bold solid lines, and the unstable branches are shown with the dashed lines.* □

Next, we discuss a type of bifurcation known as a *pitchfork bifurcation*. The name arises from the characteristic shape of its bifurcation diagram, which resembles a pitchfork. This type of bifurcation commonly appears in problems possessing a *symmetry*, such as the buckling of a beam. In a pitchfork bifurcation, equilibrium solutions emerge or disappear in a symmetric fashion, distinguishing this behavior from the bifurcations discussed previously.

Pitchfork bifurcations occur in two forms: the *supercritical bifurcation* and the *subcritical bifurcation*. A supercritical bifurcation is typically represented by equations of the form

$$x' = \mu x - x^3,$$

which we will discuss in the next example. In addition, subcritical bifurcation is typified by equations of the form

$$x' = \mu x + x^3.$$

In such a case, for $\mu < 0$, the origin is stable, and the other two equilibrium solutions $\pm\sqrt{\mu}$ are unstable. On the other hand, for $\mu > 0$, the origin is the only equilibrium solution, and it is unstable.

Example 8.4 *Consider the scalar differential equation*

$$x' = \mu x - x^3.$$

The equilibrium solutions are obtained from

$$\mu x - x^3 = 0 \quad \Longleftrightarrow \quad x(\mu - x^2) = 0.$$

Thus,

$$x^* = 0 \quad \text{for all } \mu \in \mathbb{R}, \qquad x^* = \pm\sqrt{\mu} \quad \text{for } \mu \geq 0.$$

Observe that the function $f(x) = \mu x - x^3$ is odd in x. Consequently, the system is symmetric with respect to the origin, and nontrivial equilibria appear or disappear in symmetric pairs.

To determine stability, we compute

$$f'(x) = \mu - 3x^2.$$

At the trivial equilibrium $x^* = 0$,

$$f'(0) = \mu.$$

Hence, $x^* = 0$ is asymptotically stable for $\mu < 0$ and unstable for $\mu > 0$. At $\mu = 0$, the linearization is inconclusive.

For the nontrivial equilibria $x^* = \pm\sqrt{\mu}$ (when $\mu > 0$),

$$f'(\pm\sqrt{\mu}) = \mu - 3\mu = -2\mu < 0,$$

so both equilibria are asymptotically stable.

The bifurcation diagram is shown in Fig. 8.6. It illustrates that as μ passes through zero, the trivial equilibrium loses stability and two stable equilibrium branches emerge symmetrically. This is the classical supercritical pitchfork bifurcation.

A similar but reversed stability behavior occurs in the case of a subcritical pitchfork bifurcation. □

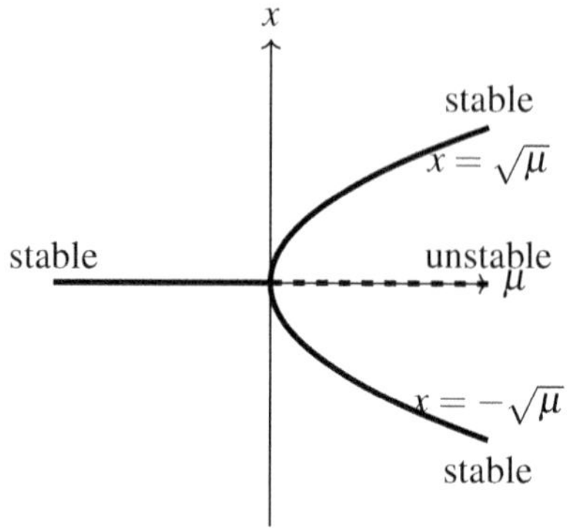

FIGURE 8.6: Bifurcation diagram determined for $x' = \mu x - x^3$; supercritical bifurcation at $(0,0)$.

The next theorem is the scalar version of the *implicit function theorem.*

Theorem 8.1 *For scalar system* (8.5)*, suppose that* $f : \mathbb{R} \times \mathbb{R} \to \mathbb{R}$ *with continuous first partial derivatives. Suppose* x_0 *is an equilibrium solution at* μ_0*, meaning that* $f(x_0, \mu_0) = 0$*. Assume*

$$f(x_0, \mu_0) = 0, \quad \frac{\partial f}{\partial x}(x_0, \mu_0) \neq 0.$$

Then there exists $\delta, \varepsilon > 0$ *and a* C^1*-function*

$$\tilde{x} : (\mu_0 - \varepsilon, \mu_0 + \varepsilon) \to \mathbb{R},$$

such that $x = \tilde{x}$ *is the unique solution of*

$$f(x, \mu) = 0,$$

with $|x - x_0| < \delta$ *and* $|\mu - \mu_0| < \varepsilon$*.*

For the purpose of bifurcation theory, we have the following important result.

Corollary 4 *Suppose that* $f : \mathbb{R} \times \mathbb{R} \to \mathbb{R}$ *has continuous first partial derivatives. Suppose* x_0 *is an equilibrium solution at* μ_0*, for* (8.5)*. Then, a necessary condition for a solution* (x_0, μ_0) *of* (8.5) *to be a bifurcation point of equilibria is that*

$$\frac{\partial f}{\partial x}(x_0, \mu_0) = 0. \tag{8.7}$$

Remark 27 *Condition* (8.7) *indicates that bifurcation points can occur only at equilibria where the linearization loses its effectiveness, that is, where*

$$\frac{\partial f}{\partial x}(x_0, \mu_0) = 0.$$

If instead $\frac{\partial f}{\partial x}(x_0, \mu_0) \neq 0$, *then by the Implicit Function Theorem, there exists a unique smooth function* $x = x(\mu)$ *near* μ_0 *such that* $f(x(\mu), \mu) = 0$. *In this case, the equilibrium persists smoothly, and no bifurcation occurs.*

Thus, bifurcations can occur only at nonhyperbolic equilibria.

The next theorem provides necessary conditions that guarantee the different types of bifurcation points that we have discussed so far.

Theorem 8.2 *Suppose that* $f : \mathbb{R} \times \mathbb{R} \to \mathbb{R}$ *is continuous in both arguments and has continuous first and second partial derivatives. Suppose at* (x_0, μ_0), f *satisfies the necessary bifurcation conditions*

$$f(x_0, \mu_0) = 0, \quad \frac{\partial f}{\partial x}(x_0, \mu_0) = 0.$$

1. If

$$\frac{\partial f}{\partial \mu}(x_0, \mu_0) \neq 0, \quad \frac{\partial^2 f}{\partial^2 x}(x_0, \mu_0) \neq 0,$$

then a saddle node bifurcation occurs at (x_0, μ_0).
2. If

$$\frac{\partial f}{\partial \mu}(x_0, \mu_0) = 0, \quad \frac{\partial^2 f}{\partial x \partial \mu}(x_0, \mu_0) \neq 0, \quad \frac{\partial^2 f}{\partial^2 x}(x_0, \mu_0) \neq 0,$$

then a transcritical bifurcation occurs at (x_0, μ_0).
3. If

$$\frac{\partial f}{\partial \mu}(x_0, \mu_0) = 0, \quad \frac{\partial^2 f}{\partial^2 x}(x_0, \mu_0) = 0,$$

$$\frac{\partial^2 f}{\partial x \partial \mu}(x_0, \mu_0) \neq 0, \quad \frac{\partial^3 f}{\partial^3 x}(x_0, \mu_0) \neq 0,$$

then a pitchfork bifurcation occurs at (x_0, μ_0). *In addition*
i) if $\frac{\partial^3 f}{\partial^3 x}(x_0, \mu_0) < 0$, *then there is a supercritical bifurcation at* (x_0, μ_0) *and*
i) if $\frac{\partial^3 f}{\partial^3 x}(x_0, \mu_0) > 0$, *then there is a subcritical bifurcation at* (x_0, μ_0).

Example 8.5 *We consider the scalar system*

$$x' = \mu x - e^x.$$

Then its bifurcation points must satisfy

$$\mu x - e^x = 0.$$

if we plot the graphs $y = \mu x$ *and* $y = e^x$ *we see that the line* $y = x$ *is tangent to the curve at* $(x, y) = (1, e)$. *In other words, if we set* $f(x, \mu) = \mu x - e^x$, *then* $f(1, e) = 0$. *This says that* $x_0 = 1$ *is an equilibrium solution for* $\mu = e$. *Moreover,* $\frac{\partial f}{\partial x}(1, e) = 0$. *In addition* $\frac{\partial f}{\partial \mu}(x_0, \mu_0) = 1 \neq 0$, *and* $\frac{\partial^2 f}{\partial x^2}(x_0, \mu_0) = -e \neq 0$, *and so by 1. of Theorem 8.2, there is a saddle node bifurcation at* $(1, e)$. □

Remark 28 *Suppose bifurcation for* (8.5) *occurs at* (x_0, μ_0) *that we may translate to the origin* $(0,0)$ *by the change of variables* $x \to x - x_0,\ \ \mu \to \mu - \mu_0$. *Moreover, if* $x = \tilde{x}(\mu)$ *is a solution branch,* $x \to x - \tilde{x}(\mu)$ *maps the branch to* $x = 0$. *Suppose we have a saddle node bifurcation. For* $(0,0)$ *to be a bifurcation point, we must have*

$$f(x_0, \mu_0) = 0, \quad \frac{\partial f}{\partial x}(x_0, \mu_0) = 0.$$

Further suppose that

$$\frac{\partial f}{\partial \mu}(0,0) = a \neq 0, \quad \frac{\partial^2 f}{\partial^2 x}(0,0) = b \neq 0.$$

Then if in the Taylor expansion of $f(x,\mu)$ *we neglect quadratic terms in* μ *and cubic terms in* x, *then near* $(x,\mu) = (0,0)$, *we obtain*

$$\begin{aligned} f(x,\mu) &= \mu \frac{\partial f}{\partial \mu}(0,0) + x \frac{\partial f}{\partial x}(0,0) + \frac{x^2}{2} \frac{\partial^2 f}{\partial^2 x}(0,0) + \dots \\ &= a\mu + bx^2 + \dots \end{aligned} \tag{8.8}$$

The form $a\mu + bx^2$ *in* (8.8) *is called a normal form or representative for scalar saddle node bifurcations. Thus, the original system may be approximated near the origin by*

$$x' = a\mu + bx^2.$$

The signs of a and b determine the type of bifurcations and the shape of the bifurcation diagrams, as shown in Example 8.2 and Remark 26.

We end the section with the following application.

Example 8.6 *(Buckling of a rod) The potential energy of the system of two rigid rods of length L connected by a torsional spring with spring constant k is given by*

$$R(x) = \frac{1}{2}kx^2 + 2\lambda L(\cos x - 1),$$

where λ *is the magnitude of the compressive force, and x is the angle of the rod with the horizontal axis.*

Equilibrium solutions satisfy $R'(x) = 0$, *which gives*

$$kx - 2\lambda L \sin x = 0.$$

Dividing by $k \neq 0$ *we obtain*

$$x - \mu \sin x = 0, \qquad \mu = \frac{2\lambda L}{k},$$

where μ *is a dimensionless parameter.*

The equation has the trivial (unbuckled) solution $x = 0$. *Let*

$$f(x,\mu) = \mu \sin x - x.$$

Note that $f(-x,\mu) = -f(x,\mu)$, *so the system is symmetric.*

For a bifurcation at $(x,\mu) = (0,\mu_0)$ *we require*

$$f(0,\mu_0) = 0, \qquad \frac{\partial f}{\partial x}(0,\mu_0) = 0.$$

Since

$$\frac{\partial f}{\partial x} = \mu \cos x - 1,$$

we obtain

$$\frac{\partial f}{\partial x}(0,\mu) = \mu - 1,$$

so the degeneracy occurs at

$$\mu = 1.$$

Expanding f *in a Taylor series about* $(0,1)$ *gives*

$$f(x,\mu) = (\mu - 1)x - \frac{1}{6}x^3 + \cdots.$$

Moreover,

$$\frac{\partial^3 f}{\partial x^3}(0,1) = -1 \neq 0,$$

so the nondegeneracy condition for a pitchfork bifurcation holds.

Therefore, there is a pitchfork bifurcation at $(x,\mu) = (0,1)$.

The bifurcating equilibria for $0 < \mu - 1 \ll 1$ *satisfy*

$$x = \pm\sqrt{6(\mu - 1)} + \cdots.$$

This illustrates the buckling of an elastic beam, one of the earliest bifurcation problems, studied by Euler (1757). □

8.1.1 Exercises

Exercise 8.1 *For the following equations, find the equilibrium solutions, sketch the phase line, and determine the type of stability of all the equilibria.*

(a) $x' = x^2(4 - x)$.

(b) $x' = 2x(1 - x) - \frac{1}{2}x$.

(c) $x' = (4-x)(2-x)^3$.

Exercise 8.2 *Find the equilibrium solutions in terms of the parameter μ and determine their stability. Show the origin is a turning point, sketch a bifurcation diagram, and label all branches as stable or unstable. When possible, identify the type of bifurcation.*

(a) $x' = \mu x(1-x)$.

(b) $x' = (x^2-\mu)(1-x)$.

Exercise 8.3 *Find the equilibrium solutions and determine their stability. Show that a bifurcation occurs at $(0,0)$ and sketch the bifurcation diagram for*

$$x' = \mu - x^4.$$

Exercise 8.4 *For the following equations, show that each one of them undergoes a transcritical bifurcation and find the bifurcation value. Sketch the bifurcation diagram.*

(a) $x' = \mu x + 2x^2$.

(a) $x' = (2+\mu)x - 5x^2$.

Exercise 8.5 *Find the equilibrium solution x_0 at μ_0 and use Theorem 8.2 to determine the type of bifurcation at (x_0, μ_0) for the scalar system*

$$x' = \mu + x - e^x + 1.$$

Exercise 8.6 *Show that*

$$x' = \mu x + x^3$$

undergoes a pitchfork bifurcation at $\mu = 0$, and sketch the bifurcation diagram.

Exercise 8.7 *(Stability via linearization for scalar systems) Consider the scalar nonlinear differential equation*

$$x' = f(x) = x - x^3.$$

(a) *Find all equilibrium solutions and classify each as stable, unstable, or semi-stable by examining the sign of $f(x)$ on intervals between equilibria (i.e., using a phase line).*

(b) *Compute $f'(x)$ and use the linearization $x' = f'(x^*)\eta$ near each equilibrium x^* to predict its stability type.*

(c) *Verify that the conclusions from the linearization in part (b) agree with the phase-line analysis in part (a).*

8.2 Stability of Systems by Linearization

For the rest of this chapter, if

$$x = \begin{pmatrix} x_1 \\ x_2 \\ \vdots \\ x_n \end{pmatrix},$$

we define the norm

$$|x| = \sum_{i=1}^{n} |x_i|,$$

and for an $n \times n$ matrix $A = (a_{ij})$,

$$|A| = \sum_{i,j=1}^{n} |a_{ij}|.$$

Consider the general system of ordinary differential equations

$$\begin{aligned} x_1' &= f_1(x_1, \ldots, x_n) \\ x_2' &= f_2(x_1, \ldots, x_n) \\ &\vdots \\ x_n' &= f_n(x_1, \ldots, x_n). \end{aligned}$$

Using vector notation,

$$x = \begin{pmatrix} x_1 \\ x_2 \\ \vdots \\ x_n \end{pmatrix}, \qquad f(x) = \begin{pmatrix} f_1(x) \\ f_2(x) \\ \vdots \\ f_n(x) \end{pmatrix},$$

the system can be written as

$$x' = f(x), \tag{8.9}$$

where $f : \mathbb{R}^n \to \mathbb{R}^n$ and each component is continuously differentiable.

Assume that $x^* = (x_1^*, \ldots, x_n^*)$ is a fixed point of (8.9), i.e., $f(x^*) = 0$.

Define the linear part of f at x^* by $J(x^*)$, where

$$J(x^*) := \begin{pmatrix} \frac{\partial f_1}{\partial x_1}(x^*) & \frac{\partial f_1}{\partial x_2}(x^*) & \cdots & \frac{\partial f_1}{\partial x_n}(x^*) \\ \frac{\partial f_2}{\partial x_1}(x^*) & \frac{\partial f_2}{\partial x_2}(x^*) & \cdots & \frac{\partial f_2}{\partial x_n}(x^*) \\ \vdots & \vdots & \ddots & \vdots \\ \frac{\partial f_n}{\partial x_1}(x^*) & \frac{\partial f_n}{\partial x_2}(x^*) & \cdots & \frac{\partial f_n}{\partial x_n}(x^*) \end{pmatrix}. \tag{8.10}$$

The matrix $J(x^*)$ is called the *Jacobian* of f at x^*.

Since $f \in C^1(\mathbb{R}^n, \mathbb{R}^n)$, Taylor's theorem yields

$$f(x) = f(x^*) + J(x^*)(x - x^*) + g(x), \tag{8.11}$$

where g satisfies

$$\lim_{x \to x^*} \frac{|g(x)|}{|x - x^*|} = 0. \tag{8.12}$$

Since $f(x^*) = 0$, (8.11) reduces to

$$f(x) = J(x^*)(x - x^*) + g(x),$$

and in particular $g(x^*) = 0$.

Our goal is to show that if all eigenvalues λ of $J(x^*)$ satisfy

$$\text{Re}(\lambda) < 0,$$

that is, if the origin of the linear system

$$y' = J(x^*)y$$

is asymptotically stable, then the equilibrium point x^* of (8.9) is asymptotically stable.

By introducing the transformation $y = x - x^*$, we may assume without loss of generality that the equilibrium is at the origin.

To work in a more general setting, the perturbation term g in (8.11) may depend explicitly on time; that is, the original system is non-autonomous. Therefore, we have the following general theorem. First, we make the following definitions regarding a function being Lipschitz.

Definition 8.1 *Let D be a subset of $\mathbb{R} \times \mathbb{R}$. The function $f : D \to \mathbb{R}$ is said to satisfy a global Lipschitz condition in x if there exists a Lipschitz constant $k > 0$ such that*

$$|f(t,x) - f(t,y)| \leq k|x - y|, \ \textit{for} \ (t,x),\ (t,y) \in D. \tag{8.13}$$

Definition 8.2 *Let D be a subset of $\mathbb{R} \times \mathbb{R}$. The function $f : D \to \mathbb{R}$ is said to satisfy a local Lipschitz condition in x if for any $(t_1, x_1) \in D$ there exists a domain $(t_1, x_1) \in D_1 \subset D$ and such that $f(t,x)$ satisfies a Lipschitz condition in x on D_1. That is there exists a positive constant K_1 such that*

$$|f(t,x) - f(t,y)| \leq K_1|x - y|, \ \textit{for} \ (t,x),\ (t,y) \in D_1. \tag{8.14}$$

Definition (8.1) can be easily extended to functions $f : D \to \mathbb{R}^n$, where $D \subset \mathbb{R} \times \mathbb{R}^n$ under a proper norm. Let R be any rectangle in D such that $R = \{(t,x) : |t| \leq a,\ |x| \leq b\}$. If we assume that f and $\frac{\partial f}{\partial x}$ are continuous on R, which is the case in this chapter, then f and $\frac{\partial f}{\partial x}$ are bounded on R. Therefore, there exists positive constants $\mathscr{M}$ and $\mathscr{K}$ such that

$$|f(t,x)| \leq \mathscr{M}, \text{ and } |\frac{\partial f}{\partial x}| \leq \mathscr{K} \tag{8.15}$$

for all points (t,x) in R. Now, for any two points $(t,x_1),(t,x_2)$ in R, by the mean value theorem, there exists a constant $\eta \in (x_1,x_2)$ such that

$$f(t,x_1)-f(t,x_2) = \frac{\partial f}{\partial x}(t,\eta)(x_1-x_2),$$

form which it follows that

$$\begin{aligned} |f(t,x_1)-f(t,x_2 &\leq |\frac{\partial f}{\partial x}(t,\eta)||x_1-x_2| \\ &\leq \mathscr{K}|x_1-x_2|. \end{aligned} \tag{8.16}$$

We have shown that if f and $\frac{\partial f}{\partial y}$ are continuous on R, then f satisfies a global Lipschitz condition on R.

Example 8.7 *Consider $f(t,x) = x^{2/3}$ in the rectangle $R = \{(t,x) : |t| \leq 1, \ |x| \leq 1\}$. We claim f does not satisfy Lipchitz condition on R. Consider the pair of points (t,x_1) and $(t,0)$ in R where $x_1 > 0$. Then*

$$\frac{|f(t,x_1)-f(t,0)}{x_1-0}| = x_1^{-1/3} \to \infty, \text{ as } x_1 \to 0^+,$$

and hence there exists no $\mathscr{K}$ such that (8.16) *holds.* □

In order to prove results regarding the (UAS) of the non-autonomous system, we state the followings.

Definition 8.3 *A (first-order) non-autonomous system is a differential equation of the form*

$$x' = f(t,x),$$

where $t \in I \subset \mathbb{R}$ (an interval), $x \in \mathbb{R}^n$, and $f : I \times \mathbb{R}^n \to \mathbb{R}^n$ is a given function.

(i) $f(t,x)$ is continuous in (t,x) on $I \times \mathbb{R}^n$;

(ii) $f(t,x)$ is locally Lipschitz continuous in x, uniformly in t on compact subsets of I, i.e., for every compact set $K \subset \mathbb{R}^n$ and every compact interval $J \subset I$, there exists $L > 0$ such that

$$|f(t,x)-f(t,y)| \leq L|x-y| \quad \text{for all } t \in J, \ x,y \in K.$$

Theorem 8.3 (Existence and uniqueness) *Under the above assumptions, for every $(t_0,x_0) \in I \times \mathbb{R}^n$, there exists a unique solution $x(t)$ of*

$$x' = f(t,x), \qquad x(t_0) = x_0, \tag{8.17}$$

defined on some interval containing t_0.

Theorem 8.4 *Assume $Re(\sigma(A)) < 0$ where A is an $n \times n$ constant matrix (or the zero solution of the linear part of (8.17) is (AS)). Assume the continuous function $g(t,x)$ satisfies Lipschitz or local Lipschitz condition in the second argument for $x \in D$. Suppose that*

$$\lim_{x\to 0} \frac{|g(t,x)|}{|x|} = 0, \text{ uniformly for } t \in [0,\infty) \tag{8.18}$$

with $g(t,0) = 0$. Then the origin of the perturbed nonlinear system

$$x' = Ax + g(t,x)$$

is uniformly asymptotically stable.

Proof *Since $\mathrm{Re}\,\lambda < 0$ for every $\lambda \in \sigma(A)$, there exist positive constants $K \geq 1$ and $\alpha > 0$ such that the matrix exponential satisfies*

$$\|e^{At}\| \leq Ke^{-\alpha t}, \quad \text{for all } t \geq 0. \tag{8.19}$$

This follows from the spectral properties of A and standard results on linear systems (see, e.g., the discussion preceding Lemma 4 in the linear theory). The uniform limit (8.18) means that for every $\varepsilon > 0$, there exists $\delta_0 > 0$ such that

$$|x| < \delta_0 \text{ implies that } \frac{|g(t,x)|}{|x|} < \varepsilon \quad \text{for all } t \geq 0,$$

or equivalently

$$|x| < \delta_0 \text{ implies that } |g(t,x)| \leq \varepsilon|x| \quad \text{for all } t \geq 0. \tag{8.20}$$

In particular, we will later choose ε small (relative to α) to ensure that g is a "small" perturbation of the linear term in a neighborhood of the origin. Because g is locally Lipschitz in x on D, for any bounded neighborhood of the origin, there exists a Lipschitz constant $L > 0$ such that

$$|g(t,x_1) - g(t,x_2)| \leq L|x_1 - x_2|,$$

for all x_1, x_2 in that neighborhood and all $t \geq 0$; this guarantees the existence and uniqueness of solutions for initial data sufficiently small. The variation of parameters formula gives

$$x(t) = e^{A(t-t_0)}x_0 + \int_{t_0}^{t} e^{A(t-s)}g(s,x(s))\,ds, \qquad t \geq t_0. \tag{8.21}$$

Taking norms in (8.21) and using (8.19) we obtain

$$\begin{aligned} |x(t)| &\leq \|e^{A(t-t_0)}\|\,|x_0| + \int_{t_0}^{t} \|e^{A(t-s)}\|\,|g(s,x(s))|\,ds \\ &\leq Ke^{-\alpha(t-t_0)}|x_0| + K\int_{t_0}^{t} e^{-\alpha(t-s)}|g(s,x(s))|\,ds. \end{aligned} \tag{8.22}$$

Next, we utilize the local smallness of g along small solutions. Fix a radius $r > 0$ *with* $r \le \delta_0$*, where* δ_0 *is given by* (8.20) *for some* $\varepsilon > 0$ *to be chosen. We now argue that if* $|x_0|$ *is small enough, then* $|x(t)|$ *remains less than r for all* $t \ge t_0$*. This is a standard continuation argument. Let*

$$\tau := \sup\{\, t \ge t_0 : |x(s)| < r \text{ for all } s \in [t_0, t]\,\}.$$

By continuity of $x(\cdot)$*, we have* $\tau > t_0$*. If* $\tau < \infty$ *then* $|x(\tau)| = r$*.*

On $[t_0, \tau)$*, we have* $|x(s)| < r \le \delta_0$*, hence by* (8.20)

$$|g(s, x(s))| \le \varepsilon |x(s)|, \qquad s \in [t_0, \tau). \tag{8.23}$$

Using (8.22) *and* (8.23) *for* $t \in [t_0, \tau)$*, we get*

$$|x(t)| \le Ke^{-\alpha(t-t_0)}|x_0| + K\varepsilon \int_{t_0}^{t} e^{-\alpha(t-s)}|x(s)|\,ds. \tag{8.24}$$

Define

$$y(t) := e^{\alpha(t-t_0)}|x(t)|, \qquad t \in [t_0, \tau).$$

Multiplying (8.24) *by* $e^{\alpha(t-t_0)}$ *yields*

$$\begin{aligned} y(t) &\le K|x_0| + K\varepsilon \int_{t_0}^{t} e^{\alpha(t-t_0)} e^{-\alpha(t-s)}|x(s)|\,ds \\ &= K|x_0| + K\varepsilon \int_{t_0}^{t} e^{\alpha(s-t_0)}|x(s)|\,ds \\ &= K|x_0| + K\varepsilon \int_{t_0}^{t} y(s)\,ds. \end{aligned}$$

Thus, y satisfies the integral inequality

$$y(t) \le K|x_0| + K\varepsilon \int_{t_0}^{t} y(s)\,ds, \qquad t \in [t_0, \tau).$$

By Gronwall's inequality,

$$y(t) \le K|x_0| \exp\big(K\varepsilon(t-t_0)\big), \qquad t \in [t_0, \tau).$$

Therefore

$$|x(t)| \le K|x_0| \exp\big(-(\alpha - K\varepsilon)(t-t_0)\big), \qquad t \in [t_0, \tau). \tag{8.25}$$

Choose $\varepsilon > 0$ *such that*

$$\alpha - K\varepsilon > 0.$$

This is always possible. Then (8.25) *shows that* $|x(t)|$ *is bounded by a decaying exponential on* $[t_0, \tau)$*. In particular,*

$$|x(t)| \le K|x_0| \quad \text{for all } t \in [t_0, \tau).$$

Now choose $\delta > 0$ such that

$$K\delta < r.$$

If $|x_0| < \delta$, then by the above bound we have

$$|x(t)| \le K|x_0| < r \quad \text{for all } t \in [t_0, \tau).$$

This contradicts the definition of τ unless $\tau = \infty$. Hence $\tau = \infty$ and

$$|x(t)| < r \le \delta_0 \quad \text{for all } t \ge t_0.$$

Thus, starting sufficiently close to the origin, the solution stays in the region where (8.20) holds for all time. Since for $|x_0| < \delta$ the solution never leaves the ball of radius $r \le \delta_0$, (8.23) (and hence (8.25)) actually hold for all $t \ge t_0$. Therefore, for all $t \ge t_0$,

$$|x(t)| \le K|x_0| \exp\big(-(\alpha - K\varepsilon)(t - t_0)\big). \tag{8.26}$$

As for the uniform stability given any $\eta > 0$, choose $\delta > 0$ so small that

$$K\delta \le \eta.$$

Then for any $t_0 \ge 0$ and any $|x_0| < \delta$, (8.26) gives

$$|x(t)| \le K|x_0| \le K\delta \le \eta \quad \text{for all } t \ge t_0.$$

The choice of δ depends only on η and not on t_0, so the origin is uniformly stable.

To finish showing the zero solution is uniformly asymptotically stable, we fix $0 < \mu < r$. From (8.26), if $|x_0| < \delta$ then

$$|x(t)| \le K|x_0| e^{-(\alpha - K\varepsilon)(t - t_0)}.$$

Given $\mu > 0$, choose

$$T(\mu) := \max\left\{0, \frac{1}{\alpha - K\varepsilon} \log\left(\frac{K\delta}{\mu}\right)\right\}.$$

Then for all $t \ge t_0 + T(\mu)$,

$$|x(t)| \le K|x_0| e^{-(\alpha - K\varepsilon)(t - t_0)} \le K\delta e^{-(\alpha - K\varepsilon)T(\mu)} = \mu.$$

Again, $T(\mu)$ is independent of t_0. Hence, the origin is uniformly attractive. That is, for every $\mu > 0$, there exists $T(\mu) > 0$ such that

$$|x_0| < \delta \text{ implies } |x(t)| < \mu \quad \text{for all } t \ge t_0 + T(\mu), \quad \text{for all } t_0 \ge 0.$$

Combining uniform stability and uniform attractivity , we conclude that the origin is uniformly asymptotically stable.

An immediate consequence of Theorem 8.4, we have the following theorem regarding autonomous systems.

Theorem 8.5 *Assume $Re(\sigma(A)) < 0$ and suppose $g(x)$ is locally Lipschitz near the origin with*

$$\lim_{x \to 0} \frac{|g(x)|}{|x|} = 0, \qquad g(0) = 0.$$

Then the origin of

$$x' = Ax + g(x)$$

is uniformly asymptotically stable.

Remark 29 *(1) If the origin is stable but not asymptotically stable for the linear system*

$$x' = Ax,$$

then no general conclusion can be drawn about the stability of the origin for the associated nonlinear system

$$x' = Ax + g(x).$$

In such marginal cases (e.g., when A has eigenvalues on the imaginary axis), nonlinear terms may destroy stability or create more complicated dynamics, so linear information alone is insufficient to draw a conclusion.

(2) If the origin is unstable for the linear system $x' = Ax$ (i.e., A has at least one eigenvalue with positive real part), then the origin is also unstable for the nonlinear system $x' = Ax + g(x)$ provided g is sufficiently small near the origin in the sense of Theorem 8.4. In this case, linear instability cannot be compensated by higher-order nonlinear terms.

Remark 30 *Consider the autonomous system*

$$x' = f(x), \tag{8.27}$$

where $f : \mathbb{R}^n \to \mathbb{R}^n$ is continuously differentiable and $f(0) = 0$. Let $A = J(0)$ denote the Jacobian matrix of f at $x = 0$. Then f admits the decomposition

$$x' = Ax + g(x), \tag{8.28}$$

where

$$g(x) := f(x) - Ax.$$

Since f is differentiable at 0 with derivative A, we have

$$\lim_{x \to 0} \frac{|f(x) - Ax|}{|x|} = \lim_{x \to 0} \frac{|g(x)|}{|x|} = 0,$$

which means precisely that $g(x) = o(|x|)$ as $x \to 0$. In particular, g satisfies the smallness condition

$$\lim_{x \to 0} \frac{|g(x)|}{|x|} = 0,$$

as required in Theorem 8.4. Thus, since f is differentiable at 0 with derivative $A = J(0)$, it follows that

$$g(x) = o(|x|) \quad \text{as } x \to 0.$$

We provide the following examples.

Example 8.8 *The zero solution of the nonlinear nonautonomous system*

$$x' = -x + \frac{t}{t+1}x^2$$

is (UAS). *It is obvious that the zero solution of the linear part* $x' = -x$ *is (AS). We must show that condition* (8.18) *holds. Let* $g(t,x) = \frac{t}{t+1}x^2$. *Then* g *is continuous and for any open subset* $D = (\alpha, \beta) \subset \mathbb{R}$ *containing the origin we have*

$$|g(t,x) - g(t,y)| \leq 2\beta |x-y|, \text{ for all } (t,x),(t,y) \in [0,\infty) \times D.$$

Moreover,

$$\lim_{x\to 0} \frac{|g(t,x)|}{|x|} \leq \lim_{x\to 0} \frac{x^2}{|x|} = 0, \text{ uniformly for } t \in [0,\infty).$$

Thus, by Theorem 8.4 , the zero solution is (UAS).

□

Example 8.9 *Consider the perturbed linear system in* $\mathbb{R}^2$

$$x'(t) = Ax(t) + g(t,x(t)), \tag{8.29}$$

where

$$A = \begin{pmatrix} -2 & 0 \\ 0 & -1 \end{pmatrix}, \qquad g(t,x) = \begin{pmatrix} x_1^2 \\ \dfrac{\sin t}{1+t^2} x_1 x_2 \end{pmatrix}, \qquad x = \begin{pmatrix} x_1 \\ x_2 \end{pmatrix}.$$

The eigenvalues of A are

$$\lambda_1 = -2, \qquad \lambda_2 = -1,$$

so $\text{Re}(\sigma(A)) = \max\{-2,-1\} = -1 < 0$. *Thus, the origin of the linear system*

$$x' = Ax$$

is exponentially (hence uniformly asymptotically) stable. Next, we look at the properties of the perturbation $g(t,x)$. *For each fixed* $t \geq 0$,

$$g(t,0) = \begin{pmatrix} 0 \\ 0 \end{pmatrix},$$

so the origin is also an equilibrium of the full system (8.29).

Moreover, $g(t,x)$ *is a polynomial in* x_1, x_2 *with coefficients depending continuously on* t, *hence it is globally Lipschitz in* x *on any bounded set (locally Lipschitz in* x *on* $\mathbb{R}^2$*). In particular, the local existence and uniqueness theorem applies.*

We must check that

$$\lim_{x\to 0} \frac{|g(t,x)|}{|x|} = 0 \quad \text{uniformly in } t \in [0,\infty),$$

where $|\cdot|$ *is any norm on* $\mathbb{R}^2$ *(all norms are equivalent, so we may use the Euclidean norm). We estimate*

$$|g(t,x)| \leq |x_1|^2 + \left|\frac{\sin t}{1+t^2}\right| |x_1|\,|x_2|.$$

Since

$$\left|\frac{\sin t}{1+t^2}\right| \leq \frac{1}{1+t^2} \leq 1 \qquad \textit{for all } t \geq 0,$$

we obtain

$$|g(t,x)| \leq |x_1|^2 + |x_1|\,|x_2| \leq |x|^2 + |x|^2 = 2|x|^2,$$

where we used $|x_1| \leq |x|$ *and* $|x_2| \leq |x|$. *Hence*

$$\frac{|g(t,x)|}{|x|} \leq 2|x| \quad \textit{for all } t \geq 0,\ x \neq 0.$$

Therefore

$$\lim_{x\to 0} \frac{|g(t,x)|}{|x|} = 0 \quad \textit{uniformly in } t \in [0,\infty),$$

which is exactly condition (8.18) *of Theorem 8.4.*

All assumptions of Theorem 8.4 are satisfied. Hence, by Theorem 8.4, the origin of the nonlinear perturbed system (8.29) *is* uniformly asymptotically stable. *In particular, there exist constants* $c > 0$, $\beta > 0$, *and a radius* $r > 0$ *such that for all* $t_0 \geq 0$ *and all initial conditions* $|x_0| < r$, *the corresponding solution satisfies*

$$|x(t)| \leq c\,|x_0|\,e^{-\beta(t-t_0)}, \qquad \forall t \geq t_0,$$

so every trajectory starting sufficiently close to the origin converges to 0 *exponentially fast and uniformly with respect to the initial time* t_0. □

The next example shows the need for a different approach when the eigenvalues of the matrix A fail to have negative real parts.

Example 8.10 *Consider the planar autonomous system*

$$x' = -y + x\sqrt{x^2+y^2}, \quad y' = x + y\sqrt{x^2+y^2}.$$

It is clear that $(0,0)$ *is the only equilibrium point. We rewrite the system in the form*

$$\begin{pmatrix} x \\ y \end{pmatrix}' = \begin{pmatrix} 0 & -1 \\ 1 & 0 \end{pmatrix} \begin{pmatrix} x \\ y \end{pmatrix} + \begin{pmatrix} x\sqrt{x^2+y^2} \\ y\sqrt{x^2+y^2} \end{pmatrix}.$$

The eigenvalues of

$$A = \begin{pmatrix} 0 & -1 \\ 1 & 0 \end{pmatrix}$$

are $\pm i$. *Thus Theorem 8.5 does not apply. Instead, we resort to the use of polar coordinates. Namely,*

$$x = r\cos(\theta),\;\; y = r\sin(\theta),\;\; \textit{where}\;\; r^2 = x^2 + y^2,$$

and $r = r(t)$ *and* $\theta = \theta(t) = \tan^{-1}(\frac{y}{x})$. *Thus*

$$\frac{d}{dt}(x^2 + y^2) = \frac{d}{dt}r^2,$$

or

$$xx' + yy' = rr'.$$

A substitution of x' *and* y' *yields,*

$$\begin{aligned} rr' &= x(-y + x\sqrt{x^2+y^2}) + y(x + y\sqrt{x^2+y^2}) \\ &= (x^2+y^2)\sqrt{x^2+y^2} \\ &= r^2 r. \end{aligned}$$

Thus, we have the first order differential equation

$$r' = r^2 \tag{8.30}$$

which implies that the function r *is increasing. Thus, trajectories are spiraling away from the origin, and hence the origin is unstable. In addition,*

$$\theta' = \frac{xy' - x'y}{x^2 + y^2} = 1.$$

Choosing $\theta(0) = 0$, *we have* $\theta(t) = t$, *which indicates trajectories spiral away from the origin counterclockwise. Note that, using separation of variables,* (8.30) *can be easily solved to find* $r(t) = \frac{1}{k-t}$ *for some suitable constant* k. *Notice that* $r(t) \to \infty$ *as* $t \to k^-$. □

Example 8.11 *In Example 6.1 we consider a pendulum with friction that has the equation*

$$mL\theta'' + mg\sin(\theta) + bL\theta' = 0,$$

by letting $x_1 = \theta(t)$, $x_2 = \theta'(t)$, *we arrive at the planar autonomous system*

$$\begin{aligned} x_1' &= x_2 \\ x_2' &= -\frac{g}{L}\sin(x_1) - \frac{b}{m}x_2. \end{aligned} \tag{8.31}$$

It is clear that system (8.31) *has infinitely many equilibrium points that are given by* $(n\pi, 0), n = 0, 1, \ldots$ *which correspond to vertical positions with zero velocity. We will only analyze the two equilibrium points* $(0,0)$, $(\pi, 0)$. *For the equilibrium point* $(0,0)$, *we put system* (8.31) *in the form*

$$\begin{pmatrix} x_1 \\ x_2 \end{pmatrix}' = \begin{pmatrix} 0 & 1 \\ -\frac{g}{L} & -\frac{b}{m} \end{pmatrix}\begin{pmatrix} x_1 \\ x_2 \end{pmatrix} + \frac{g}{L}\begin{pmatrix} 0 \\ x_1 - sin(x_1) \end{pmatrix}.$$

The linear part has the eigenvalues

$$\lambda_1 = -\frac{b}{2m} + \frac{1}{2}\left(\frac{b^2}{m^2} - 4\frac{g}{L}\right)^{1/2},$$

and

$$\lambda_1 = -\frac{b}{2m} - \frac{1}{2}\left(\frac{b^2}{m^2} - 4\frac{g}{L}\right)^{1/2}.$$

If $\frac{b^2}{m^2} - 4\frac{g}{L} < 0$ *(when damping is weak), then the eigenvalues are complex with negative real parts and the origin is a stable spiral. If* $\frac{b^2}{m^2} - 4\frac{g}{L} \geq 0$ *(when the damping is strong),*

$$\frac{b^2}{m^2} > \frac{b^2}{m^2} - 4\frac{g}{L}$$

we have that $\lambda_2 \leq \lambda_1 < 0$, *and hence the equilibrium solution* $(0,0)$ *for the system* (8.31) *is asymptotically stable by Theorem 8.5.*
The physical and geometrical interpretation for $(x_1, x_2') = (0,0)$ *being a stable spiral corresponds to the pendulum hanging vertically downward* $(x_1 = 0)$ *with zero velocity* $(x_2' = 0)$, *thus, when the damping is weak, a perturbation will cause the pendulum to spiral and approach the vertically downward position.*
For the equilibrium point $(\pi, 0)$, *we use the transformation*

$$\begin{pmatrix} x_1 \\ x_2 \end{pmatrix} = \begin{pmatrix} y_1 \\ y_2 \end{pmatrix} + \begin{pmatrix} \pi \\ 0 \end{pmatrix}$$

in (8.31) *obtaining*

$$\begin{pmatrix} y_1 \\ y_2 \end{pmatrix}' = \begin{pmatrix} y_2 \\ \frac{g}{L}\sin(y_1) - \frac{b}{m}y_2 \end{pmatrix}.$$

To examine this system near $(0,0)$, *we write the above new nonlinear system in y in the form*

$$\begin{pmatrix} y_1 \\ y_2 \end{pmatrix}' = \begin{pmatrix} 0 & 1 \\ \frac{g}{L} & -\frac{b}{m} \end{pmatrix}\begin{pmatrix} y_1 \\ y_2 \end{pmatrix} - \frac{g}{L}\begin{pmatrix} 0 \\ y_1 - \sin(y_1) \end{pmatrix}.$$

The linear parts have the eigenvalues

$$\lambda_1 = -\frac{b}{2m} + \left(\frac{b^2}{m^2} + 4\frac{g}{L}\right)^{1/2},$$

$$\lambda_1 = -\frac{b}{2m} - \left(\frac{b^2}{m^2} + 4\frac{g}{L}\right)^{1/2}.$$

It is clear that $\lambda_2 < 0 < \lambda_1$ *and the equilibrium point* $(0,0)$, *and hence* $(\pi,0)$ *is unstable.*

Note that we could have calculated the Jacobian given by (8.10) *at* $(\pi,0)$ *and obtained*

$$J((\pi,0)) = \begin{pmatrix} 0 & 1 \\ \frac{g}{L} & -\frac{b}{m} \end{pmatrix}.$$

instead of using a transformation.
If we compute $J((\pm n\pi,0))$*, we will see that* $(\pm n\pi,0), n = 2,4,\ldots$ *are stable and unstable for* $n = 1,3,\ldots$ □

Example 8.12 *(Lotka–Volterra Predator–Prey Model) In Example 7.1, we considered the Lotka–Volterra system*

$$\frac{dx}{dt} = ax - bxy$$

$$\frac{dy}{dt} = -cy + dxy \tag{8.32}$$

and found that the equilibrium solutions are

$$x_1^* = (0,0) \text{ and } x_2^* = (c/d, a/b).$$

The first equilibrium solution x_1^* *effectively represents the extinction of both species. If both populations are at* 0*, then they will continue to be so indefinitely. The second equilibrium solution* x_2^* *represents a fixed point at which both populations sustain their current, nonzero numbers, and do so indefinitely. The levels of population at which this equilibrium is achieved depend on the chosen values of the parameters* $a,b,c,$ *and* d*. Now, we will further discuss the stability of both equilibrium solutions,* x_1^* *and* x_2^* *and hope that* x_1^* *is unstable and* x_2^* *is stable. The Jacobian matrix is*

$$J(x,y) = \begin{pmatrix} a - by & -bx \\ dy & -c + dx \end{pmatrix}.$$

For $x_1^* = (0,0)$*, we have*

$$J(0,0) = \begin{pmatrix} a & 0 \\ 0 & -c \end{pmatrix}.$$

with eigenvalues $a > 0$ *and* $-c < 0$*. Thus, the equilibrium point* $(0,0)$ *is a saddle and hence unstable.*
For $x_2^* = (c/d, a/b)$

$$J(c/d, a/b) = \begin{pmatrix} 0 & -\frac{bc}{d} \\ \frac{da}{b} & 0 \end{pmatrix}$$

which has purely imaginary eigenvalues. Hence, linearization is inconclusive. Note that the use of polar coordinates will not work here in the sense that it is hard, if not impossible, to obtain any information from it. □

We end this section by considering the system obtained in Section 2.3.3 concerning bacterial growth in a chemostat.

Example 8.13 *(Bacterial growth in a chemostat). In Section 2.3.3, we arrived at the dimensionless systems*

$$\frac{dc}{d\tau} = 1 - c - \beta cn,$$
$$\frac{dn}{d\tau} = \beta cn - n = n(\beta c - 1),$$

with parameter $\beta > 0$. When finding the equilibria, we set

$$1 - c - \beta cn = 0, \quad n(\beta c - 1) = 0.$$

The equation implies that $c = \frac{1}{1+\beta n}$, and from the second equation we have

$$n = 0 \quad or \quad c = \frac{1}{\beta}.$$

For $n = 0$, we obtain $c = 1$, hence the equilibrium $(c,n) = (1,0)$.
On the other hand, for $c = \dfrac{1}{\beta}$, we get $n = 1 - \dfrac{1}{\beta}$ and we obtain a second equilibrium

$$(c^*, n^*) = \left(\frac{1}{\beta}, 1 - \frac{1}{\beta}\right),$$

which lies in $n \geq 0$ only when $\beta > 1$. Next, we construct the general Jacobin matrix in order to study the stability of the equilibria. The Jacobian matrix is

$$J(c,n) = \begin{pmatrix} -1 - \beta n & -\beta c \\ \beta n & \beta c - 1 \end{pmatrix}.$$

At $(1,0)$ we have

$$J(1,0) = \begin{pmatrix} -1 & -\beta \\ 0 & \beta - 1 \end{pmatrix},$$

whose eigenvalues are -1 and $\beta - 1$. Thus:

- $0 < \beta < 1$*: both eigenvalues are negative, so $(1,0)$ is a stable node.*
- $\beta > 1$*: the eigenvalues have opposite signs, so $(1,0)$ is a saddle.*

At $(c^*, n^*) = \left(\frac{1}{\beta}, 1 - \frac{1}{\beta}\right)$ (for $\beta > 1$) we have

$$J(c^*, n^*) = \begin{pmatrix} -\beta & -1 \\ \beta - 1 & 0 \end{pmatrix}.$$

The trace and determinant are

$$T = -\beta < 0, \qquad \Delta = \beta - 1 > 0,$$

so both eigenvalues are real and negative, and (c^*, n^*) *is a stable node. To study trajectories, one can eliminate* τ *and write*

$$\frac{dn}{dc} = \frac{\frac{dn}{d\tau}}{\frac{dc}{d\tau}} = \frac{n(\beta c - 1)}{1 - c - \beta cn}.$$

This first-order equation can be separated and (in principle) integrated to obtain an implicit first integral, whose level curves describe the orbits in the (c,n)*-plane. The nullclines and the signs of* $\dot{c}, \dot{n}$ *in each region give a qualitative phase portrait showing convergence to the stable equilibrium and the role of the saddle point. In this nondimensional chemostat model, the stability of the equilibria determines whether the chemostat ultimately washes out or sustains a nonzero bacterial population, depending on the parameter* β.

Meaning of β

Recall that

$$\beta = \frac{KC_{in}}{D},$$

which is the growth rate at feed concentration divided by the dilution rate. Biologically,

- *If* $\beta < 1$, *even at the highest available nutrient level* $C = C_{in}$, *the per-capita growth rate* KC_{in} *is smaller than the dilution rate* D.
- *If* $\beta > 1$, *growth at feed concentration exceeds dilution, so the bacterial population can in principle, maintain itself.*

Thus, β *is a control parameter that can be tuned by changing either* D *(the flow rate) or* C_{in} *(the feed concentration).*

Washout equilibrium $(c,n) = (1,0)$

In dimensionless variables, the washout equilibrium is

$$(c,n) = (1,0),$$

which corresponds to

$$C = C_{in}, \qquad N = 0.$$

From the linear stability analysis, this equilibrium is

- *a stable node for* $0 < \beta < 1$,
- *a saddle point for* $\beta > 1$.

Biologically,

- *For* $0 < \beta < 1$, *any small inoculum decays and is washed out of the reactor. The only long-time state is a sterile reactor filled with fresh medium at concentration* C_{in}.
- *For* $\beta > 1$, *the washout state is unstable: once bacteria are present, a small population tends to grow away from zero, so washout is not the expected long-time outcome.*

Positive equilibrium for $\beta > 1$

For $\beta > 1$, *there is a positive equilibrium*

$$(c^*, n^*) = \left(\frac{1}{\beta}, 1 - \frac{1}{\beta}\right),$$

which corresponds in dimensional variables to

$$C^* = \frac{C_{in}}{\beta}, \qquad N^* = N_c\left(1 - \frac{1}{\beta}\right) = C_{in}Y\left(1 - \frac{1}{\beta}\right).$$

Our analysis shows that this equilibrium is a stable node when $\beta > 1$. *Biologically, this means:*

- *The chemostat settles to a steady operating point with constant bacterial density and reduced nutrient concentration.*
- *Nutrient inflow is exactly balanced by consumption and outflow.*
- *Small perturbations in* N *or* C *decay in time: if the biomass or nutrient concentration is disturbed, the system relaxes back to* (c^*, n^*).

This is the non-washout operating regime used in practice for continuous cultures.

Practical conclusions for reactor operation

The condition for avoiding washout is

$$\beta > 1 \quad \Longleftrightarrow \quad KC_{in} > D.$$

For a fixed organism (fixed K*) and feed concentration* C_{in}, *this places an upper bound on the dilution rate* D *needed to sustain a culture. If* D *is increased past the critical value*

$$D_{crit} = KC_{in},$$

then β *drops below* 1, *the positive equilibrium disappears, and the washout state* $(c, n) = (1, 0)$ *becomes globally attracting: the culture collapses, and the reactor runs clear.*

Mathematically, the change of stability at $\beta = 1$ *is a transcritical bifurcation between the washout equilibrium and the positive equilibrium. Biologically, it marks the transition between a functioning continuous culture and complete washout of bacteria.* □

8.3 An SIR Epidemic Model

The SIR model is a classical compartmental model in mathematical epidemiology that describes the spread of an infectious disease in a homogeneous population. The total population is divided into three compartments: *susceptible* individuals (S), who can contract the disease; *infected* individuals (I), who are currently infectious; and *recovered* individuals (R), who have acquired immunity (or have been removed) after infection. The model is governed by a system of nonlinear ordinary differential equations that track how individuals move between these compartments due to infection and recovery. Despite its simplicity, the SIR framework captures key qualitative features of epidemic dynamics, such as outbreak thresholds, peak infection levels, and the eventual decline of an epidemic, and it serves as a foundation for more sophisticated epidemiological models. We begin by introducing our model by letting,

$S = S(t)$ is the number of susceptible individuals, (not yet having the disease but capable of getting it)
$I = I(t)$ is the number of infected individuals, and
$R = R(t)$ is the number of recovered or removed individuals.
We assume the total population is constant and normalized to one. Then

$$S + I + R = N \text{ (constant)}.$$

That is

$$S + I + R = 1 \text{ (by normalization)}.$$

Let

$$\begin{aligned} \mu &= \text{ birth rate} = \text{death rate (because the number of people is constant)},\\ \gamma &= \text{ removal rate}, \end{aligned}$$

and

$$\lambda = \text{ infection rate}.$$

We assume all newborns are susceptible and arrive at the epidemic model

$$\begin{cases} S'(t) &= (S+I+R)\mu - \mu S - (SI)\lambda \\ I'(t) &= -\mu I - \gamma I + \lambda SI \\ R'(t) &= -\mu R + \gamma I \end{cases}$$

The term $(S+I+R)\mu = \mu$, since the population is constant and normalized at 1. The term of μS represents the death of susceptible, and the negative implies loss or removal from being susceptible. On the other hand, the term $(SI)\lambda$ denotes those who got infected and the minus represents that they are infected and not susceptible anymore. Similarly, the term μI represents those who die from the infection, and the minus indicates a loss for those who are infected already. In addition, the term

γI represents recovery/removal of infected individuals. Finally, the term μR denotes those who are removed due to death. Keep in mind that *minus* means loss from a certain class for S, I, and R and *plus* implies gain for the same classes. Thus, by only analyzing $S \geq 0$ and $I \geq 0$ we arrive at the 2×2-planar system

$$\begin{cases} S' = \mu - \mu S - \lambda SI \\ I' = -(\mu+\gamma)I + \lambda SI \end{cases} \tag{8.33}$$

Note that system (8.33) is nonlinear and $(0,0)$ is not an equilibrium solution. Also, the system is valid in the region $0 \leq S+I \leq 1$ since we are neglecting R. It is important to note that the system does not allow those who recover from the epidemic to become susceptible again. Suppose $S(t_0) = S_0$ and $I(t_0) = I_0$. Since the right side of (8.33) is continuous with continuous first partial derivatives, we know the solutions $S(t, t_0, S_0)$ and $I(t, t_0, I_0)$ exist and are unique. Moreover, $S(t, t_0, S_0)$ and $I(t, t_0, I_0)$ are continuous and differentiable with respect to all variables. Next, we examine how big the interval is on which the solutions exist. Due to our constrains we are interested in the triangular region of existence

$$T = \{(S,I) : S \geq 0, I \geq 0,\ S+I \leq 1\}.$$

Along $S = 0$, we have $S' = \mu > 0, 0 \leq I < 1$ and $I' = -(\mu+\gamma)I < 0$, if $I > 0$. Moreover,

$$\frac{dI}{dS} = \frac{I'}{S'} < 0.$$

Thus, any trajectory that starts on the I-axis moves downward inside T in the positive direction of S.

Now, along $I = 0$, we have $S' = \mu(1-S) > 0, 0 \leq I < 1$ and $I' = 0$, Thus, any trajectory that starts on the S-axis moves toward the point $(1,0)$ at which the line $S+I = 1$ intersects the S-axis. Note that $(1,0)$ is an equilibrium solution.

Along the line $S+I = 1$, we have

$$(S+I)' = S' + I' = \mu - \mu(S+I) - \gamma I,$$

and since $S+I = 1$, the above expression reduces to

$$(S+I)' = -\gamma I < 0.$$

This tells us that any trajectory starting at the line $S+I = 1$ moves to the left of the line toward the inside of the triangle T. We conclude that the triangular region T is positively invariant. That is, if $(S_0, I_0) \in T$, then $S(t, t_0, S_0)$, $I(t, t_0, I_0) \in T$, for all $t \geq t_0$. Thus, the maximum interval of existence for solutions in T is $[t_0, \infty)$. From now on we refer to the triangular region T as the feasible region. Next, we search for the equilibrium solutions (S^*, I^*) in the feasible region. To do so, we must solve for S and I by setting the right sides of (8.33) to zero. That is, $S' = 0$ and $I' = 0$.

$$\mu - \mu S - \lambda SI = 0, \text{ and } -(\mu+\gamma)I + \lambda SI = 0.$$

From the second equation, we obtain

$$(-\mu-\gamma+\lambda S)I=0.$$

Thus, if $I\neq 0$, then we arrive at

$$S^*=\frac{\mu+\gamma}{\lambda}.$$

Clearly $S^*\geq 0$, and in order for S^* to be in the feasible region, we must have

$$\mu+\gamma<\lambda.$$

It is convenient to introduce the quantity

$$\mathscr{R}_0=\frac{\lambda}{\mu+\gamma},$$

called the *basic reproduction number*. Biologically, $\mathscr{R}_0$ represents the average number of secondary infections generated by a single infected individual in a completely susceptible population. Hence, the condition

$$\mu+\gamma<\lambda$$

can be rewritten as

$$\mathscr{R}_0>1.$$

This threshold condition plays a fundamental role in epidemiology. If $\mathscr{R}_0<1$, then the disease eventually dies out, whereas if $\mathscr{R}_0>1$, the infection can persist in the population, and an endemic equilibrium may exist. If we substitute $S^*=\frac{\mu+\gamma}{\lambda}$ into the equation $\mu-\mu S^*-\lambda S^*I^*=0$, we arrive at

$$I^*=\frac{\mu}{\mu+\gamma}-\frac{\mu}{\lambda}.$$

It is clear $I^*>0$, since

$$I^*=\frac{\mu}{\mu+\gamma}-\frac{\mu}{\lambda}>\frac{\mu}{\lambda}-\frac{\mu}{\lambda}=0.$$

Next we show (S^*,I^*), belongs to the feasible region.

$$\begin{aligned}0<S^*+I^* &= \frac{\mu+\gamma}{\lambda}+\frac{\mu}{\mu+\gamma}-\frac{\mu}{\lambda}\\ &= \frac{\gamma}{\lambda}+\frac{\mu}{\mu+\gamma}\\ &< \frac{\gamma}{\mu+\gamma}+\frac{\mu}{\mu+\gamma}\\ &= 1.\end{aligned}$$

Thus, the two equilibrium solutions $(1,0)$ and (S^*,I^*) belong to the feasible region T. Next, we use Theorem 8.5 to study the stability of the two equilibrium solutions. We do not want solutions to approach $(1,0)$ and stay there. Instead, we rather have a solutions approach (S^*,I^*). That is, we hope for $(1,0)$ to be unstable. We begin with $(1,0)$. The Jacobian matrix is

$$J(S,I) = \begin{pmatrix} -\mu - \lambda I & -\lambda S \\ \lambda I & -(\mu+\gamma)+\lambda S \end{pmatrix}.$$

To study the stability of the endemic equilibrium (S^*,I^*), it is more convenient to use the trace and determinant of the Jacobian matrix rather than the explicit eigenvalue formula. Recall that

$$J(S^*,I^*) = \begin{pmatrix} -\mu - \lambda I^* & -\lambda S^* \\ \lambda I^* & 0 \end{pmatrix}.$$

Hence,

$$\mathrm{tr}(J(S^*,I^*)) = -(\mu + \lambda I^*) < 0,$$

and

$$\det(J(S^*,I^*)) = \lambda^2 S^* I^* > 0.$$

For a 2×2 matrix, if the trace is negative and the determinant is positive, then both eigenvalues have negative real parts. Therefore, both eigenvalues of $J(S^*,I^*)$ have negative real parts.

Consequently, by Theorem 8.5, the endemic equilibrium solution

$$(S^*,I^*)$$

is asymptotically stable. The asymptotic stability of the endemic equilibrium solution

$$(S^*,I^*)$$

has an important epidemiological interpretation. It implies that if the initial population levels of susceptible and infected individuals are slightly perturbed from (S^*,I^*), then the corresponding solution trajectory will eventually return to (S^*,I^*) as $t \to \infty$.

In other words, the disease persists in the population at a steady long-term level rather than disappearing. The number of susceptible individuals approaches the constant level S^*, while the number of infected individuals approaches the positive constant level I^*. Thus, the epidemic becomes endemic in the population.

Biologically, this means that although the disease does not infect the entire population, it continues to circulate indefinitely at a stable equilibrium level. Small fluctuations in the numbers of susceptible or infected individuals will eventually damp out, and the system will return to the endemic state.

We now examine the disease-free equilibrium point $(1,0)$. The Jacobian matrix is

$$J(S,I) = \begin{pmatrix} -\mu - \lambda I & -\lambda S \\ \lambda I & -(\mu+\gamma) + \lambda S \end{pmatrix}.$$

Evaluating at $(1,0)$, we obtain

$$J(1,0) = \begin{pmatrix} -\mu & -\lambda \\ 0 & -(\mu+\gamma) + \lambda \end{pmatrix}.$$

Since this matrix is triangular, its eigenvalues are

$$d_1 = -\mu, \qquad d_2 = \lambda - (\mu+\gamma).$$

Clearly,

$$d_1 < 0.$$

Moreover, if

$$\lambda > \mu + \gamma,$$

then

$$d_2 > 0.$$

Hence, the Jacobian matrix at $(1,0)$ has one negative eigenvalue and one positive eigenvalue. Therefore, $(1,0)$ is a saddle point and is consequently unstable by Theorem 8.5. This instability has an important epidemiological interpretation: when

$$\lambda > \mu + \gamma,$$

the infection rate exceeds the combined removal and death rates, allowing the disease to invade the population rather than die out.

8.3.1 Exercises

Exercise 8.8 *Investigate the stability of each of the equilibrium points for the given systems:*

(a) $x' = x - 3y + 2xy,\ y' = 4x - 6y - xy.$

(b) $x' = 6x - 5y + x^2,\ y' = 2x - y + y^2.$

(c) $x' = x - 2y + 3xy,\ y' = 2x - 3y - x^2 - y^2.$

(d) $x' = x - y,\ y' = x^2 - y.$

(e) $x' = y^2 - 1,\ y' = x^3 - y.$

(f)) $x' = 8x - y^2,\ y' = -y + x^2.$

Exercise 8.9 *Consider the system*

$$x' = -x + hy$$
$$y' = x - y.$$

Show that

(a) $(0,0)$ *is asymptotically stable if* $h = 0$,

(b) $(0,0)$ *is an asymptotically stable spiral if* $h < 0$,

(c) $(0,0)$ *asymptotically stable node if* $0 < h < 1$

Exercise 8.10 *Consider the nonlinear non-autonomous scalar system*

$$x' = -2x + \frac{1}{1+t^2}x^2.$$

(a) Identify the linear part and the perturbation term by writing the system in the form

$$x' = Ax + g(t,x),$$

and state the matrix (scalar) A and the function $g(t,x)$.

(b) Show that the zero solution of the linear part $x' = -2x$ *is asymptotically stable.*

(c) Prove that $g(t,x)$ *is (locally) Lipschitz in x on any bounded interval* $|x| \le \beta$, *uniformly in* $t \ge 0$.

(d) Verify that

$$\lim_{x \to 0} \frac{|g(t,x)|}{|x|} = 0$$

uniformly for $t \in [0,\infty)$, *and conclude from Theorem 8.4 that the zero solution is uniformly asymptotically stable.*

Exercise 8.11 *Consider the nonlinear non-autonomous scalar system*

$$x' = -x + \frac{\sin t}{1+t^2}x^3.$$

(a) Write the system as $x' = Ax + g(t,x)$ *and specify A and* $g(t,x)$.

(b) Show that the linear system $x' = -x$ *has an exponentially stable (and hence asymptotically stable) zero solution.*

(c) Prove that $g(t,x)$ *is continuous and locally Lipschitz in x on any bounded set* $|x| \le \beta$, *uniformly in* $t \ge 0$.

(d) Show that

$$|g(t,x)| = \left| \frac{\sin t}{1+t^2}x^3 \right| \le |x|^3$$

for all $t \geq 0$*, and use this estimate to verify that*

$$\lim_{x\to 0} \frac{|g(t,x)|}{|x|} = \lim_{x\to 0} |x|^2 = 0$$

uniformly for $t \in [0,\infty)$.

(e) *Deduce from Theorem 8.4 that the origin is uniformly asymptotically stable.*

Exercise 8.12 *Consider the nonlinear non-autonomous scalar system*

$$x' = -3x + \frac{e^{-t}}{1+|x|}x^2.$$

(a) *Decompose the equation as* $x' = Ax + g(t,x)$ *and identify* A *and* $g(t,x)$.

(b) *Show that the linear system* $x' = -3x$ *has an exponentially stable zero solution.*

(c) *Show that for* $|x| \leq \beta$ *with any fixed* $\beta > 0$,

$$|g(t,x) - g(t,y)| \leq L_\beta |x-y|$$

for some constant $L_\beta > 0$ *independent of* $t \geq 0$*; that is,* g *is locally Lipschitz in* x*, uniformly in* t.

(d) *Prove that*

$$\lim_{x\to 0} \frac{|g(t,x)|}{|x|} \leq \lim_{x\to 0} |x| = 0,$$

uniformly in $t \in [0,\infty)$.

(e) *Conclude, using Theorem 8.4, that the zero solution of the nonlinear system is uniformly asymptotically stable.*

Exercise 8.13 *Consider the perturbed linear system in* $\mathbb{R}^2$

$$x'(t) = Ax(t) + g(t,x(t)),$$

where

$$A = \begin{pmatrix} -1 & 0 \\ 0 & -3 \end{pmatrix}, \qquad g(t,x) = \begin{pmatrix} \dfrac{1}{1+t^2}x_1^2 \\ e^{-t}x_1x_2 \end{pmatrix}, \qquad x = \begin{pmatrix} x_1 \\ x_2 \end{pmatrix}.$$

(a) *Compute the eigenvalues of* A *and show that* $\mathrm{Re}(\rho(A)) < 0$.

(b) *Verify that* $g(t,0) = 0$ *for all* $t \geq 0$ *and that* g *is (locally) Lipschitz in* x *on any bounded subset of* $\mathbb{R}^2$*, uniformly in* $t \geq 0$.

(c) *Show that for some constant* $C > 0$,

$$|g(t,x)| \leq C|x|^2 \qquad \text{for all } t \geq 0,\ x \in \mathbb{R}^2,$$

and use this to prove that

$$\lim_{x\to 0} \frac{|g(t,x)|}{|x|} = 0 \quad \text{uniformly in } t \in [0,\infty).$$

(d) Conclude, by applying Theorem 8.4, that the origin of this system is uniformly asymptotically stable.

Exercise 8.14 *Consider the perturbed linear system in* $\mathbb{R}^2$

$$x'(t) = Ax(t) + g(t,x(t)),$$

where

$$A = \begin{pmatrix} -2 & 1 \\ 0 & -2 \end{pmatrix}, \qquad g(t,x) = \begin{pmatrix} \dfrac{\sin t}{1+t^2} x_1^3 \\ \dfrac{t}{1+t^2} x_1 x_2^2 \end{pmatrix}, \qquad x = \begin{pmatrix} x_1 \\ x_2 \end{pmatrix}.$$

(a) Find the eigenvalues of A and show that all of them have strictly negative real part.

(b) Argue that $g(t,x)$ *is continuous in* (t,x) *and locally Lipschitz in x on any bounded subset of* $\mathbb{R}^2$, *uniformly for* $t \geq 0$.

(c) Prove that there exists a constant $C > 0$ *such that*

$$|g(t,x)| \leq C|x|^3 \qquad \text{for all } t \geq 0,\ x \in \mathbb{R}^2,$$

and deduce that

$$\frac{|g(t,x)|}{|x|} \leq C|x|^2 \longrightarrow 0 \quad \text{as } x \to 0,$$

uniformly in $t \in [0,\infty)$.

(d) Using Theorem 8.4, conclude that the zero solution of this perturbed system is uniformly asymptotically stable.

Exercise 8.15 *Consider the perturbed linear system in* $\mathbb{R}^2$

$$x'(t) = Ax(t) + g(t,x(t)),$$

where

$$A = \begin{pmatrix} -2 & 1 \\ 0 & -1 \end{pmatrix}, \qquad g(t,x) = \begin{pmatrix} x_1^2 + x_1 x_2 \\ \dfrac{1-\sin t}{1+t^2} x_1 x_2 \end{pmatrix}, \qquad x = \begin{pmatrix} x_1 \\ x_2 \end{pmatrix}.$$

(a) Compute the eigenvalues of A and show that

$$\mathrm{Re}(\rho(A)) < 0,$$

so that the zero solution of the linear system $x' = Ax$ *is asymptotically stable.*

(b) Verify that $g(t,0) = 0$ *for all* $t \geq 0$ *and that* $g(t,x)$ *is (locally) Lipschitz in x on any bounded subset of* $\mathbb{R}^2$, *uniformly in* $t \geq 0$.

(c) *Show that there exists a constant $C > 0$ such that*

$$|g(t,x)| \leq C|x|^2 \qquad \text{for all } t \geq 0,\ x \in \mathbb{R}^2,$$

and use this to prove that

$$\lim_{x \to 0} \frac{|g(t,x)|}{|x|} = 0 \quad \text{uniformly in } t \in [0, \infty).$$

(d) *Conclude, by applying Theorem 8.4, that the origin of this perturbed system is uniformly asymptotically stable, even though A is not diagonal.*

Exercise 8.16 *Use polar coordinates to study the stability of the origin for the system*

$$x' = -y - x\sqrt{x^2+y^2}, \ y' = x - y\sqrt{x^2+y^2}.$$

Exercise 8.17 *Use polar coordinates to study the stability of the origin for the system*

$$x' = -x - \frac{y}{Ln(\sqrt{x^2+y^2})}, \quad y' = -y + \frac{x}{Ln(\sqrt{x^2+y^2})}.$$

Exercise 8.18 *(A Predator–Prey Model with Harvesting) Consider a predator–prey system in which the prey population is subject to constant harvesting. Let $x(t)$ denote the prey population and $y(t)$ denote the predator population at time t. The model is given by*

$$\begin{cases} x' = rx\left(1 - \dfrac{x}{K}\right) - axy - h, \\ y' = -dy + bxy, \end{cases} \tag{8.34}$$

where all parameters are positive constants. Here,

- *r is the intrinsic growth rate of the prey,*
- *K is the carrying capacity of the prey,*
- *a is the predation rate,*
- *b is the conversion efficiency of prey into predators,*
- *d is the predator death rate, and*
- *h is a constant harvesting rate of the prey.*

(a) *Show that system (8.34) is well-defined for $x \geq 0$ and $y \geq 0$, and explain why only the first quadrant of the phase plane is biologically meaningful.*

(b) *Determine conditions on h under which the region*

$$D = \{(x,y) : x \geq 0,\ y \geq 0,\ x \leq K\}$$

is positively invariant.

(c) *Find all equilibrium solutions of system* (8.34). *Identify which equilibria correspond to extinction states and which correspond to the coexistence of the two species.*

(d) *Compute the Jacobian matrix of system* (8.34) *and evaluate it at each equilibrium point.*

(e) *Determine conditions on the parameters under which the coexistence equilibrium is locally asymptotically stable.*

(f) *Give a biological interpretation of the stability result obtained in part (e), focusing on the role of the harvesting rate h.*

8.4 Limit Cycle

In many nonlinear dynamical systems, long-term behavior is not governed solely by equilibrium points but also by closed, isolated trajectories known as *limit cycles*. A limit cycle is a periodic orbit that attracts or repels nearby trajectories, and it often represents a self-sustained oscillation whose amplitude and period are determined by the system dynamics rather than by initial conditions. Such behavior arises naturally in a wide range of applications, including electrical circuits (Van der Pol oscillator), biological rhythms (heartbeat, neuronal firing), chemical reactions (Belousov–Zhabotinsky oscillations), and mechanical systems with friction or impact. In this section, we develop the basic theory of limit cycles and their role in planar autonomous systems. We introduce precise definitions of periodic orbits, stability of limit cycles, and surrounding phase portraits, and we discuss analytical tools such as Bendixson-type criteria for the existence or nonexistence of closed orbits. Throughout, examples from classical models illustrate how limit cycles emerge, how they can bifurcate from equilibria, and how they organize the qualitative dynamics of nonlinear systems.

We consider the two-dimensional autonomous system

$$x' = P(x,y), \quad y' = Q(x,y) \tag{8.35}$$

where P and Q are continuous on some subset of $\mathbb{R}^2$. Systems given by (8.35) may admit solutions that commonly tend to a limiting finite periodic solution. This limiting closed curve in the phase plane is called a *limit cycle*.

Definition 8.4 *A limit cycle of system* (8.35) *is a closed curve, and no other solution that is a closed curve exists in its neighborhood.*

Note that every neighboring trajectory spirals and approaches the limit cycle from the inside or from the outside as $t \to \infty$ or as $t \to -\infty$.

Definition 8.5 *If every neighboring trajectory approaches a limit cycle as $t \to \infty$ or as $t \to -\infty$, then the limit cycle is said to be stable.*

It is hard if not impossible to determine if a nonlinear system admits a limit cycle. However, the next result, known as Bendixson–Dulac criteria should provide a criterion for the *nonexistence* of limit cycles, or closed trajectories of the system given by (8.35).

Theorem 8.6 *(Bendixson–Dulac Theorem) Suppose the first partial derivatives of $P(x,y)$ and $Q(x,y)$ are continuous on some connected subset U in $\mathbb{R}^2$. If*

$$\frac{\partial P}{\partial x} + \frac{\partial Q}{\partial y}$$

have the same sign in any open set of U, then the system (8.35) has no closed trajectory in U.

Proof *Suppose the contrary. That is, there is a closed curve C in U. Then by Green's theorem, we have*

$$\int_C P(x,y)dy - Q(x,y)dx = \int\int_R (\frac{\partial P}{\partial x} + \frac{\partial Q}{\partial y})dxdy, \tag{8.36}$$

where R is the region enclosed by C. Let $x = x(t), y = y(t)$ be a parametrization of C. Then

$$\int_C P(x,y)dy - Q(x,y)dx = \int_0^T (P\frac{dy}{dt} - Q\frac{dx}{dt})dt,$$

where T is the period of C. Using (8.35), we arrive at

$$\int_C P(x,y)dy - Q(x,y)dx = \int_0^T (PQ - QP)dt = 0.$$

Thus, from (8.36) we have

$$\int\int_R (\frac{\partial P}{\partial x} + \frac{\partial Q}{\partial y})dxdy = 0$$

which could only hold true if $\frac{\partial P}{\partial x} + \frac{\partial Q}{\partial y}$ changes sign. This is a contradiction to the fact that C is a closed curve. Thus, C cannot be close and the system does not admit a limit cycle. This completes the proof.

Example 8.14 *Let g and h be differentiable functions, and consider the two-dimensional system*

$$x' = g(y), \quad y' = (1+x^2)y + h(x).$$

Then

$$\frac{\partial P}{\partial x} + \frac{\partial Q}{\partial y} = 1 + x^2 > 0, \ x \in \mathbb{R},$$

and by Theorem 8.6, the system does not admit a periodic solution. □

The next theorem generalizes Theorem 8.6. For its proof, we refer you to Exercise 8.28

Theorem 8.7 *(Generalization of Bendixson–Dulac Theorem) Suppose the first partial derivatives of $P(x,y), Q(x,y)$, and $h(x,y)$ are continuous on some connected subset U in $\mathbb{R}^2$. If*

$$\frac{\partial(hP)}{\partial x} + \frac{\partial(hQ)}{\partial y}$$

is of one sign in any open set of U in $\mathbb{R}^2$, then system (8.35) cannot have a closed trajectory in U.

We have the following example as an application.

Example 8.15 *For constants a,b,c,k, consider the two-dimensional system*

$$x' = (y-a)x, \quad y' = bx + cy + ky^2.$$

Then

$$\frac{\partial P}{\partial x} + \frac{\partial Q}{\partial y} = y - a + c + 2ky$$

and hence Theorem 8.6 is not applicable. Let us try Theorem 8.7 with $h(x,y) = x^{-(2k+1)}$. Then

$$\frac{\partial(hP)}{\partial x} + \frac{\partial(hQ)}{\partial y} = (c + 2ka)x^{-(2k+1)} \neq 0,$$

if $c + 2ka \neq 0$, and by Theorem 8.7, the system does not admit a periodic solution in the half plane $x > 0$ or $x < 0$.
Actually, we can do better. If $x = 0$, then starting from any point on the y-axis, the equation has a solution that lies on the y-axis. Thus, we conclude that the equation has no closed cycle or trajectory and hence a periodic solution in $\mathbb{R}^2$, otherwise uniqueness will be violated. □

So far, we have discussed results concerning the nonexistence of periodic solutions, or periodic trajectories, or limit cycles. Next, we look at the existence of limit cycles.

Theorem 8.8 *A closed orbit or limit cycle of (8.35) surrounds at least one equilibrium point of the system. In other words, if no equilibrium point of the system exists in a region, then there can be no periodic orbits in that region.*

Let $(0,0)$ be the only equilibrium solution of (8.35). Moreover, suppose any solution of (8.35) that starts near the origin will spiral away from the origin. Since there are no equilibrium solutions, our first thought might be that all solutions of (8.35) have trajectories that spiral out to infinity. However, the next theorem presents us with a different scenario. The next theorem is due to Poincaré and Bendixson that shows, under certain conditions, the existence of a limit cycle. Its proof requires high-level arguments, and we omit it.

Theorem 8.9 *(Poincaré–Bendixson) Suppose R is a closed and bounded region of the plane that is enclosed by two simple curves μ_1 and μ_2. If R is free of equilibrium solutions of* (8.35) *and there is a trajectory C that is confined to R, then either C is a limit cycle, or it spirals toward a limit cycle.*

If a solution stays in a bounded region that contains no equilibrium points, then its long-term behavior approaches a periodic orbit. In particular, the trajectory either is a limit cycle or approaches one asymptotically.

The main implication of the theorem is that if a solution exists for all sufficiently large t and remains in a bounded region that contains no equilibrium points, then its long-term behavior must approach a periodic orbit. Thus, asymptotically, the solution behaves like a periodic function.

The theorem is primarily qualitative rather than computational. In practice, it is often difficult to verify that a solution remains in a bounded region for all time. Moreover, the result is specific to planar systems; in higher dimensions, the dynamics can be significantly more complicated, and such a conclusion no longer holds.

Example 8.16 *Consider the system*

$$x' = y + x(1 - x^2 - y^2), \quad y' = -x + y(1 - x^2 - y^2). \tag{8.37}$$

It is easy to verify that $(0,0)$ *is the only equilibrium solution.*

We introduce polar coordinates

$$x = r\cos\theta, \quad y = r\sin\theta, \quad r^2 = x^2 + y^2,$$

where $r = r(t)$ and $\theta = \theta(t)$. Then

$$xx' + yy' = rr'.$$

Substituting from (8.37), *we obtain*

$$r' = r(1 - r^2), \qquad \theta' = \frac{xy' - x'y}{x^2 + y^2} = -1.$$

Solving these equations gives

$$r(t) = \frac{1}{\sqrt{1 + ce^{-2t}}}, \quad c > -1, \qquad \theta(t) = -t + k.$$

If $c = 0$, then $r(t) = 1$, and the corresponding solution lies on the unit circle

$$x^2 + y^2 = 1,$$

which is a periodic orbit, and parametrically is given by $(x(t), y(t)) = (\cos(t), -\sin(t))$.

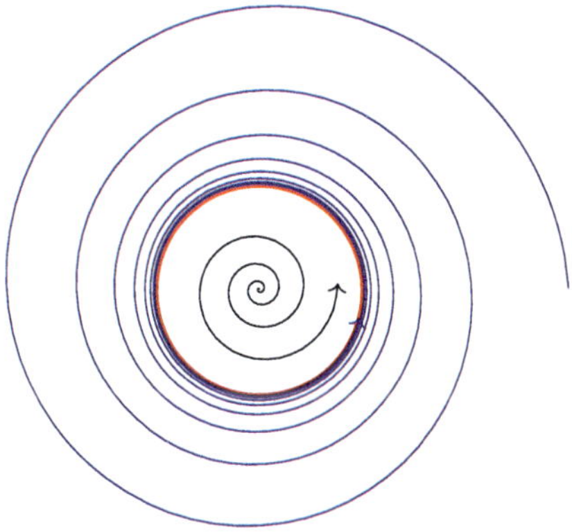

FIGURE 8.7: Limit cycle.

If $c > 0$, then $0 < r(t) < 1$ and $r(t) \nearrow 1$ as $t \to \infty$, so the trajectory spirals outward toward the unit circle.

If $-1 < c < 0$, then $r(t) > 1$ and $r(t) \searrow 1$ as $t \to \infty$, so the trajectory spirals inward toward the unit circle.

Thus, the unit circle is a limit cycle, and it is asymptotically stable (see Fig. 8.7).

To relate this to the Poincaré–Bendixson Theorem, consider the annular region

$$R = \{(x,y) : \mu_1 < x^2 + y^2 < \mu_2\}, \quad 0 < \mu_1 < 1 < \mu_2.$$

This region contains no equilibrium points. Moreover, trajectories that enter R remain in R for all sufficiently large t. Hence, by Theorem 8.9, the trajectories spiral toward the limit cycle, which in this case is the unit circle.

8.5 Lotka–Volterra Competition Model

If two populations $x_1(t)$ and $x_2(t)$ were growing logistically and not affecting each other, their growth could be described by the two uncoupled logistic equations

$$\begin{cases} x_1'(t) = r_1 x_1(t)\left(1 - \frac{x_1(t)}{K_1}\right) \\ x_2'(t) = r_2 x_2(t)\left(1 - \frac{x_2(t)}{K_2}\right) \\ x_1(0) > 0,\ x_2(0) > 0 \end{cases}$$

where $r_i, i = 1,2$ are the growth rates (birth - death), and $K_i, i = 1,2$ are positive constants. If the two populations were competing for the same resources, then the growth rate of each population is affected by the other. In that case, the model becomes

$$\begin{cases} x_1'(t) = r_1x_1(t)\left(1 - \frac{x_1(t)}{K_1} - \lambda_1 x_2(t)\right) \\ x_2'(t) = r_2x_2(t)\left(1 - \frac{x_2(t)}{K_2} - \lambda_2 x_1(t)\right) \\ x_1(0) > 0,\ x_2(0) > 0 \end{cases}$$

where $\lambda_i, i = 1,2$, are positive constants. To simplify the calculations, we convert the system by using nondimensional variables by measuring x_1 in units of K_1, x_2 in units of K_2, and time in units of $\frac{1}{r_1}$. We let

$$x_1(t) = K_1x(t), \quad x_2(t) = K_2y(t), \quad \text{and} \quad t = \frac{s}{r_1}.$$

It follows from using the chain rule that

$$K_1\frac{dx}{ds} = \frac{dx_1}{dt}\frac{dt}{ds} = \frac{dx_1}{dt}\frac{1}{r_1}.$$

Or,

$$x_1'(t) = r_1K_1\frac{dx}{ds}.$$

Substituting back into the first equation of the model, we arrive at

$$r_1K_1\frac{dx}{ds} = r_1K_1x(s)\left(1 - x(s) - \lambda_1K_2y(s)\right),$$

from which we get

$$\frac{dx}{ds} = x(s)\left(1 - x(s) - \lambda_1K_2y(s)\right).$$

Similarly we obtain

$$\frac{dy}{ds} = ry(s)\left(1 - y(s) - \lambda_2K_1x(s)\right),$$

where $r = \frac{r_2}{r_1}$.

Interchanging s with t and by letting $\mu_1 = \lambda_1K_2$, and $\mu_2 = \lambda_2K_1$, we arrive at the dimensionless Lotka–Volterra competition model

$$\begin{cases} x'(t) = x(t)\left(1 - x(t) - \mu_1y(t)\right) \\ y'(t) = ry(t)\left(1 - y(t) - \mu_2x(t)\right) \\ x(0) = x_0 > 0,\ y(0) = y_0 > 0 \end{cases} \tag{8.38}$$

Next, we show that the solutions of (8.38) are positive and remain bounded for all time, $t > 0$.

Proposition 1 *If $x(t), y(t)$ are solutions of* (8.38) *with positive initial conditions x_0 and y_0, then they are positive and bounded for all $t > 0$.*

Proof *Let $x(t), y(t)$ be solution of* (8.38) *with positive initial conditions x_0 and y_0. Dividing the first equation by $x(t)$ and the second equation by $y(t)$ of* (8.38), *respectively, and integrating, we get the integral relations*

$$x(t) = x_0 e^{\int_0^t \left(1-x(s)-\mu_1 y(s)\right) ds},$$

and

$$y(t) = y_0 e^{r\int_0^t \left(1-y(s)-\mu_2 x(s)\right) ds}.$$

We see if the initial conditions are positive, then the components of the solution are positive. Left to show solutions remain bounded for all positive time. Since x and y are positive, we have that

$$x(t)\big(1-x(t)-\mu_1 y(t)\big) \le x(t)\big(1-x(t)\big),$$

and

$$ry(t)\big(1-y(t)-\mu_2 x(t)\big) \le ry(t)\big(1-y(t)\big).$$

It follows that

$$x'(t) \le x(t)\big(1-x(t)\big) \text{ and } y'(t) \le ry(t)\big(1-y(t)\big).$$

From now on, we write x for $x(t)$ and y for $y(t)$. Consider the differential equation

$$z' = z(1-z),\ z(0) = x(0) = x_0.$$

Separating the variables and then integrating to get $\int \frac{dz}{z(1-z)} = \int dt$. Using partial fractions on the left side integral, we obtain $\text{Ln}|\frac{z}{1-z}| = t + C$. Taking exponential on both sides, solving for z and then applying the initial condition, we arrive at the unique solution

$$z(t) = \frac{x_0 e^t}{1+x_0(e^t-1)} \to 1,\ as\ t \to \infty.$$

By the Comparison Theorem,

$$x(t) \le z(t), \qquad t \ge 0.$$

Since $z(t)$ is bounded and approaches 1 as $t \to \infty$, it follows that $x(t)$ is bounded for all positive times. In fact,

$$0 < x(t) \le \max\{1, x_0\}.$$

Similar arguments apply to $y(t)$. This concludes the proof.

Setting the right sides of (8.38) equal to zero, we obtain the equilibrium solutions

$$E_0 = (0,0), \quad E_1 = (0,1), \quad E_2 = (1,0),$$

and

E_3 : which consists of the solutions, if they exist, of $x + \mu_1 y = 1,\ \mu_2 x + y = 1.$

Next, we use Theorem 8.5 to analyze the stability of the equilibrium solutions. The Jacobian matrix is found to be

$$J(x,y) = \begin{pmatrix} 1-2x-\mu_1 y & -\mu_1 x \\ -r\mu_2 y & r(1-2y-\mu_2 x) \end{pmatrix}.$$

For $E_0 = (0,0)$, we have

$$J(0,0) = \begin{pmatrix} 1 & 0 \\ 0 & r \end{pmatrix}$$

with eigenvalues $\lambda = 1, r$, and hence by Theorem 8.5 and Remark 29 E_0 is unstable or repeller. At E_1,

$$J(0,1) = \begin{pmatrix} 1-\mu_1 & 0 \\ -r\mu_2 & -r \end{pmatrix}$$

with eigenvalues $\lambda = 1-\mu_1, -r$ and hence by Theorem 8.5 and Remark 29 E_1 is either saddle or asymptotically stable. Similarly, at E_2,

$$J(1,0) = \begin{pmatrix} -1 & -\mu_1 \\ 0 & r(1-\mu_2) \end{pmatrix}$$

with eignevalues $\lambda = -1, r(1-\mu_2)$ and hence by Theorem 8.5 and Remark 29 E_2 is either saddle or asymptotically stable. For, E_3 we solve the system

$$x + \mu_1 y = 1, \quad \mu_2 x + y = 1.$$

After some calculations, we obtain

$$E_3 = \left(\frac{1-\mu_1}{1-\mu_1\mu_2}, \frac{1-\mu_2}{1-\mu_1\mu_2}\right).$$

The stability of the equilibrium solution of E_3 is important to biologists since it shows the coexistence of both species. We will analyze the stability by considering different cases that are based on the eigenvalues. Biologically, the parameters μ_1 and μ_2 measure the competitive effect of one species on the other.

When

$$\mu_1 < 1, \quad \mu_2 < 1,$$

competition between species is relatively weak, and coexistence becomes possible.

On the other hand, when

$$\mu_1 > 1, \quad \mu_2 > 1,$$

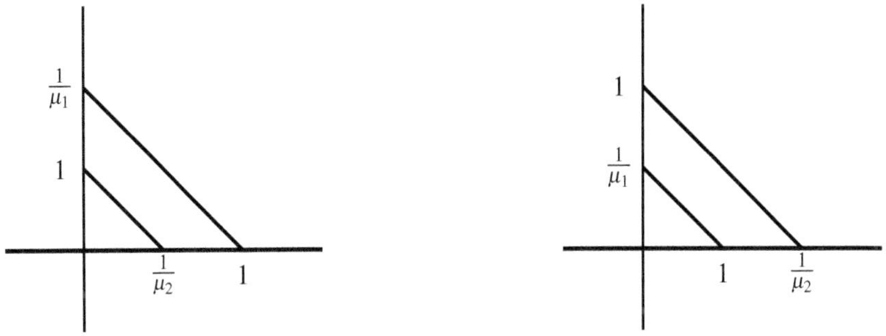

Case : $(a)\ \mu_1 < 1,\ \mu_2 > 1$ *Case* : $(b)\ \mu_1 > 1,\ \mu_2 < 1$

FIGURE 8.8: No equilibrium points in the first quadrant.

the interspecific competition is strong, and the eventual winner depends on the initial population levels.

Case (a): $\mu_1 < 1,\ \mu_2 > 1.$
In this case, the two lines $x + \mu_1 y = 1,\ \mu_2 x + y = 1$ do not intersect in the first quadrant, and it follows that E_3 is not an equilibrium solution as depicted in Fig. 8.8. In addition, in Case (a), E_2 is asymptotically stable and E_1 is a saddle point. Since the interior of the first quadrant does not contain any equilibrium solution, it follows from Theorem 8.8 that there is no periodic solution in the first quadrant. Recall Theorem 8.8 says a closed orbit or limit cycle surrounds at least one equilibrium point. Thus, from Theorem 8.8, the omega limit set of any trajectory is the equilibrium solution E_2, and hence E_2 is globally asymptotically stable (see Fig. 8.9). The biological meaning is that population x wins the competition. In other words, population y becomes extinct.

Case (b): $\mu_1 > 1,\ \mu_2 < 1.$
In this case, the roles of E_1 and E_2 are reversed, but still only one competitor survives.

Next, we consider the cases where E_3 is in the first quadrant. That is, E_3 is an equilibrium solution. It turned out that E_3 is an equilibrium solution in the two cases:

Case (c): $\mu_1 < 1,\ \ \mu_2 < 1$ and **Case (d):** $\mu_1 > 1,\ \ \mu_2 > 1.$
Next, we make use of Theorem 8.7 (Generalization of Bendixson–Dulac Theorem) to study the possibilities of the existence of a periodic solution that surrounds E_3. Let $h(x,y) = \frac{1}{xy}$. Then h is continuously differentiable function in $G = \{(x,y) : x > 0, y > 0\}$.

$$\frac{\partial}{\partial x}\Big[\frac{1}{xy}x\big(1 - x - \mu_1 y\big)\Big] = -\frac{1}{y},$$

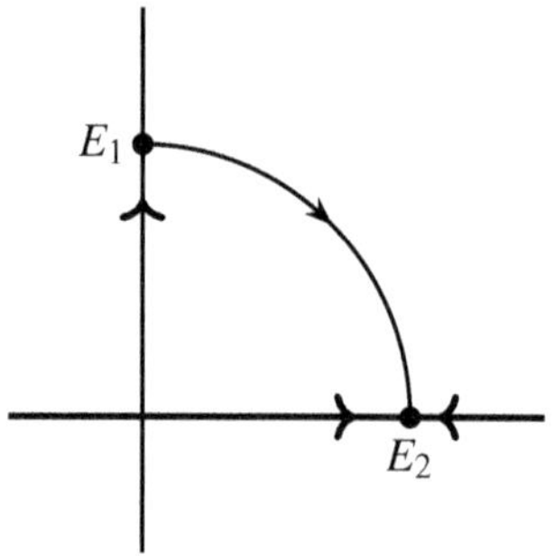

FIGURE 8.9: E_2 is globally asymptotically stable.

and

$$\frac{\partial}{\partial y}\left[\frac{1}{xy}ry(1-y-\mu_2 x)\right] = -\frac{r}{x}.$$

Thus,

$$\frac{\partial}{\partial x}\left[\frac{1}{xy}x(1-x-\mu_1 y)\right] + \frac{\partial}{\partial y}\left[\frac{1}{xy}ry(1-y-\mu_2 x)\right] = -\left(\frac{1}{y}+\frac{r}{x}\right),$$

which is negative in G. Therefore, there is no periodic solution in the first quadrant in either Case (c) or Case (d).
Let (x^*,y^*) be a solution of $x+\mu_1 y = 1, \mu_2 x + y = 1$. That is $(x^*,y^*) = \left(\frac{1-\mu_1}{1-\mu_1\mu_2}, \frac{1-\mu_2}{1-\mu_1\mu_2}\right)$. Then

$$\begin{aligned} J(x^*,y^*) &= \begin{pmatrix} 1-2x^*-\mu_1 y^* & -\mu_1 x^* \\ -r\mu_2 y^* & r(1-2y^*-\mu_2 x^*) \end{pmatrix} \\ &= \begin{pmatrix} 1-2x^*-\mu_1\frac{1-\mu_2}{1-\mu_1\mu_2} & -\mu_1 x^* \\ -r\mu_2 y^* & r(1-2y^*-\mu_2\frac{1-\mu_1}{1-\mu_1\mu_2}) \end{pmatrix} \\ &= \begin{pmatrix} -x^* & -\mu_1 x^* \\ -r\mu_2 y^* & -ry^* \end{pmatrix}. \end{aligned}$$

$$\operatorname{tra}(J(x^*,y^*)) = -x^* - ry^* < 0, \text{ for } x^*,y^* > 0,$$

implies the sum of the eigenvalues is negative. Moreover,

$$\det(J(x^*,y^*)) = rx^*y^* - r\mu_1\mu_2 x^*y^* = rx^*y^*(1-\mu_1\mu_2).$$

As

$$\det(J(x^*,y^*)-\lambda I) = \lambda^2 - \operatorname{tra}(J(x^*,y^*))\lambda + \det(J(x^*,y^*)),$$

we have that

$$\lambda^2 + (x^* + ry^*)\lambda + rx^*y^*(1 - \mu_1\mu_2) = 0.$$

It follows that eigenvalues are given by

$$\lambda = \frac{-x^* - ry^* \pm \sqrt{(x^* + ry^*)^2 - 4rx^*y^*(1 - \mu_1\mu_2)}}{2}.$$

E_3 is asymptotically stable if $1 - \mu_1\mu_2 > 0$ and a saddle if $1 - \mu_1\mu_2 < 0$.
If $\mu_1\mu_2 = 1$, the two lines are parallel or coincident. If coincident, there is a line of equilibrium solutions, and if parallel, the cases, Case (a) and Case (b) apply. In Case (c), $1 - \mu_1 > 0$ and $1 - \mu_2 > 0$, so E_1 and E_2 are saddle unstable points. Since $\mu_1\mu_2 < 1$, E_3 is an attractor. That is all the solutions tend to E_3. In that case, the two populations coexist. See Fig. 8.10 case(c).
In Case (d), $\mu_1 - 1 > 0$, and $\mu_2 - 1 > 0$, and hence E_1 and E_2 are asymptotically stable. Also in Case (d) $\mu_1\mu_2 > 1$, and so E_3 is a saddle point. See Fig. 8.10 (f). Here, the initial conditions determine the winner.

8.6 Bifurcation in Planar Systems

We consider the two-dimensional autonomous system

$$x' = P(x, y, \mu), \quad y' = Q(x, y, \mu) \tag{8.39}$$

depending upon a real parameter μ. We assume P and Q are continuous in all arguments on some subset of $\mathbb{R}^2$. As μ changes values or signs, the behavior of solutions changes and may result in bifurcation points. For example, as we move across a particular equilibrium solution, the stability may change from stable to unstable and vice versa. We begin by considering the following example.

Example 8.17 *Consider the two-dimensional linear system*

$$x' = x + \mu y, \quad y' = x - y,$$

for parameter μ. The matrix

$$\begin{pmatrix} 1 & \mu \\ 1 & -1 \end{pmatrix}$$

has the eigenvalues

$$\lambda = \pm\sqrt{1+\mu}.$$

If $\mu > -1$, the eigenvalues are real with opposite signs, and therefore the origin $(0,0)$ is a saddle point (unstable). For the special value $\mu = -1$, the system becomes

$$x' = x - y, \quad y' = x - y,$$

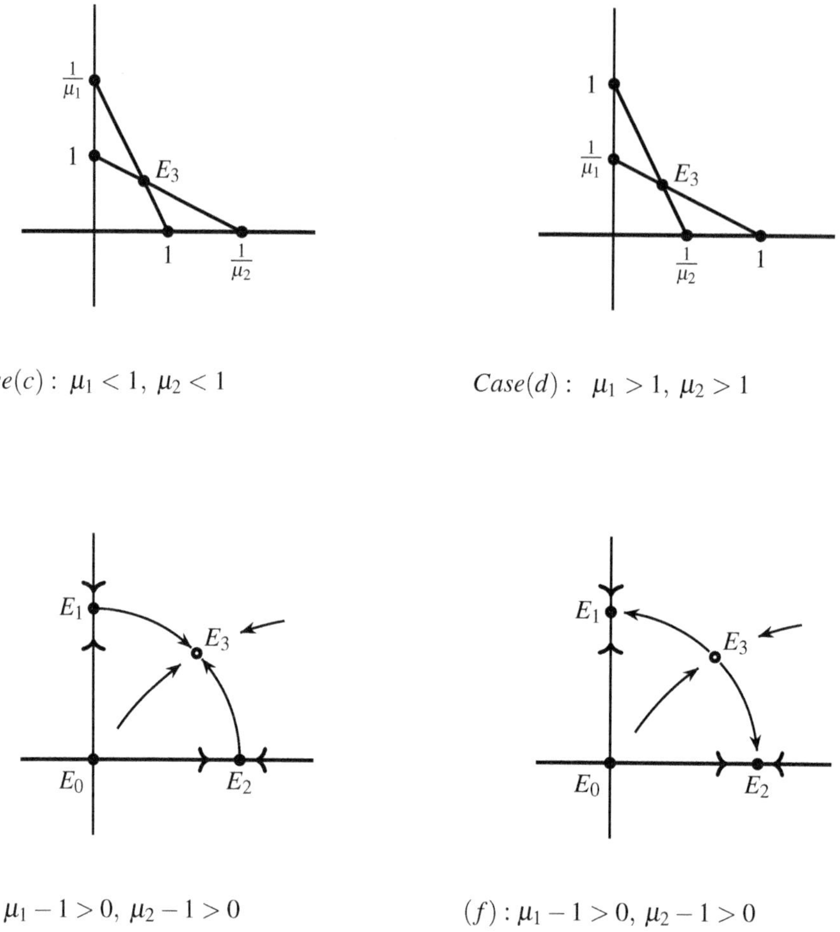

FIGURE 8.10: Phase portraits for coexistence and competitive exclusion.

and results in the whole line $y = x$ *of fixed points. If* $\mu < -1$*, the eigenvalues are purely imaginary and the origin* $(0,0)$ *is a center (stable). As* $\mu > -1$ *decreases to* $\mu < -1$*, the nature of the solution changes at* $\mu = -1$*; and as a result, the equilibrium solution* $(0,0)$ *changes from an unstable saddle to a stable center as* μ *passes the critical value* -1*. Thus, we have a bifurcation at* $\mu = -1$*.* □

Example 8.18 *Consider the planar autonomous system*

$$x' = -y - x(x^2 + y^2 - \mu), \quad y' = x - y(x^2 + y^2 - \mu).$$

It is clear that $(0,0)$ *is the only equilibrium point. We rewrite the system in the form*

$$\begin{pmatrix} x \\ y \end{pmatrix}' = \begin{pmatrix} \mu & -1 \\ 1 & \mu \end{pmatrix} \begin{pmatrix} x \\ y \end{pmatrix} + \begin{pmatrix} -x(x^2+y^2) \\ -y(x^2+y^2) \end{pmatrix}.$$

The eigenvalues of

$$A = \begin{pmatrix} \mu & -1 \\ 1 & \mu \end{pmatrix}$$

are $\mu \pm i$.
(1) If $\mu < 0$, *then trajectories spiral toward the origin and the origin is asymptotically stable by Theorem 8.5.*
(2) If $\mu > 0$, *then the origin is unstable for the nonlinear system.*
(3) If $\mu = 0$, *then Theorem 8.5 says nothing about the stability or instability of the origin of the nonlinear system.*
Next, we examine in more detail the trajectories of the system by using polar coordinates as we did in Example 8.10. By doing so, we arrive at

$$r' = r(\mu - r^2) \text{ and } \theta' = 1. \tag{8.40}$$

Separating the variables and using partial fractions, we arrive at

$$\int \frac{dr}{r(r-\sqrt{\mu})(r+\sqrt{\mu})} = t + k, \text{ for some constant } k.$$

This leads to the solution

$$r(t) = \frac{\sqrt{\mu}}{\sqrt{1+ce^{-2\mu t}}}. \tag{8.41}$$

Similarly, if we choose $\theta(0) = 0$, *we have* $\theta(t) = t$.
Assume $\mu > 0$ *and consider the following three cases.*
(3a) When $c = 0$, *we have from* (8.41) $r(t) = \sqrt{\mu}$, *which represents a periodic solution with radius* $\sqrt{\mu}$
(3b) If $c < 0$, *then* (8.41) *represents an orbit or trajectory that spirals counterclockwise toward the circle* $r(t) = \sqrt{\mu}$ *from the outside.*
(3c) If $c > 0$, *then* (8.41) *represents an orbit or trajectory that spirals counterclockwise toward the circle* $r(t) = \sqrt{\mu}$ *from the inside.*
Let $0 < \mu_1 < \mu < \mu_2$ *and define the annulus region R by*

$$R = \{(x,y) : \mu_1 < x^2 + y^2 < \mu_2\}.$$

Clearly, R contains no equilibrium solution, and then cases (3a)-(3c) prove to us that the circle given by $r^2(t) = \mu$ *or* $x^2 + y^2 = \mu$ *is a stable limit cycle, and the origin is unstable.*
Next, we examine the stability of the origin when $\mu = 0$.
In this case, (8.40) *gives*

$$r' = -r^3,$$

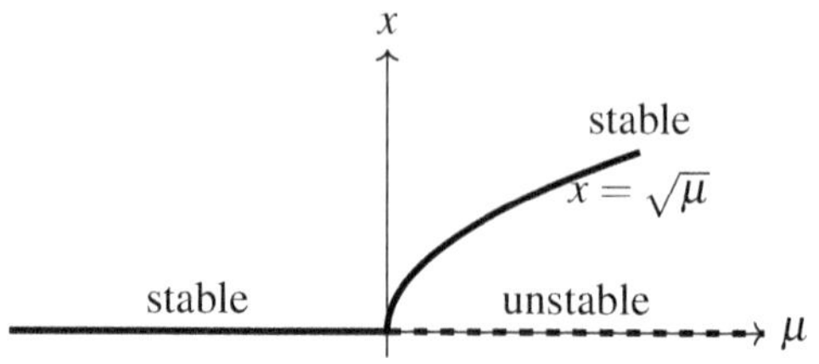

FIGURE 8.11: Bifurcation diagram for Example 8.18.

which has the solution

$$r(t) = \frac{1}{\sqrt{2t+K}}$$

for a suitable constant K. *Clearly,* $r(t) \to 0$, *as* $t \to \infty$ *and hence the zero solution, or the origin, is asymptotically stable (spirally stable).*
Finally, for $\mu < 0$, *we have from* (8.40)

$$r' = r(r^2 + l^2) \text{ and } \theta' = 1,$$

where $l^2 = -\mu > 0$. *Separating the variables and using partial fractions, we arrive at*

$$\int \frac{dr}{r(r^2+l^2)} = -dt, \text{ for some constant } k.$$

This leads to the solution

$$r(t) = \frac{cl^2 e^{-2l^2 t}}{1 - cl^2 e^{-2l^2 t}} \text{ for some constant } c.$$

Clearly $r(t) \to 0$, *as* $t \to \infty$ *and hence the zero solution, or the origin is asymptotically stable (spirally stable). In conclusion, for* $\mu \leq 0$, *the origin* $(0,0)$ *is stable, and for* $\mu > 0$, *(0,0) is unstable, and* $r = \sqrt{\mu}$ *is stable. Thus, as* μ *passes through the critical value* $\mu = 0$ *from negative to positive, the origin changes from stable spiral to unstable spiral and there appears a new limit cycle that is a periodic solution. This type of bifurcation is common and known as the Hopf bifurcation.*

The bifurcation diagram is shown in Fig. 8.11. □

The next theorem provides a necessary condition for the existence of a Hopf bifurcation. The proof is complicated, and hence we omit it.

Theorem 8.10 *Consider the two-dimensional autonomous system*

$$x' = P(x,y,\mu), \quad y' = Q(x,y,\mu) \tag{8.42}$$

depending upon a real parameter μ, *where* P *and* Q *are differentiable in the neighborhood of* $(0,0,0)$. *Assume for small* μ,

$$P(0,0,\mu) = Q(0,0,\mu) = 0. \tag{8.43}$$

Let the Jacobian

$$J(0,0,0)=\begin{pmatrix} P_x(0,0,0) & P_y(0,0,0) \\ Q_x(0,0,0) & Q_y(0,0,0) \end{pmatrix}=\begin{pmatrix} 0 & -w \\ w & 0 \end{pmatrix}, \text{ for some } w\neq 0. \tag{8.44}$$

Let

$$\begin{aligned} a &= \frac{1}{16}\big(P_{xxx}+Q_{xxy}+P_{xyy}+Q_{yyy}\big) \\ &+ \frac{1}{16w}\big(P_{xy}(P_{xx}+P_{yy})-Q_{xy}(Q_{xx}+Q_{yy})-P_{xx}Q_{xx}+P_{yy}Q_{yy}\big) \end{aligned} \tag{8.45}$$

evaluated at $(0,0,0)$ *with* $a\neq 0$. *Suppose that*

$$a\big(P_{\mu x}+Q_{\mu y}\big)>0. \tag{8.46}$$

Then, for sufficiently small $\mu<0$, *there is a unique periodic solution with amplitude* $A(\mu)$ *such that*

$$\lim_{\mu\to 0^-}\frac{A(\mu)}{\sqrt{|\mu|}}=c>0.$$

If

$$a\big(P_{\mu x}+Q_{\mu y}\big)<0, \tag{8.47}$$

then the periodic solution exists for $\mu>0$. *The stability of the periodic solution is determined by the sign of*

$$P_{\mu x}+Q_{\mu y}. \tag{8.48}$$

and is the opposite of the stability of the zero solution in the range of μ *where the periodic solution exists.*

Example 8.19 *As an application of Theorem 8.10, we consider the system given in Example 8.17. It readily follows that the Jacobian*

$$J(0,0,0)=\begin{pmatrix} \mu & -1 \\ 1 & \mu \end{pmatrix}=\begin{pmatrix} 0 & -1 \\ 1 & 0 \end{pmatrix}.$$

Thus, condition (8.44) *is satisfied with* $w=1$. *After simple calculations, it follows that, evaluated at* $(0,0,0)$, $a=-1$. *Moreover,*

$$a\big(P_{\mu x}+Q_{\mu y}\big)=-1(1+1)<0.$$

Thus, by Theorem 8.10, the system possesses a periodic solution for $\mu>0$ *that is stable (spirally stable) since the origin is unstable in the region where* μ *is positive.*
□

We end this section with an application to Euler's buckling beam.

Example 8.20 *(Euler's Buckling Beam) One of the classical applications of bifurcation theory is Euler's buckling beam. Consider the different beam configurations shown in Fig. 8.12, where increasing compressive pressure is applied to a beam. For*

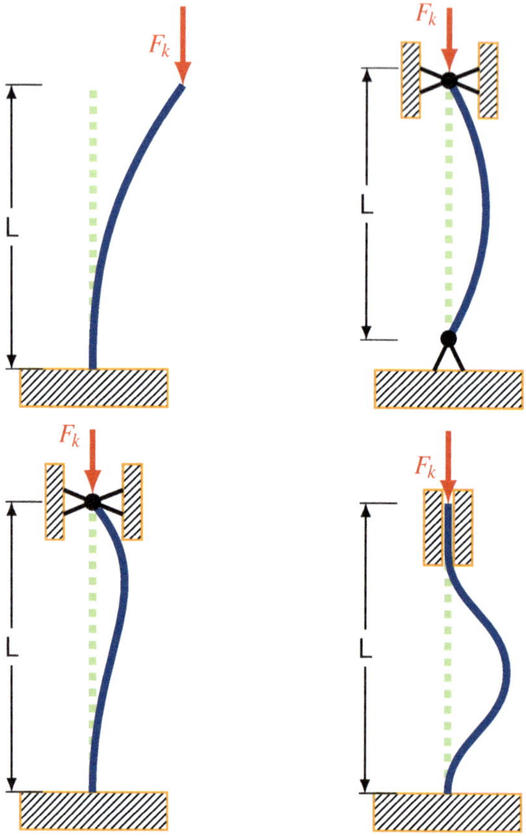

FIGURE 8.12: Buckling beam.

relatively small applied forces, the beam maintains its straight vertical configuration. However, as the applied force increases, the beam eventually undergoes a qualitative change in shape. Beyond a certain critical force, the straight configuration loses stability, and the beam bends into a new configuration, a phenomenon known as buckling.

Suppose the beam has length one unit and lies along the horizontal axis. One end of the beam is supported by a fixed wall, and a compressive force $\mathbb{F} \geq 0$ *is applied along its axis. Let* $\delta > 0$ *denote the viscous damping coefficient and let* $K > 0$ *represent the stiffness constant of the beam. The small transverse deflection of the beam is denoted by* $v(x,t)$*, where* x *represents the spatial position along the beam and* t *denotes time.*

Using a first-mode approximation of the form

$$v(x,t) \approx u(t)\sin(\pi x),$$

where $u(t)$ represents the time-dependent amplitude of the beam deflection, the governing partial differential equation can be reduced to an ordinary differential equation describing the evolution of the dominant buckling mode.

It can be shown that the amplitude $u(t)$ satisfies the second-order differential equation

$$u''(t)+\delta u'(t)-\pi^2(\mathbb{F}-\pi^2)u(t)+\frac{1}{2}K\pi^4u^3(t)=0. \tag{8.49}$$

To better study the differential equation, we put it in the system form

$$\begin{cases} x' = y \\ y' = \pi^2(\mathbb{F}-\pi^2)x-\delta y-\frac{1}{2}K\pi^4x^3 \end{cases} \tag{8.50}$$

where the force $\mathbb{F}$ is the only parameter. The system has the three equilibrium solutions

$$(0,0),\ (\sqrt{\frac{2(\mathbb{F}-\pi^2)}{K\pi^2}},0),\ \text{ and } (-\sqrt{\frac{2(\mathbb{F}-\pi^2)}{K\pi^2}},0).$$

Note that the equilibrium solutions

$$(\sqrt{\frac{2(\mathbb{F}-\pi^2)}{K\pi^2}},0),\ \text{ and } (-\sqrt{\frac{2(\mathbb{F}-\pi^2)}{K\pi^2}},0)$$

are only valid for $\mathbb{F}>\pi^2$.

Now

$$J(0,0)=\begin{pmatrix} 0 & 1 \\ \pi^2(\mathbb{F}-\pi^2) & -\delta \end{pmatrix},$$

with eigenvalues

$$\lambda=\frac{-\delta\pm\sqrt{\delta^2+4\pi^2(\mathbb{F}-\pi^2)}}{2}$$

and as a consequence, $(0,0)$ is stable if $\mathbb{F}<\pi^2$ and unstable if $\mathbb{F}>\pi^2$. The physical implication is that the beam will buckle if $\mathbb{F}>\pi^2$.
Now we consider the equilibrium solution $(\sqrt{\frac{2(\mathbb{F}-\pi^2)}{K\pi^2}},0)$ for $\mathbb{F}>\pi^2$.

We translate this equilibrium to the origin by setting

$$x_1=x-\sqrt{\frac{2(\mathbb{F}-\pi^2)}{K\pi^2}}, \qquad y_1=y.$$

The linearization at the translated origin is

$$J(0,0)=\begin{pmatrix} 0 & 1 \\ -2\pi^2(\mathbb{F}-\pi^2) & -\delta \end{pmatrix}.$$

Hence, its eigenvalues are

$$\lambda=\frac{-\delta\pm\sqrt{\delta^2-8\pi^2(\mathbb{F}-\pi^2)}}{2}.$$

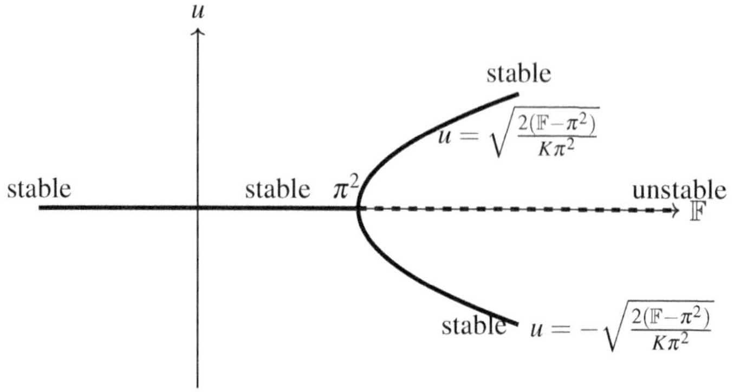

FIGURE 8.13: Bifurcation diagram for $u = \pm\sqrt{\frac{2(\mathbb{F}-\pi^2)}{K\pi^2}}$.

Since $\delta > 0$ and $2\pi^2(\mathbb{F}-\pi^2) > 0$, both eigenvalues have negative real parts. Therefore, the translated origin is asymptotically stable, and consequently the equilibrium

$$\left(\sqrt{\frac{2(\mathbb{F}-\pi^2)}{K\pi^2}}, 0\right)$$

is asymptotically stable for $\mathbb{F} > \pi^2$. Similarly, the equilibrium

$$\left(-\sqrt{\frac{2(\mathbb{F}-\pi^2)}{K\pi^2}}, 0\right)$$

is also asymptotically stable for $\mathbb{F} > \pi^2$. See Fig. 8.13.

Thus, the trivial straight configuration is asymptotically stable for $\mathbb{F} < \pi^2$ and loses stability at $\mathbb{F} = \pi^2$. For $\mathbb{F} > \pi^2$, two symmetric stable buckled equilibria appear. Hence, the system undergoes a supercritical pitchfork bifurcation at $\mathbb{F} = \pi^2$. □

8.6.1 Exercises

Exercise 8.19 *Consider the planar autonomous system*

$$\begin{cases} x' = y - x(1 - x^2 - y^2), \\ y' = -x - y(1 - x^2 - y^2). \end{cases}$$

(a) Show that $(0,0)$ is the only equilibrium point of the system.

(b) Rewrite the system in polar coordinates

$$x = r\cos\theta, \qquad y = r\sin\theta,$$

and derive the differential equations for $r'(t)$ and $\theta'(t)$.

(c) *Determine the equilibrium points of the radial equation $r' = r'(r)$ and analyze their stability.*

(d) *Show that there exists a unique radius $r_0 > 0$ such that $r(t) \to r_0$ as $t \to \infty$ for all solutions with $r(0) > 0$, and conclude that the circle $r = r_0$ defines a stable limit cycle in the phase plane.*

Exercise 8.20 *Let $P(x,y) = x - y + x^3$ and $Q(x,y) = x + y^3$, and consider the autonomous system*

$$\begin{cases} x' = P(x,y), \\ y' = Q(x,y). \end{cases}$$

(a) *Compute*

$$\frac{\partial P}{\partial x}(x,y) + \frac{\partial Q}{\partial y}(x,y),$$

and determine where in $\mathbb{R}^2$ this quantity has a constant sign.

(b) *Use the Bendixson–Dulac Theorem (Theorem 8.6) to show that the system has no closed trajectories in any region where*

$$\frac{\partial P}{\partial x} + \frac{\partial Q}{\partial y}$$

does not change sign.

(c) *Now let $h(x,y) = \dfrac{1}{1+x^2}$ and consider the Dulac function hP, hQ. Compute*

$$\frac{\partial(hP)}{\partial x}(x,y) + \frac{\partial(hQ)}{\partial y}(x,y),$$

and identify an open region $U \subset \mathbb{R}^2$ where this sum has a constant sign.

(d) *Apply the generalized Bendixson–Dulac Theorem (Theorem 8.7) to conclude that there are no limit cycles contained entirely in U.*

Exercise 8.21 *Use polar coordinates to study the stability of the origin for the system*

$$x' = -y - x\sqrt{x^2+y^2}, \; y' = x - y\sqrt{x^2+y^2}.$$

Exercise 8.22 *Show that the system*

$$x' = y + \frac{x(1-x^2-y^2)}{\sqrt{x^2+y^2}}, \; y' = -x + \frac{y(1-x^2-y^2)}{\sqrt{x^2+y^2}}$$

has a limit cycle.

Exercise 8.23 *Show that*

$$x' = y + x(1-x^2-y^2), \; y' = -x + y(1-x^2-y^2)$$

has a limit cycle.

Exercise 8.24 *Consider the system*

$$x' = 4x + 4y - x(x^2 + y^2)$$
$$y' = -4x + 4y - y(x^2 + y^2)$$

(a) Show that there is a closed path in the region $1 \le r \le 3$, *where* $r^2 = x^2 + y^2$.
(b) Find the general solution.

Exercise 8.25 *Show that* $x'' + x' + f(x) = 0$ *has no periodic solution.*

Exercise 8.26 *Determine the region on which there is no periodic solution for*

$$x'' + \beta(x^2 - 1)x' + x = 0.$$

Exercise 8.27 *Consider the predator–prey model*

$$x' = x((1-x) - y), \quad y' = y(x - a), \; a > 0$$

where $x(t)$ *and* $y(t)$ *are the numbers of prey and predators at time* t, *respectively.*
(a) Sketch the nullclines and indicate the direction field.
(b) Find the equilibrium points and analyze them.
(c) What can you say for each of the cases; $a > 1$; $a = 1/2$; *and* $0 < a < 1$.

Exercise 8.28 *(a) Prove Theorem 8.7*
(b) Use (a) to show that the system

$$x' = x(A - ax + by), \quad y' = y(B - cy + dx)$$

where a, b, c, d, A, *and* B *are constants with* $a,\ c > 0$, *has no closed trajectories or periodic orbits.*
(Hint: let $h(x,y) = \frac{1}{xy}$.*)*

Exercise 8.29 *Consider the nonlinear system*

$$x' = x - y - x^3, \quad y' = x + y - y^3$$

Given that $(0,0)$ *is the only critical point, use polar coordinates to show that there is a limit cycle in the region*

$$R = \{(r, \theta) : 1 < r < \sqrt{2}\}.$$

Exercise 8.30 *Use Theorem 8.10 to show the system*

$$x' = x\mu - y - x^3 - xy^2 - y, \quad y' = x + y\mu - y^3 - x^2 y$$

undergoes a Hopf bifurcation. That is, the system has a stable periodic solution with radius $\sqrt{\mu}$.

Exercise 8.31 *Decide if Theorem 8.10 applies to the van der Pol equation given in system form by*

$$x' = y, \quad y' = -x - \mu(x^2 - 1)y$$

8.7 Lyapunov Functions

Lyapunov function is named after Aleksandr Mikhailovich Lyapunov, a Russian mathematician who defended the thesis *The General Problem of Stability of Motion* at Kharkov University in 1892. Lyapunov was a pioneer in creating the global approach to the analysis of the stability of nonlinear dynamical systems. His work, initially published in Russian and then translated to French, received little attention for many years. The current use of Lyapunov functions has proved that Lyapunov was ahead of his time in developing his theory that is being used in many different areas of sciences. In the theory, Lyapunov functions are scalar functions that may be used to prove the stability of the equilibrium of a given dynamical system. Since the inception of Lyapunov functions, their successful usage has been extended to integral equations, integro-differential equations, and functional differential equations.

8.8 Lyapunov Method

The existence of Lyapunov functions is a necessary and sufficient condition for stability in differential equations. There are no concrete procedures on how to find Lyapunov functions, but in some cases, the construction of Lyapunov functions is known. In the particular case of homogeneous autonomous systems with constant coefficients, the Lyapunov function can be found as a quadratic form. As we shall see later, Lyapunov functions allow us to study the stability of nonlinear dynamical systems when the equilibrium solution of an associated linear differential equation is a center.

Let D be an open set in $\mathbb{R}^n$ containing 0 and t_0 be any initial time. Define

$$V : [t_0, \infty) \times D \to [0, \infty),$$

where V is any differentiable *scalar* function. If $\psi : [t_0, \infty) \times D \to D$ is any differentiable function, then $V(t) := V(t, \psi(t))$ is a scalar function of t, and using the chain rule, we can compute its derivative,

$$V'(t) = \frac{\partial V}{\partial x_1}\psi_1'(t) + \ldots + \frac{\partial V}{\partial x_n}\psi_n'(t) + \frac{\partial V}{\partial t}.$$

Let $f : [t_0, \infty) \times D \to \mathbb{R}^n$ with $f(t, 0) = 0$. Assume the existence of the unknown solution $x : [t_0, \infty) \to D$ on each $(t^*, x_0) \in [t_0, \infty) \times D$ and consider the system

$$x' = f(t, x), \tag{8.51}$$

where

$$f(t,x) = \begin{pmatrix} f_1(t,x) \\ f_2(t,x) \\ \vdots \\ f_n(t,x) \end{pmatrix}.$$

Then it follows from the above argument that

$$\begin{aligned} V'(t,x(t)) &= \frac{\partial V(t,x(t))}{\partial x_1} f_1(t,x(t)) + \dots \\ &+ \frac{\partial V(t,x(t))}{\partial x_n} f_n(t,x(t)) + \frac{\partial V}{\partial t}. \end{aligned} \tag{8.52}$$

Thus, expression (8.52) defines the derivative of the function $V(t,x)$ along the unknown solutions of (8.56). Note that if $\nabla V = gradV = (\frac{\partial V}{\partial x_1}, \dots, \frac{\partial V}{\partial x_n})$ then expression (8.52) takes the form

$$V'(t,x(t)) = gradV \cdot f(t,x) + \frac{\partial V}{\partial t},$$

where "$\cdot$" represents the dot product.
Suppose we are able to find V so that for any $(t,x) \in [t_0, \infty) \times D$, we have

$$\frac{\partial V(t,x(t))}{\partial x_1} f_1(t,x(t)) + \dots + \frac{\partial V(t,x(t))}{\partial x_n} f_n(t,x(t)) + \frac{\partial V}{\partial t} \leq 0.$$

Then along the solutions of , we have $V'(t, \psi(t)) \leq 0$.Since V is nonincreasing along the solutions and V is increasing in $|x|$, then with some work, we are able to find boundedness of solutions and stability of equilibrium solutions.
The situation is considerably simplified when V is independent of t. In this case, (8.52) simplifies to

$$V'(x(t)) = \frac{\partial V(x(t))}{\partial x_1} f_1(t,x(t)) + \dots + \frac{\partial V(x(t))}{\partial x_n} f_n(t,x(t)). \tag{8.53}$$

Recall, if x^* is an equilibrium point of (8.56) then the change of variables $x = y + x^*$ translates every nonzero equilibrium solution to the origin.

Definition 8.6 *Let D be an open set in $\mathbb{R}^n$ containing 0. A continuous autonomous function $V : D \to [0, \infty)$ is positive definite if*

$$V(0) = 0 \ \ and \ \ V(x) > 0 \ \ if\, x \neq 0.$$

V is said to be negative definite if $-V$ is positive definite.

Definition 8.7 *Let D be an open set in $\mathbb{R}^n$ containing 0. Let*

$$V : D \to [0, \infty),$$

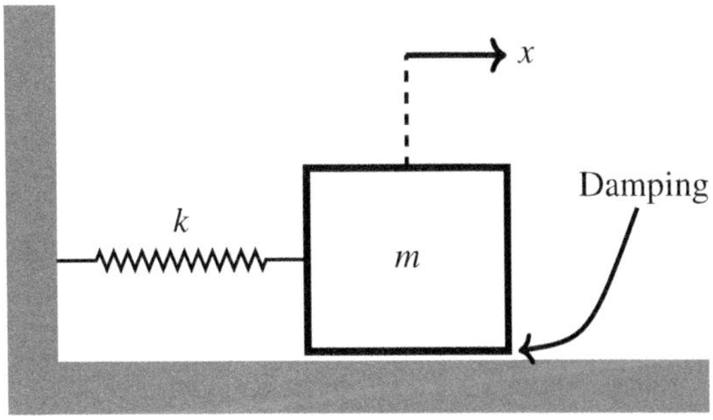

FIGURE 8.14: Damping harmonic motion.

have continuous first partial derivatives. If V is positive definite and

$$V'(x(t)) = \frac{\partial V(x(t))}{\partial x_1} f_1(t,x(t)) + \ldots + \frac{\partial V(x(t))}{\partial x_n} f_n(t,x(t)) \leq 0,$$

for $(t,x) \in [t_0,\infty) \times D$, and $x \neq 0$, then V is called Lyapunov function for the system (8.56).

If the inequality is strict, that is $V'(x(t)) < 0$, then V is said to be a *strict Lyapunov function.*

Example 8.21 *Consider the damped harmonic motion depicted in Fig. 8.14 where m,k are positive constants and $b \geq 0$. If x is the displacement of the mass from its equilibrium position, then applying Newton's second law to all forces acting on the system, we arrive at the second order differential equation*

$$mx'' + bx' + kx = 0.$$

We introduce the transformation $x_1 = x$ and $x_2 = x'$ and arrive at the system

$$x_1' = x_2, \quad x_2' = -\frac{k}{m}x_1 - \frac{b}{m}x_2. \tag{8.54}$$

Clearly, if $b = 0$, then linearization gives no information regarding the stability of the origin. Next, we try to construct a Lyapunov function that overcomes such difficulties. Since we are dealing with a physical system, we expect that the system energy will decrease over time to zero. Thus, a logical choice for the Lyapunov function would be the sum of kinetic energy and spring potential energy. That is

$$V(x) = \frac{1}{2}m(x')^2 + \frac{1}{2}kx^2.$$

Clearly, V is positive definite and

$$V' = mx'x'' + kxx'$$

$$\begin{aligned} &= x'(mx'' + kx) \\ &= x'(-bx') \\ &= -b(x')^2. \end{aligned}$$

Later on, we show that this is enough to demonstrate the origin is stable. We make the following definitions regarding stability. □

Definition 8.8 *Consider the system*

$$x' = f(t,x), \tag{8.55}$$

where $x(t) \in \mathbb{R}^n$, and assume that f is continuous and locally Lipschitz in x.

Let $x(t,t_0,x_0)$ denote the maximal solution of (8.55) *satisfying*

$$x(t_0) = x_0.$$

Let

$$\phi : [t_0,\infty) \to \mathbb{R}^n$$

be a solution of (8.55).

(a) The solution ϕ is called stable *(S) on $[t_0,\infty)$ if for every $\varepsilon > 0$ there exists*

$$\delta = \delta(t_0,\varepsilon) > 0$$

such that

$$|\phi(t_0) - x_0| < \delta$$

implies that the solution $x(t,t_0,x_0)$ exists for all $t \geq t_0$ and

$$|\phi(t) - x(t,t_0,x_0)| < \varepsilon, \qquad t \geq t_0.$$

(b) The solution ϕ is called uniformly stable *(US) if it is stable and the number δ in part (a) depends only on ε and is independent of t_0.*

(c) The solution ϕ is called asymptotically stable *(AS) if it is stable and there exists*

$$\delta_1 = \delta_1(t_0,\varepsilon) > 0$$

such that

$$|\phi(t_0) - x_0| < \delta_1$$

implies

$$\lim_{t\to\infty} |\phi(t) - x(t,t_0,x_0)| = 0.$$

(d) *If ϕ is not stable, then it is called* unstable. *That is, there exists $\varepsilon > 0$ such that for every $\delta > 0$ there exists x_0 satisfying*

$$|\phi(t_0) - x_0| < \delta,$$

but

$$|\phi(t_1) - x(t_1, t_0, x_0)| \geq \varepsilon$$

for some $t_1 \geq t_0$.

(e) *The solution ϕ is called* uniformly asymptotically stable *(UAS) if it is uniformly stable and there exists $\gamma > 0$ such that for every $\mu > 0$, there exists $T = T(\mu) > 0$ satisfying*

$$|\phi(t_0) - x_0| < \gamma$$

implies

$$|x(t, t_0, x_0) - \phi(t)| < \mu, \qquad t \geq t_0 + T.$$

Since any equilibrium solution can be translated to the origin, the previous definitions reduce to the following definitions for the zero solution.

Assume

$$f : [t_0, \infty) \times D \to \mathbb{R}^n,$$

where $D \subset \mathbb{R}^n$ contains the origin, and consider

$$x' = f(t, x), \qquad f(t, 0) = 0. \tag{8.56}$$

Let $x(t, t_0, x_0)$ denote the solution satisfying $x(t_0) = x_0$.

Definition 8.9 *The zero solution $x = 0$ of (8.56) is:*

(a) stable *(S) if for every $\varepsilon > 0$ and $t_0 \geq 0$, there exists $\delta = \delta(t_0, \varepsilon) > 0$ such that*

$$|x_0| < \delta \quad \text{implies} \quad |x(t, t_0, x_0)| < \varepsilon, \qquad t \geq t_0.$$

(b) uniformly stable *(US) if it is stable and δ is independent of t_0.*

(c) unstable *if it is not stable.*

(d) asymptotically stable *(AS) if it is stable and*

$$|x_0| < \delta \quad \text{implies} \quad \lim_{t \to \infty} x(t, t_0, x_0) = 0.$$

(e) uniformly asymptotically stable *(UAS) if it is uniformly stable and there exists $\gamma > 0$ such that for every $\mu > 0$ there exists $T = T(\mu) > 0$ satisfying*

$$|x_0| < \gamma \quad \text{implies} \quad |x(t, t_0, x_0)| < \mu, \qquad t \geq t_0 + T.$$

Thus, $x = 0$ is stable if solutions starting near $x = 0$ do not wander too far from $x = 0$ in future time. Now, $x = 0$ is asymptotically stable, which requires that it is stable, and solutions near x_0 must approach 0 as $t \to \infty$. The next result is useful in constructing positive definite or negative definite functions, and is stated in the following theorem.

Theorem 8.11 *For constants $a, b,$ and c, the function*

$$V(x,y) = ax^2 + bxy + cy^2$$

is positive definite if, and only if,

$$a > 0, \text{ and } 4ac - b^2 > 0,$$

and is negative definite if, and only if,

$$a < 0, \text{ and } 4ac - b^2 > 0.$$

Example 8.22 *Consider the planner system*

$$x' = -x - xy^2$$
$$y' = -y - yx^2.$$

Set

$$V(x,y) = ax^2 + bxy + cy^2.$$

Computing the derivative of V along the solutions of the system, we obtain

$$V'(x,y) = -2[2a(x^2 + x^2y^2) + b(2xy + xy^3 + yx^3) + 2c(y^2 + x^2y^2)].$$

If we choose $b = 0$, and a and c to be any positive numbers, then V' is negative definite and V is positive definite by Theorem 8.11. □

8.8.1 Stability of Autonomous Systems

Let $D \subset \mathbb{R}^n$ be an open set containing the origin, and consider the autonomous system

$$x' = f(x), \qquad x(0) = x_0, \tag{8.57}$$

where $f : D \to \mathbb{R}^n$ is continuous, locally Lipschitz, and satisfies

$$f(0) = 0.$$

Thus, the origin is an equilibrium point of (8.57). Note that Definitions 8.6, 8.7, and 8.9 carry over to the autonomous system (8.57). We make the following definitions regarding an open ball so that we may state the stability notion in terms of such balls.

Definition 8.10 *Let $x^* \in \mathbb{R}^n$ and $r > 0$. Then $B_r(x^*)$ denotes the open ball centered at x^* with radius r. That is*

$$B_r(x^*) = \{x \in \mathbb{R}^n : ||x - x^*|| < r\}.$$

Definition 8.11 *An equilibrium point x^* of* (8.57) *is:*

(a) stable *if for every $\varepsilon > 0$ there exists $\delta > 0$ such that*

$$x_0 \in B_\delta(x^*) \quad \longrightarrow \quad x(t, x_0) \in B_\varepsilon(x^*), \qquad t \geq 0.$$

(b) asymptotically stable *if it is stable and there exists $c > 0$ such that*

$$x_0 \in B_c(x^*) \quad \longrightarrow \quad x(t, x_0) \to x^* \quad \text{as } t \to \infty.$$

We warm up with the following example.

Example 8.23 *The origin $(0,0)$ is the only equilibrium point and is a node for the system*

$$x' = -x,$$
$$y' = -y.$$

We apply Definition 8.11. The solution is easily computed and given by

$$\begin{pmatrix} x(t) \\ y(t) \end{pmatrix} = \begin{pmatrix} c_1 e^{-t} \\ c_2 e^{-t} \end{pmatrix}$$

for $x(0) = c_1$ and $y(0) = c_2$.

For all $\varepsilon > 0$, let $\delta = \varepsilon$. Then for all $c = (c_1, c_2) \in B_\delta(0) = \{(x,y) \in \mathbb{R}^2 : \sqrt{x^2 + y^2} < \delta\}$ and $t \geq 0$, we have

$$||(x(t), y(t))|| = \sqrt{x^2(t) + y^2(t)} = \sqrt{(c_1^2 + c_2^2)e^{-2t}} = \sqrt{c_1^2 + c_2^2}e^{-t} \leq \sqrt{c_1^2 + c_2^2} < \delta = \varepsilon.$$

It follows that

$$(x(t), y(t)) \in B_\varepsilon(0),$$

and hence $(0,0)$ is stable. Moreover,

$$(x(t), y(t)) \to (0,0) \quad \text{as } t \to \infty,$$

which shows that $(0,0)$ is asymptotically stable. □

Theorem 8.12 *Suppose there exists a continuously differentiable function*

$$V : D \to [0, \infty)$$

such that

$$V(0) = 0, \qquad V(x) > 0 \text{ for } x \neq 0,$$

and

$$V'(x) \leq 0$$

along solutions of (8.57)*. Then the zero solution of* (8.57) *is stable.*

Proof *Pick an* $\varepsilon > 0$ *so that the sphere* $B_\varepsilon(0)$ *is in* D. *Let* S_ε *be the boundary of* $B_\varepsilon(0)$. *Due to the continuity of* V *at the origin and* $V(0) = 0$, *there must be a point on* S_ε *where* V *attains its minimum that we denote by* $m = \min_{x \in S_\varepsilon} V(x)$, *where* $m > 0$. *Let* $B_\delta(0)$ *be another sphere in* D. *Since* $V(0) = 0$ *and* V *is continuous, we can choose* $\delta > 0$ *such that* $V(x) < m$ *for all* $x \in B_\delta(0)$. *Our goal is to show that if* $x_0 \in B_\delta(0)$, *then* $x(t,0,x_0) \in B_\varepsilon(0)$ *for all* $t > 0$. *Since* $V' \leq 0$, *then* $V(x_0) < m$ *implies that* $V(x(t,0,x_0)) < m$ *for all* $t > 0$, *since* V *is nonincreasing along trajectories, with the initial condition* x_0. *This, in turn, implies that* $x(t,0,x_0)$ *cannot cross* S_ε *since* $V(x) \geq m$ *for all* $x \in S_\varepsilon$. *This implies the origin is stable. This completes the proof.*

Example 8.24 *Consider the planar system*

$$x' = -y + xy$$
$$y' = x - x^2.$$

Clearly, $(0,0)$ *is an equilibrium point of the system. The linearization theorem, Theorem 8.5 is inconclusive since* $(0,0)$ *is a center for the linear part*

$$x' = -y$$
$$y' = x.$$

Moreover, using the method of polar coordinates gives no information about the stability of the origin. Let's try to construct a Lyapunov function. Let a *and* b *be positive constants and set*

$$V(x,y) = ax^2 + by^2,$$

where $a, b > 0$ *are to be determined. It is clear that* $V(0,0) = 0$ *and* $V(x,y) > 0$ *for* $(x,y) \neq (0,0)$.

Computing the derivative of V *along the solutions of the system using* (8.53) *we obtain*

$$\begin{aligned} V'(x,y) &= 2ax' + 2by' \\ &= 2ax(-y + xy) + 2by(x - x^2) \\ &= 0 \end{aligned}$$

by setting $a = b > 0$. *The simplest choice would be* $a = \frac{1}{2} = b$. *In this case*

$$V(x,y) = \frac{1}{2}(x^2 + y^2).$$

Thus, by Theorem 8.12 the origin is stable. □

Here is another example.

Example 8.25 *Consider the three dimensional system*

$$x' = -2y + yz$$
$$y' = x - xz$$
$$z' = xy$$

Clearly $(0,0,0)$ *is an equilibrium solution of the system and*

$$J(0,0,0) = \begin{pmatrix} 0 & -2 & 0 \\ 1 & 0 & 0 \\ 0 & 0 & 0 \end{pmatrix}.$$

Thus, $J(0,0,0)$ *has the eigenvalues* $\lambda_1 = 0$, $\lambda_2 = i\sqrt{2}$, $\lambda_3 = -i\sqrt{2}$. *The linearization theorem, Theorem 8.5 cannot be applied. Let's try to construct a Lyapunov function. Let* a, b, *and* c *be positive constants and set*

$$V(x,y,z) = ax^2 + by^2 + cz^2,$$

where $a,b,c > 0$ *are to be determined. It is clear that* $V(0,0) = 0$ *and* $V(x,y,z) > 0$ *for* $(x,y,z) \neq (0,0,0)$. *Computing the derivative of* V *along the solutions of the system using* (8.53), *we obtain*

$$\begin{aligned} V'(x,y) &= 2axx' + 2byy' + 2czz' \\ &= 2ax(-2y + yz) + 2by(x - xz) + 2cz(xy) \\ &= 2[(a - b + c)xyz + (-2a + b)xy]. \end{aligned}$$

Hence, if $b = 2a$ *and* $c = a > 0$, *we have* $V'(x,y,z) = 0$ *for all* $(x,y,z) \in \mathbb{R}^3$ *and therefore by Theorem 8.12, the origin* $(0,0,0)$ *is stable. Furthermore, choosing* $a = c = 1$, *and* $b = 2$, *we see that the trajectories of this system lie on the ellipsoids*

$$x^2 + 2y^2 + z^2 = k^2$$

for a positive constant k. □

Example 8.24 reveals the advantage of the use of Lyapunov function over linearization since it shows that a center for a linear system is stable for the accompanying nonlinear system.

The next theorem provides a criterion for asymptotic stability.

Theorem 8.13 *Suppose there exists a strict Lyapunov function*

$$V : D \to [0, \infty)$$

such that

$$V(0) = 0, \qquad V(x) > 0, \qquad V'(x) < 0 \quad \text{for } x \neq 0.$$

Then the zero solution of (8.57) *is asymptotically stable.*

Proof *By Theorem 8.12, the origin is stable. Since*

$$V'(x(t)) < 0$$

for all nontrivial solutions, the function $V(x(t))$ *is strictly decreasing along trajectories. Moreover,*

$$V(x(t)) \geq 0,$$

so $V(x(t))$ converges to a limit $c \geq 0$ as $t \to \infty$.

We claim that $c = 0$. Suppose, to the contrary, that $c > 0$. Choose $\varepsilon > 0$ so that $B_\varepsilon(0) \subset D$, and let

$$m = \min\{-V'(x) : c \leq V(x) \leq V(x_0)\}.$$

Since $V'(x) < 0$ for $x \neq 0$, we have $m > 0$. Hence

$$V'(x(t)) \leq -m$$

along the trajectory, and integrating from 0 to t gives

$$V(x(t)) - V(x_0) \leq -mt.$$

Thus $V(x(t)) \to -\infty$ as $t \to \infty$, contradicting $V(x(t)) \geq 0$. Therefore $c = 0$.

Finally, suppose $x(t) \not\to 0$. Then there exist $\alpha > 0$ and a sequence $t_k \to \infty$ such that

$$|x(t_k)| \geq \alpha.$$

By continuity and positive definiteness of V, there exists $\eta > 0$ such that

$$|x| \geq \alpha \quad \Longrightarrow \quad V(x) \geq \eta.$$

Hence

$$V(x(t_k)) \geq \eta > 0,$$

contradicting the fact that

$$V(x(t)) \to 0.$$

Therefore,

$$x(t) \to 0 \qquad \text{as } t \to \infty.$$

Thus, the origin is asymptotically stable.

Example 8.26 *Consider the planar system*

$$\begin{aligned} x' &= -x - xy^2 \\ y' &= -y + 3x^2 y. \end{aligned}$$

Clearly, $(0,0)$ is an equilibrium point of the system. The linearization theorem, Theorem 8.5, tells us that $(0,0)$ is (AS). We verify the result by constructing a strict Lyapunov function. Let a and b be positive constants and set

$$V(x,y) = ax^2 + by^2,$$

where a and b are to be precisely determined. Clearly $V(0,0) = 0$ and $V(x,y) > 0$ for $(x,y) \neq (0,0)$. Computing the derivative of V along the solutions of the system using (8.53), we obtain

$$V'(x,y) \quad = \quad 2axx' + 2byy'$$

$$\begin{aligned} &= 2ax(-x-xy^2)+2by(-y+3x^2y) \\ &= -2ax^2-2ax^2y^2-2by^2+6bx^2y^2 \\ &= -(6x^2+2y^2)<0 \end{aligned}$$

unless $x=y=0$, *where we choose* $a=3, b=1$ *in order to eliminate the mixed terms. Thus, by Theorem 8.13, the origin is asymptotically stable.* □

Note that the origin $(0,0)$ of the system in Example 8.22 is asymptotically stable by Theorem 8.13.

First integrals are useful because they identify quantities that remain constant along solution trajectories. They provide geometric information about the phase plane and are often used to study stability, boundedness, and invariant sets.

Definition 8.12 *A first integral function of an autonomous system is a scalar function that remains constant along every solution trajectory. For a planar autonomous system*

$$x' = P(x,y), \qquad y' = Q(x,y),$$

a differentiable function $H(x,y)$ *is called a first integral if, for every solution* $(x(t),y(t))$,

$$\frac{d}{dt}H(x(t),y(t)) = 0.$$

Equivalently,

$$H_x(x,y)P(x,y)+H_y(x,y)Q(x,y)=0$$

for all (x,y) *in the region of interest. Thus,* H *is constant along trajectories, so solutions remain on level curves*

$$H(x,y)=C.$$

For example, for the system

$$x' = y, \qquad y' = -x,$$

the function

$$H(x,y) = x^2+y^2$$

is a first integral because

$$\frac{d}{dt}(x^2+y^2) = 2xx'+2yy' = 2xy+2y(-x) = 0.$$

Hence, the trajectories are circles that satisfy $x^2+y^2=C$.

Example 8.27 *Recall that in Example 8.12, linearization was inconclusive regarding the stability of the equilibrium point*

$$x^* = \left(\frac{c}{d}, \frac{a}{b}\right).$$

In this example, we show that a first integral *establishes the stability of* x^*.

Consider the Lotka–Volterra system

$$\begin{cases} x' = ax - bxy, \\ y' = -cy + dxy, \end{cases} \qquad a,b,c,d > 0,$$

which has the coexistence equilibrium

$$x^* = \frac{c}{d}, \qquad y^* = \frac{a}{b}.$$

The Jacobian matrix at (x^*, y^*) *is*

$$J(x^*,y^*) = \begin{pmatrix} 0 & -\frac{bc}{d} \\ \frac{ad}{b} & 0 \end{pmatrix},$$

with eigenvalues

$$\lambda = \pm i\sqrt{ac}.$$

Thus, linearization is inconclusive.

Define

$$H(x,y) = dx - c\ln x + by - a\ln y, \qquad x > 0,\ y > 0.$$

Then along solutions,

$$\frac{d}{dt}H(x(t),y(t)) = \left(d - \frac{c}{x}\right)x' + \left(b - \frac{a}{y}\right)y'.$$

Substituting the system shows that all terms cancel, hence

$$\frac{d}{dt}H(x(t),y(t)) = 0.$$

Thus, H is a first integral, and trajectories lie on level sets

$$H(x,y) = constant.$$

At the equilibrium,

$$H_x(x^*,y^*) = 0, \qquad H_y(x^*,y^*) = 0,$$

so (x^*,y^*) *is a critical point. Moreover,*

$$H_{xx} = \frac{c}{x^2} > 0, \qquad H_{yy} = \frac{a}{y^2} > 0, \qquad H_{xy} = 0.$$

Hence, the Hessian matrix

$$D^2H(x,y) = \begin{pmatrix} H_{xx} & 0 \\ 0 & H_{yy} \end{pmatrix}$$

is positive definite in the positive quadrant. Therefore, (x^*, y^*) *is a strict local minimum of* H.

It follows that, for values of H *close to* $H(x^*, y^*)$*, the level sets*

$$H(x,y) = constant$$

are closed curves surrounding (x^*, y^*)*. Since trajectories remain on these invariant level sets, any solution starting sufficiently close to* (x^*, y^*) *remains close for all time.*

Thus, the equilibrium

$$\left(\frac{c}{d}, \frac{a}{b}\right)$$

is Lyapunov stable.

It is not asymptotically stable, since solutions remain on level curves of H *and do not converge to the equilibrium unless they start there.* □

Suppose $V(x)$ is a strict Lyapunov function in D. Then the *domain of attraction* of an equilibrium consists of all points such that a solution starting at them tends to the equilibrium. In engineering applications, it is often important to get an estimate of the size of the domain of attraction. Thus, if V is a strict Lyapunov function on a set $D \in \mathbb{R}^n$ then the origin is asymptotically stable and the *domain of attraction*

$$\bar{N} = \{x \in D : V(x) \leq C\}$$

for some positive constant C.

Example 8.28 *Consider the planar system*

$$\begin{aligned} x' &= -x - y, \\ y' &= 2x - y + y^3. \end{aligned}$$

Clearly, $(0,0)$ *is an equilibrium point of the system. Set*

$$V(x,y) = x^2 + \frac{y^2}{2}.$$

Clearly $V(0,0) = 0$ *and* $V(x,y) > 0$ *for* $(x,y) \neq (0,0)$*. Computing the derivative of* V *along the solutions of the system we obtain*

$$\begin{aligned} V'(x,y) &= 2xx' + yy' \\ &= 2x(-x-y) + y(2x - y + y^3) \\ &= -2x^2 - 2xy + 2xy - y^2 + y^4 \\ &= -2x^2 \quad y^2 + y^4. \end{aligned}$$

To keep $V'(x,y) < 0$*, we must restrict* y *so that*

$$-y^2 + y^4 = y^2(-1 + y^2) < 0,$$

that is, $-1 < y < 1$. *On the other hand, to keep the ellipses*

$$V(x,y) = x^2 + \frac{y^2}{2} = C$$

inside $D = \{(x,y) : x \in \mathbb{R},\ -1 < y < 1\}$, *we must take* $0 < C < \frac{1}{2}$. *Thus, for any such value of* C, *the origin is asymptotically stable, and a Lyapunov sublevel set contained in the domain of attraction is*

$$\bar{N} = \{(x,y) \in D : V(x,y) \le C < \tfrac{1}{2}\}.$$

□

The next theorem only asks for the existence of a Lyapunov function to obtain asymptotic stability, but with a catch.

Theorem 8.14 *Let* $V(x)$ *be a Lyapunov function in D. Let*

$$Z = \{x : V'(x) = 0, x \in D\}.$$

Suppose that the only bounded solution of (8.57) *that remains inside Z for all time is the equilibrium solution* 0. *Then the origin of* (8.57) *is asymptotically stable.*

Proof *We will omit the proof of Theorem 8.14.*

As an application of Theorem 8.14 we provide the following example.

Example 8.29 *Consider the planar system*

$$x' = -x - 2y$$
$$y' = 3x - 2y.$$

Clearly, $(0,0)$ *is an equilibrium point of the system. Set*

$$V(x,y) = \frac{x^2}{2} + y^2.$$

Clearly $V(0,0) = 0$ *and* $V(x,y) > 0$ *for* $(x,y) \neq (0,0)$. *Computing the derivative of* V *along the solutions of the system, we obtain*

$$\begin{aligned} V'(x,y) &= xx' + 2yy' \\ &= x(-x-2y) + y(3x-2y) \\ &= -(x-2y)^2. \end{aligned}$$

It is clear that $V'(x,y) = 0$, *for* $x = 2y$ *for all* x *and* y. *Thus,* $V(x,y)$ *is a Lyapunov function (not strict) on the set*

$$Z = \{(x,y) \in \mathbb{R}^2 : x = 2y\}.$$

Let (x,y) be a bounded solution (trajectory) in Z. That is, there exists a positive constant K such that $|x|,\ |y| \leq K$. Substitute $x = 2y$ in the original system to get $x' = -2x$ and $y' = 4y$. By a straight forward calculations we obtain the solutions,

$$x(t) = x(0)e^{-2t} \text{ and } y(t) = y(0)e^{4t},$$

respectively, for $t \in \mathbb{R}$. These functions cannot possibly satisfy $|x|,\ |y| \leq K$, unless $x(0) = y(0) = 0$. To have you more convinced, the solutions must also satisfy $x = 2y$, or

$$x(0)e^{-2t} = 2y(0)e^{4t},$$

which can only holds if $x(0) = y(0) = 0$. Thus, the only bounded solution in Z is the origin and hence by Theorem 8.14, the origin is asymptotically stable. □

Lyapunov-type functions can also be used to detect limit cycles, as the next example demonstrates.

Example 8.30 *In Example 8.18, we considered the planar autonomous system*

$$x' = -y - x(x^2+y^2-\mu), \qquad y' = x - y(x^2+y^2-\mu),$$

with $\mu > 0$, and used polar coordinates to show the existence of a stable limit cycle. Here, we use a Lyapunov-type function to arrive at the same result.

Let

$$V(x,y) = x^2+y^2.$$

Then along solutions of the system,

$$\begin{aligned} V'(t) &= 2xx' + 2yy' \\ &= 2x[-y - x(x^2+y^2-\mu)] + 2y[x - y(x^2+y^2-\mu)] \\ &= -2(x^2+y^2-\mu)(x^2+y^2) \\ &= 2(\mu - V(x,y))V(x,y). \end{aligned}$$

Since $V > 0$ away from the origin, the sign of V' depends on $\mu - V(x,y)$.

We have $V' = 0$ when $V = \mu$, that is,

$$x^2+y^2 = \mu,$$

which is a circle centered at the origin with radius $\sqrt{\mu}$.

If $0 < V < \mu$, then $V' > 0$, so $V(t)$ is increasing and trajectories move outward toward the circle.

If $V > \mu$, then $V' < 0$, so $V(t)$ is decreasing, and trajectories move inward toward the circle.

On the circle $x^2+y^2 = \mu$, we have $V' = 0$, so it is invariant. By the uniqueness of solutions, trajectories cannot cross this circle.

Therefore, all nearby trajectories approach the circle $x^2+y^2 = \mu$, and we conclude that it is a stable limit cycle. □

8.8.2 Instability

Lyapunov functions are also powerful tools for proving that an equilibrium point is *unstable*. In this context, one seeks a continuously differentiable scalar function V whose behavior near the equilibrium contradicts the conditions required for stability. For example, if V is chosen so that it attains a local *maximum* at the equilibrium and its derivative $\dot{V}$ along trajectories is nonpositive but not identically zero, then any sufficiently small perturbation must drive the state away from the equilibrium, since V cannot increase to remain near its maximal value. Alternatively, if V is negative definite near the equilibrium while $\dot{V}$ is positive definite, trajectories are forced to move in directions where V increases, again implying that solutions cannot stay arbitrarily close to the equilibrium. In both cases, Lyapunov's direct method provides a coordinate-free, non-perturbative argument that establishes instability without explicitly solving the system. The next theorem provides a criterion for the instability of the zero solution of (8.57).

Theorem 8.15 *Let $D \subset \mathbb{R}^n$ be an open set containing the origin, and let*

$$V : D \to [0, \infty)$$

be continuous on D, continuously differentiable on $D \setminus \{0\}$, and positive definite. If

$$V'(x) > 0, \qquad x \neq 0,$$

where V' denotes the derivative of V along solutions of (8.57)*, then the origin of* (8.57) *is unstable.*

Proof *Suppose, to the contrary, that the origin is stable. Then, for $\varepsilon = 1$ and $t_0 = 0$ there exists $\delta > 0$ such that $|x_0| < \delta$ implies that $|x(t,0,x_0)| \leq 1$ for all $t \geq 0$. Hence for $x(t) = x(t,0,x_0)$, $V(x(t))$ is bounded for $t \geq 0$ whenever $|x_0| \leq \delta$.*

Choose x_0 with $0 < |x_0| < \delta$. Then $V(x_0) > 0$. Since $V'(x) > 0$ for $x \neq 0$, V is strictly increasing along any nontrivial solution. Therefore,

$$V(x(t)) \geq V(x_0) > 0, \quad t \geq 0.$$

By continuity of V and $V(0) = 0$, there exists $p \in (0,1)$ such that

$$|x| < p \quad \text{implies} \quad V(x) < \tfrac{1}{2}V(x_0).$$

Because $V(x(t)) \geq V(x_0)$ for all $t \geq 0$, the trajectory can never enter the ball $\{x : |x| < p\}$, so for all $t \geq 0$ we have

$$p \leq |x(t)| \leq 1.$$

On the closed and bounded set

$$A := \{x \in D : p \leq |x| \leq 1\},$$

V' is continuous and positive, so

$$l := \min_{x \in A} V'(x) > 0.$$

Thus, along the solution,

$$\frac{d}{dt}V(x(t)) = V'(x(t)) \geq l, \quad t \geq 0,$$

and integrating from 0 to t yields

$$V(x(t)) - V(x_0) = \int_0^t V'(x(s))\,ds \geq \int_0^t l\,ds = lt \underset{t\to\infty}{\longrightarrow} \infty.$$

Since $|x(t)| \leq 1$ for all $t \geq 0$ and V is continuous, $V(x(t))$ must remain bounded. This contradicts

$$V(x(t)) \to \infty.$$

It follows that the zero solution $x(t) = 0$ is unstable.

Example 8.31 *Consider the planar system*

$$x' = 2xy + x^3$$
$$y' = -x^2 + y^5.$$

Clearly, $(0,0)$ is an equilibrium point of the system. Linearization will not work here since there are no linear terms at $(0,0)$. Set

$$V(x,y) = x^2 + 2y^2.$$

Clearly $V(0,0) = 0$ and $V(x,y) > 0$ for $(x,y) \neq (0,0)$. Computing the derivative of V along the solutions of the system we obtain

$$\begin{aligned} V'(x,y) &= 2xx' + 4yy' \\ &= 2x(2xy + x^3) + 4y(-x^2 + y^5) \\ &= 2x^4 + 4y^6 > 0, \ \ (x,y) \neq (0,0). \end{aligned}$$

It follows from Theorem 8.15 that the origin is unstable.

□

Example 8.32 *We examine the stability and asymptotic stability for different values of a for the planner system*

$$x' = ax(x^2 + y^2) - xy^2$$
$$y' = ay(x^2 + y^2) + x^2 y.$$

Clearly $(0,0)$ is an equilibrium point of the system. Linearization will not work here since there are no linear terms in the neighborhood of $(0,0)$. Set

$$V(x,y) = x^2 + y^2.$$

Clearly $V(0,0)=0$ *and* $V(x,y)>0$ *for* $(x,y)\neq(0,0)$. *Set* $V(t):=V(x(t),y(t))$. *Then, along the solutions of the system we have that*

$$\begin{aligned} V'(t) &= 2xx'+2yy' \\ &= 2x[ax(x^2+y^2)-xy^2]+2y[ay(x^2+y^2)+x^2y] \\ &= 2a(x^2+y^2)^2. \end{aligned} \tag{8.58}$$

It is clear from relation (8.58) *that the origin is stable if* $a=0$, *asymptotically stable if* $a<0$, *and unstable if* $a>0$. *In the following, we will use inequality* (8.58) *and try to understand the behavior of the Lyapunov function along the solutions.*

We observe that (8.58) *is equivalent to*

$$V'(t)=2aV^2(t),$$

which is a separable first-order differential equation in $V(t)$. *Thus, separating the variables and then integrating from* 0 *to* t *gives*

$$V(t)=\frac{V(0)}{1-2aV(0)t},\ t\geq 0. \tag{8.59}$$

Case 1. *If* $a=0$, *then from* (8.59) *we have that* $V(t)=V(0)>0$. *In this case, we have a periodic solution surrounding the origin*

$$x^2+y^2=k^2,$$

for a nonzero constant k. *Hence, the origin is stable.*
Case 2. *If* $a<0$, *then from* (8.59) *we have that* $V(t)\to 0$, *as* $t\to\infty$. *That is* $x^2+y^2\to 0$, *as* $t\to\infty$, *which means solutions spiral toward the origin, and we conclude the origin is asymptotically stable.*
Case 3. *If* $a>0$, *then from* (8.59) $V(t)$ *is not defined for all times* $t\geq 0$. *In fact* $V(t)\to\infty$ *as* $t\to\frac{1}{2aV(0)}$ *from the left. We conclude that the origin is unstable.* □

The next theorem provides a criterion for instability of the zero solution that differs from Theorem 8.15.

Theorem 8.16 *Let* x^* *be an equilibrium solution of* (8.57). *Let* U *be a neighborhood of* x^*. *Let* $V:U\to\mathbb{R}$ *be a continuously differentiable function on* $U\setminus\{x^*\}$. *Assume that*

- $V(x^*)=0$ *and for every neighborhood of* x^*, *there exists a point* x *such that* $V(x)>0$, *and*
- $V'(x)>0$ *for all* $x\in U\setminus\{x^*\}$,

where $V'(x)$ *denotes the derivative of* V *along solutions of* (8.57). *Then* x^* *is unstable for* (8.57).

Proof *Without loss of generality, we take* $x^*=0$. *Let* $B_\beta(0)\subseteq U$, $\beta>0$. *Because* V *takes positive values arbitrarily close to* 0, *we may choose a point* $x_0\in B_\beta(0)\setminus\{0\}$

such that $V(x_0) > 0$. Let (a,b) be the interval of existence of the solution $x(t) = x(t,x_0)$ with $x(0) = x_0$.

To prove instability, it is enough to show that this solution eventually leaves the closed ball $\bar{B}_\beta(0)$. Assume the contrary, i.e., that

$$x(t) \in B_\beta(0) \quad \textit{for all } t \geq 0.$$

Since $\bar{B}_\beta(0)$ is compact and V is continuous on $\bar{B}_\beta(0)$, V attains its maximum on $\bar{B}_\beta(0)$; denote

$$V_0 := \max_{x \in \bar{B}_\beta(0)} V(x).$$

Thus

$$V(x(t)) \leq V_0, \quad t \geq 0. \tag{8.60}$$

Because $V'(x) > 0$ for all $x \in U \setminus \{0\}$, V is strictly increasing along any nontrivial solution that remains in U. In particular,

$$V(x(t)) > V(x_0) > 0, \quad t > 0.$$

Moreover, V' is continuous on the compact annulus

$$A := \{x \in \mathbb{R}^n : \rho \leq \|x\| \leq \beta\} \subset U$$

for any $0 < \rho < \beta$. Since $V'(x) > 0$ for $x \neq 0$, we can choose $0 < \rho < \beta$ so small that

$$V(x) < V(x_0) \quad \textit{for all } x \in B_\rho(0),$$

so that the trajectory cannot enter $B_\rho(0)$ once $V(x(t)) \geq V(x_0)$. Hence, for all $t \geq 0$ such that the solution remains in $B_\beta(0)$, we have

$$\rho \leq \|x(t)\| \leq \beta,$$

i.e., $x(t) \in A$.

By continuity of V' and compactness of A, the minimum

$$m := \min\{V'(x) : x \in A\}$$

exists. Since $V'(x) > 0$ on A, we have $m > 0$. For all $t \geq 0$ with $x(t) \in A$ we then have

$$V'(x(t)) \geq m.$$

Integrating from 0 to t yields

$$V(x(t,x_0)) - V(x_0) = \int_0^t V'(x(s,x_0))\, ds \geq mt,$$

or

$$V(x(t,x_0)) \geq V(x_0) + mt.$$

As $t \to \infty$, the right-hand side tends to $+\infty$, contradicting (8.60). Thus, our assumption that the solution $x(t)$ stays in $B_\beta(0)$ for all $t \geq 0$ is false, and the solution must leave $B_\beta(0)$ in finite time.

Therefore, for every neighborhood of the origin, there exist initial data arbitrarily close to 0 for which the corresponding solution eventually leaves that neighborhood, and hence the origin is unstable. This completes the proof.

The next example shows the importance of Theorem 8.16 and illustrates its usage, as Theorem 8.15 is not applicable.

Example 8.33 *Consider the planar autonomous system*

$$x' = 3x + y^2, \quad y' = -2y + x^3.$$

Clearly, the origin is the only equilibrium point of the system. Linearization already shows that the origin is unstable. We now verify this conclusion using Theorem 8.16.

We consider the Lyapunov function

$$V(x,y) = \frac{x^2}{2} - \frac{y^2}{2}.$$

There are points in the neighborhood of the origin where V is positive, for example, points on the x-axis. Since we are only interested in what happens in the neighborhood of $(0,0)$, we limit ourselves to $|x| < 1$ and $|y| < 1$. To better understand the derivative of V, we will utilize the following inequality: for any two real numbers a and b,

$$0 < (a-b)^2 = a^2 - 2ab + b^2,$$

from which it follows that

$$ab \leq \frac{a^2}{2} + \frac{b^2}{2}.$$

It readily follows that along the solutions of the system

$$\begin{aligned} V'(x,y) &= xx' - yy' \\ &= x(3x + y^2) - y(-2y + x^3) \\ &= 3x^2 + xy^2 + 2y^2 - x^3y \\ &= (3x^2 + 2y^2) + (xy^2 - x^3y). \end{aligned} \tag{8.61}$$

Next we compare the term $(3x^2 + 2y^2)$ with $(xy^2 - x^3y)$.

$$\begin{aligned} |xy^2 - x^3y| &= |xy||y - x^2| \\ &\leq \frac{1}{2}(x^2 + (y - x^2)^2) \quad \textit{(by the inequality above)} \\ &\leq \frac{1}{2}(x^2 + 2(y^2 + x^4)) \end{aligned}$$

$$= \frac{1}{2}(x^2+2y^2+2x^4).$$

For $|x|<1$ *we have* $x^4 \le x^2$*, hence*

$$|xy^2 - x^3y| \le \frac{1}{2}(3x^2+2y^2) < 3x^2+2y^2.$$

Thus, from (8.61),

$$V'(x,y) = (3x^2+2y^2)+(xy^2-x^3y) \ge (3x^2+2y^2) - |xy^2-x^3y| > 0$$

for $(x,y) \ne (0,0)$ *with* $|x|<1,\ |y|<1$.

We conclude that $V' > 0$ *on the set* $\{(x,y): |x|<1,\ |y|<1\}$ *minus the origin, and by Theorem 8.16, the origin is unstable. Note that the set* U *of Theorem 8.16 is defined by*

$$U = \{(x,y) \in \mathbb{R}^2 : |x|<1,\ |y|<1\}.$$

□

8.8.3 Exercises

Exercise 8.32 *Use Lyapunov function to investigate the stability of the origin for each of the following systems and compare with the linearization method:*
(a) $x' = y,\ y' = -x.$
(b) $x' = -xy^2,\ y' = 3yx^2.$
(c) $x' = y-2x,\ y' = 2x-y-x^3.$
(d) $x' = x+3y,\ y' = 2x$
(e) $x' = x^3+y,\ y' = x+y^3.$
(f) $x' = y,\ y' = -x-y-x^3.$

Exercise 8.33 *Find the two equilibrium solutions and use Lyapunov functions to investigate their stability.*

$$x' = y - x^2,\ y' = x - y^2.$$

Exercise 8.34 *Consider the planar system*

$$x' = -xy^2$$
$$y' = -y - yx^2.$$

Examine the stability of the origin using:
(a) Theorem 8.5.
(b) Lyapunov function.

Exercise 8.35 *(Lyapunov function for a nonlinear scalar system) Consider the scalar nonlinear system*

$$x' = -x + x^3.$$

(a) *Determine all equilibrium solutions and their stability via linearization.*

(b) *Propose a Lyapunov function of the form*

$$V(x) = \frac{1}{2}x^2 + \alpha x^4,$$

where $\alpha > 0$ *is a constant to be chosen. Compute* $\dot{V}(x)$ *along solutions of the system.*

(c) *Show that for a suitable choice of* α, *one has* $\dot{V}(x) \leq 0$ *in a neighborhood of the origin and use this to conclude that* $x^* = 0$ *is locally stable.*

(d) *Compare the Lyapunov-based conclusion with the result from linearization and discuss any additional information provided by* $V(x)$ *about the size of the domain of attraction.*

Exercise 8.36 *Consider the planar system*

$$x' = ax(x+y) - xy^3$$
$$y' = ay(x+y) + x^2y^2.$$

Examine the stability of the origin for different values of a using:
(a) *Lyapunov function.*
(b) *Linearization method.*

Exercise 8.37 *Put the differential equation*

$$x'' + (x')^2 + x' + x(x-2) = 0,$$

into a system and find its two equilibrium solutions, and examine their stability using:
(a) *Lyapunov function.*
(b) *Linearization method.*

Exercise 8.38 *Use the indicated Lyapunov function V and Theorem 8.14 to show that the origin is asymptotically stable. Show that all solutions tend to the origin as* $t \to \infty$.
(a) $x' = 2y,\ y' = -3x - 3y^3,\ V = 3x^2 + 2y^2.$
(b) $x' = -x + 6y,\ y' = -x - 2y,\ V = x^2 + 2y^2.$
(c) $x' = -2x - y,\ y' = 6x - y,\ V = 2x^2 + y^2.$
(d) $x' = y,\ y' = -x^2y - x^3,\ V = x^4 + 2y^2.$

Exercise 8.39 *Study the stability of the origin for the planar system*

$$x' = x(y-1)$$
$$y' = -\frac{x^2}{1+x^2}.$$

by considering the Lyapunov function $V(x,y) = Ln(1+x^2) + y^2.$

Exercise 8.40 *Use Lyapunov function to show the existence of a stable limit cycle for the system*

$$x' = y + x(\mu^2 - x^2 - y^2), \;\; y' = -x + y(\mu^2 - x^2 - y^2)$$

for $\mu \neq 0$.

Exercise 8.41 *Consider the planar system*

$$x' = x + g_1(x,y), \;\; y' = -y + g_2(x,y),$$

where g_1 *and* g_2 *are continuous on a subset D of* $\mathbb{R}^2$ *that contains* $(0,0)$. *Suppose for a positive constant k,*

$$||g_i|| \leq k(x^2 + y^2), i = 1,2,$$

with $g_1(0,0) = g_2(0,0) = 0$. *Use Theorem 8.16 to show the origin is unstable.*

Exercise 8.42 *Consider the second-order differential equation*

$$x'' + g(x) = 0,$$

where g is continuously differentiable, $g(0) = 0$ *and* $xg(x) > 0$. *Write the equation into a system and use a Lyapunov function to show its zero solution is stable.*

Exercise 8.43 *Consider the second-order differential equation*

$$x'' + f(x)x' + g(x) = 0,$$

where $f(x)$ *and* $g(x)$ *are even and odd polynomials, respectively.*
(a) Show that the equation is equivalent to the system

$$x' = y - F(x), \;\; y' = g(x),$$

where $F(x) = \int_0^x f(s)ds.$

(b) Let $G(x) = \int_0^x g(s)ds$ *and suppose there are positive constants* α *and* β *such that* $g(x)F(x) > 0$ *for* $0 \leq |x| < \alpha$, *and* $G(x) < \beta$ *for* $|x| < \alpha$.
Show that

$$V(x,y) = \frac{y^2}{2} + G(x)$$

is a Lyapunov function in the region $A = \{\alpha, \beta \in \mathbb{R} : |x| < \alpha, y^2 < \beta\}$, *and that the origin is asymptotically stable.*

Exercise 8.44 *Consider the predator–prey model with carrying capacity*

$$x' = x[a - dx - by], \;\; y' = y[-c + kx],$$

where $x(t)$ *and* $y(t)$ *are the numbers of prey and predators at time t, respectively, with* a,b,c,d, *and k being positive constants.*

(a) Find the equilibrium point (x^*, y^*) that is located in the first quadrant.
(b) Use the transformation

$$u = Ln(\frac{x}{x^*}) \text{ and } v = Ln(\frac{y}{y^*})$$

to obtain a system in (u, v).
(c) Show that

$$V(u,v) = kx^{(}e^u - u - 1) + by^{(}e^v - v - 1)$$

is a Lyapunov function, and that the equilibrium solution $(x^{\cdot}y^{)}$ is stable.

Exercise 8.45 *Consider the second-order differential equation*

$$x'' + (\mu + \nu x^2)x' + x + x^3 = 0$$

(a) If $\nu = 0$, and $\mu > 0$, show that the zero solution is asymptotically stable.
(b) If $\mu = 0$, and $\nu > 0$, show that the zero solution is stable using an appropriate Lyapunov function.

Exercise 8.46 *Consider the planar system*

$$x' = (x - by)(x^2 + y^2 - 1)$$
$$y' = (ax + y)(x^2 + y^2 - 1).$$

Show that the origin is stable if $a, b > 0$.

Exercise 8.47 *Show the origin is unstable for the planar system*

$$x' = x + y$$
$$y' = x - y + xy.$$

Exercise 8.48 *Suppose f is continuously differentiable, and consider the planar system*

$$x' = y - xf(x,y)$$
$$y' = -x - yf(x,y).$$

Show that the origin is asymptotically stable if $f(x,y) > 0$ and unstable of $f(x,y) < 0$.

Exercise 8.49 *Let $V = x^2 + y^2$ and consider the system*

$$x' = (x - y)(x^2 + y^2 - 1)$$
$$y' = (x + y)(x^2 + y^2 - 1)$$

Show that the origin is asymptotically stable and find the region of attraction.

Exercise 8.50 *Let $V = x^2 + y^2$ and consider the system*

$$x' = -x + 2x^2 + y^2$$
$$y' = -y + y^2$$

Show that the origin is asymptotically stable and find the region of attraction.

Exercise 8.51 *Consider*

$$y'' + By' + y + y^3 = 0, \quad B \geq 0$$

which is a special case of mass-spring nonlinear damper.
(a) Use the transformation $v = y'$ and obtain a corresponding system in y and v.
(b) Show

$$V(y, v) = \frac{y^2 + y^4}{2} + \frac{v^2}{2}$$

is a Lyapunov function .
(c) Decide whether the origin is stable or asymptotically stable.

Appendices

A

Quick Review of Ordinary Differential Equations

1. Introduction and General Definition

An **ordinary differential equation (ODE)** is an equation involving an unknown function y of one variable (say t or x) and its derivatives. A **general nth–order ODE** has the form

$$F\left(t, y, y', y'', \dots, y^{(n)}\right) = 0.$$

If the highest derivative that appears is $y^{(n)}$, we say the ODE is of **order** n.

A **solution** of an ODE on an interval I is a function $y(t)$ that has as many derivatives as required, and when substituted into the ODE, satisfies it for all $t \in I$.

Example A.1 *The function $y(t) = Ce^t$ satisfies the equation*

$$y' - y = 0,$$

since substituting gives $y' - y = Ce^t - Ce^t = 0$. □

2. Initial Value Problem (IVP)

An **initial value problem** specifies the value of the function and possibly some of its derivatives at a point t_0. For an nth–order ODE:

$$F\left(t, y, y', \dots, y^{(n)}\right) = 0, \quad y(t_0) = y_0,\ y'(t_0) = y_1, \dots, y^{(n-1)}(t_0) = y_{n-1}.$$

A function $y(t)$ is a **solution to the IVP** if it satisfies both the differential equation and all initial conditions.

Example A.2 *Consider*

$$y' = 2y, \quad y(0) = 1.$$

DOI: 10.1201/9781042009220-A

Separating variables gives $\frac{dy}{y} = 2\,dt$. *An integration yields to*

$$\ln|y| = 2t + C \quad \Rightarrow \quad y = Ce^{2t}.$$

Using $y(0) = 1$ *gives* $C = 1$, *so* $y = e^{2t}$. □

3. First–Order Equations

(a) Separable Equations

A first–order ODE is **separable** if it can be written in the form

$$\frac{dy}{dt} = g(t)h(y),$$

which can be rearranged as

$$\frac{dy}{h(y)} = g(t)\,dt.$$

Integrate both sides and, if possible, solve for y.

Example A.3 *Consider* $\frac{dy}{dt} = ty$.

$$Then, separating, the, variables, we, get \frac{dy}{y} = t\,dt \quad \Rightarrow \quad \ln|y| = \frac{t^2}{2} + C.$$

Hence the solution is $y = Ce^{t^2/2}$. □

(b) Linear First–Order Equations

A general linear first–order ODE has the form

$$y' + P(t)y = Q(t).$$

It can be solved using an **integrating factor**:

$$\mu(t) = e^{\int P(t)\,dt}.$$

Multiplying through by $\mu(t)$ gives

$$\frac{d}{dt}[\mu(t)y] = \mu(t)Q(t),$$

so

$$y = \frac{1}{\mu(t)} \int \mu(t)Q(t)\,dt + C/\mu(t).$$

Example A.4 *Solve* $y' + y = e^{-t}$.

$$\mu(t) = e^{\int 1\,dt} = e^t.$$

Then $\frac{d}{dt}(e^t y) = 1 \Rightarrow e^t y = t + C$, *so* $y = e^{-t}(t + C)$. □

4. Second–Order Linear Equations

A **second–order linear ODE** has the general form

$$a(t)y'' + b(t)y' + c(t)y = f(t),$$

where $a(t) \neq 0$. If $f(t) = 0$, the equation is **homogeneous**; otherwise, it is **nonhomogeneous**.

(a) Constant–Coefficient Case

If a, b, and c are constants, then

$$ay'' + by' + cy = 0.$$

Assume a solution of the form $y = e^{rt}$, giving the **characteristic equation**

$$ar^2 + br + c = 0.$$

Depending on the discriminant $\Delta = b^2 - 4ac$:

$$\begin{cases} \text{Two real roots } r_1, r_2: & y = C_1 e^{r_1 t} + C_2 e^{r_2 t}, \\ \text{Repeated root } r: & y = (C_1 + C_2 t)e^{rt}, \\ \text{Complex roots } \alpha \pm i\beta: & y = e^{\alpha t}(C_1 \cos\beta t + C_2 \sin\beta t). \end{cases}$$

Example A.5 *Consider* $y'' - 3y' + 2y = 0$. *Its characteristic equation is given by* $r^2 - 3r + 2 = 0$, *and has the roots* $r = 1, 2$. *So the solution takes the form* $y = C_1 e^t + C_2 e^{2t}$.
□

(b) Nonhomogeneous Case

To solve

$$ay'' + by' + cy = f(t), \tag{A.1}$$

first find the general solution y_h of the homogeneous part. Then find a **particular solution** y_p using:

- Undetermined coefficients (when $f(t)$ is polynomial, exponential, sine, or cosine),
- Variation of parameters (for general $f(t)$).

The general solution is $y = y_h + y_p$. Below, we provide some examples to explain how to obtain the particular solution. Finding the particular solution y_p depends on two things:
(a) The type of function $f(x)$ in (A.1), and
(b) The nature of the homogeneous solution y_h of (A.1).
The method of this section only applies to functions $f(x)$ that are polynomial in x, combinations of *sine* or *cosine*, exponentials in x, or combinations of the aftermentioned forms of $f(x)$. We illustrate the idea by displaying a few examples.

Example A.6 *The differential equation*

$$y'' - 3y' + 2y = 4$$

has the homogenous solution $y_h(x) = c_1e^x + c_2e^{2x}$. *Since* $f(x) = 4$, *is a constant, we consider a particular solution of the form* $y_p = A$, *where A is to be determined. Substituting* y_p *into the differential equation and solving for A gives,* $A = 2$. *Thus, the general solution is*

$$y(x) = y_h(x) + y_p = c_1e^x + c_2e^{2x} + 2.$$

□

Example A.7 *For*

$$y'' - 3y' + 2y = 2e^{3x}$$

we have $y_h(x) = c_1e^x + c_2e^{2x}$. *Since* $f(x) = 2e^{3x}$, *we consider a particular solution of the form* $y_p = Ae^{3x}$, *where A is to be determined. Substituting* y_p *into the differential equation, we arrive at the relation* $2Ae^{3x} = 2e^{3x}$. *This gives* $A = 1$, *and hence the general solution is*

$$y(x) = y_h(x) + y_p = c_1e^x + c_2e^{2x} + e^{3x}.$$

□

Example A.8 *The equation*

$$y'' - 3y' + 2y = 2e^x$$

has $y_h(x) = c_1e^x + c_2e^{2x}$. *Since* $f(x) = e^x$, *we consider a particular solution of the form* $y_p = Ae^x$, *where A is to be determined. Substituting* y_p *into the differential equation we arrive at the relation* $(A - 3A + 2A)e^x = 2e^x$. *Or,* $0e^x = 2e^x$, *which can only imply that*

$$0 = 2.$$

So what went wrong? Well, we said in the beginning that $f(x)$ *depends on the form of* y_h *too. Let's start over. Now* $f(x) = e^x$, *and so we try a particular solution of the form* $y_p = Ae^x$. *But the term* e^x *is already present in* y_h, *and so we try to multiply* Ae^x *by* x. *Thus, we end up with the particular solution of the form* $y_p = Axe^x$. *Substituting* y_p *into the differential equation, we arrive at* $A = -2$, *and hence the general solution is*

$$y(x) = y_h(x) + y_p = c_1 e^x + c_2 e^{2x} - 2xe^x.$$

□

The table below provides guidance on how to construct y_p in the case that none of the terms present in y_h are parts of the forcing function $f(x)$.

TABLE A.1: Shows how to find y_p.

$f(x)$	y_p
Constant C	A
e^{ax}	Ae^{ax}
Cx^n, $n = 0, 1, 2, \ldots$	$A_0 + A_1 x + A_2 x^2 + \ldots + A_n x^n$
$\cos(bx)$, or $\sin(bx)$	$A_1 \cos(bx) + A_2 \sin(bx)$
$x^n \cos(bx)$, or $x^n \sin(bx)$	$(A_0 + A_1 x + A_2 x^2 + \ldots + A_n x^n) \cos(bx)$ $+(B_0 + B_1 x + B_2 x^2 + \ldots + B_n x^n) \sin(bx)$
$x^n e^{ax}$	$(A_0 + A_1 x + A_2 x^2 + \ldots + A_n x^n) e^{ax}$

5. Cauchy–Euler (Equidimensional) Equations

A **Cauchy–Euler equation** has the form

$$at^2 y'' + bty' + cy = 0, \quad t > 0.$$

We try a power–law solution $y = t^r$. Then

$$y' = rt^{r-1}, \quad y'' = r(r-1)t^{r-2}.$$

Substitute into the equation:

$$ar(r-1) + br + c = 0.$$

This gives the **auxiliary (characteristic) equation**. Solve for r and write the general solution as:

$$\begin{cases} y = C_1 t^{r_1} + C_2 t^{r_2}, & \text{if roots are real distinct;} \\ y = (C_1 + C_2 \ln t) t^r, & \text{if repeated root;} \\ y = t^{\alpha} \left(C_1 \cos(\beta \ln t) + C_2 \sin(\beta \ln t) \right), & \text{if complex roots } r = \alpha \pm i\beta. \end{cases}$$

Example A.9 *The auxiliary equation for* $t^2 y'' - 3ty' + 4y = 0$. *is* $r(r-1) - 3r + 4 = 0$, *or* $r^2 - 4r + 4 = 0$, *which has the solution* $r = 2$. *Hence* $y = (C_1 + C_2 \ln t)t^2$. □

Bibliography

[1] Arnold, V. I., *Ordinary Differential Equations,* MIT Press, Cambridge, 1973.

[2] Bellman, R., *Stability Theory of Differential Equations,* McGraw-Hill Book Company, New York, London, 1953.

[3] Bender, C. M., and Orszag, S. A., *Advanced Mathematical Methods for Scientists and Enginners,* McGraw-Hill Book Company, New York, 1978.

[4] Boyce, W. E., and DiPrima, R. C., *Elementary Differential Equations and Boundary Value Problems,* Wiley, New York, 2009.

[5] J. A. Nohel and M. Braun, *Ordinary Differential Equations: A First Course,* Dover Publications, New York, 1980.

[6] Carr, J., *Applications of Centre Manifold Theory,* Springer, New York, 1981.

[7] Coddington, E. A., and Levinson, N., *Theory of Ordinary Differential Equations,* McGraw-Hill, New York, 1955.

[8] Eckhaus, W., *Matched Asymptotic Expansions and Singular Perturbations,* North-Holland, Amsterdam, 1979.

[9] Guckenheimer, J., and Holmes, P., *Nonlinear Oscillations, Dynamical Systems, and Bifurcations of Vector Fields,* Springer, New York, 1983.

[10] Hale, J. K., *Ordinary Differential Equations,* Wiley-Interscience, New York, 1969.

[11] Hale, J. K., and Koçak, H., *Dynamics and Bifurcations,* Springer, New York, 1991.

[12] Hartman, P., *Ordinary Differential Equations,* Wiley, New York, 1964.

[13] Hinch, E. J., *Perturbation Methods,* Cambridge University Press, Cambridge, 1991.

[14] Hirsch, M. W., Smale, S., and Devaney, R. L., *Differential Equations, Dynamical Systems, and an Introduction to Chaos,* Academic Press, San Diego, 2004.

[15] Holmes, M. H., *Introduction to Perturbation Methods,* Springer, New York, 1995.

[16] Hoppensteadt, F. C., *Mathematical Methods of Population Biology,* Cambridge University Press, Cambridge, 1982.

[17] Jordan, D. W., and Smith, P., *Nonlinear Ordinary Differential Equations,* Oxford University Press, Oxford, 1999.

[18] Kevorkian, J., and Cole, J. D., *Perturbation Methods in Applied Mathematics,* Springer-Verlag, New York, 1981.

[19] Kevorkian, J., and Cole, J. D., *Multiple Scale and Singular Perturbation Methods,* Springer, New York, 1996.

[20] Khalil, H. K., *Nonlinear Systems,* Prentice Hall, Upper Saddle River, 2002.

[21] LaSalle, J., and Lefschetz, S., *Stability by Liapunov's Direct Method,* Academic Press, New York, 1961.

[22] Logan, J. D., *Applied Mathematics,* Wiley, New York, 2006.

[23] Lyapunov, A. M., *The General Problem of the Stability of Motion,* Taylor & Francis, London, 1992.

[24] Murray, J. D., *Mathematical Biology I: An Introduction,* Springer, New York, 2002.

[25] Nayfeh, A. H., *Perturbation Methods,* Wiley-Interscience, New York, 1973.

[26] Nayfeh, A. H., *Introduction to Perturbation Techniques,* Wiley, New York, 1981.

[27] Olver, F. W. J., *Asymptotics and Special Functions,* Academic Press, New York, 1974.

[28] O'Malley, R. E., *Singular Perturbation Methods for Ordinary Differential Equations,* Springer, New York, 1991.

[29] Perko, L., *Differential Equations and Dynamical Systems,* Springer, New York, 2001.

[30] Pontryagin, L. S., *Ordinary Differential Equations,* Addison-Wesley, Reading, 1962.

[31] Raffoul, Y. N., *Advanced Differential Equations,* Elsevier/Academic Press, N Y, (2022).

[32] Schlichting, H., and Gersten, K., *Boundary-Layer Theory,* Springer, Berlin, 2000.

[33] Segel, L. A., *Modeling Dynamic Phenomena in Molecular and Cellular Biology,* Cambridge University Press, Cambridge, 1984.

[34] Strogatz, S. H., *Nonlinear Dynamics and Chaos,* Westview Press, Boulder, 2015.

[35] Tenenbaum, M., and Pollard, H., *Ordinary Differential Equations,* Dover, New York, 1985.

[36] Tikhonov, A. N., and Samarskii, A. A., *Equations of Mathematical Physics,* Dover, New York, 1990.

[37] Van Dyke, M., *Perturbation Methods in Fluid Mechanics,* Academic Press, New York, 1964.

[38] Verhulst, F., *Nonlinear Differential Equations and Dynamical Systems,* Springer, Berlin, 1996.

[39] Wasow, W., *Asymptotic Expansions for Ordinary Differential Equations,* Dover Publications, New York, 1987.

[40] Whitham, G. B., *Linear and Nonlinear Waves,* Wiley, New York, 1974.

[41] Wiggins, S., *Introduction to Applied Nonlinear Dynamical Systems and Chaos,* Springer, New York, 2003.

Index